CONTEMPORARY RESEARCH IN THE FOUNDATIONS AND

PHILOSOPHY OF QUANTUM THEORY

THE UNIVERSITY OF WESTERN ONTARIO SERIES IN PHILOSOPHY OF SCIENCE

A SERIES OF BOOKS

ON PHILOSOPHY OF SCIENCE, METHODOLOGY,

AND EPISTEMOLOGY

PUBLISHED IN CONNECTION WITH

THE UNIVERSITY OF WESTERN ONTARIO

PHILOSOPHY OF SCIENCE PROGRAMME

VOLUME 2

CONTEMPORARY RESEARCH IN THE FOUNDATIONS AND PHILOSOPHY OF QUANTUM THEORY

PROCEEDINGS OF A CONFERENCE HELD AT THE UNIVERSITY OF WESTERN ONTARIO, LONDON, CANADA

Edited by

C. A. HOOKER

University of Western Ontario, Ontario, Canada

D. REIDEL PUBLISHING COMPANY

DORDRECHT-HOLLAND / BOSTON-U.S.A.

Library of Congress Catalog Card Number 72–83377

ISBN-13:978-90-277-0338-5   e-ISBN-13:978-94-010-2534-8
DOI: 10.1007/978-94-010-2534-8

Published by D. Reidel Publishing Company,
P.O. Box 17, Dordrecht, Holland

Sold and distributed in the U.S.A., Canada, and Mexico
by D. Reidel Publishing Company, Inc.
306 Dartmouth Street, Boston,
Mass. 02116, U.S.A.

TABLE OF CONTENTS

INTRODUCTION

To mathematicians, mathematics is a happy game, to scientists a mere tool and to philosophers a Platonic mystery – or so the caricature runs. The caricature reflects the alleged 'cultural gap' between the disciplines – a gap for which there too often has been, sadly, sound historical evidence. In many minds the lack of communication between philosophy and the exact disciplines is especially prominent. Yet in the past there was no separation – exact knowledge, covering both scientists and mathematicians, was known as natural *philosophy* and the business of providing a critical view of the nature of reality and an accurate mathematical description of it constituted a single task from the glorious tradition begun by the early Greek philosophers even up until Newton's day (but I am thinking of Descartes and Leibniz!).

The lack of communication between these professional groups has been particularly unfortunate, for the past half century has seen the most exciting developments in mathematical physics since Newton. These developments hinged on the introduction of vast new reaches of mathematics into physics (non-Euclidean geometries, covariant formulations, non-commutative algebras, functional analysis and so on) and conversely have challenged mathematicians to develop the appropriate mathematical fields. Equally, these developments have posed profound philosophical problems to do with the rejection of traditional conceptions concerning the nature of physical reality and physical theorising. Out of these challenges, however, have come exciting new philosophical ideas, such as the relative conventionality of geometry vis-à-vis physics, the concept of non-standard 'lattice logics' (quantum logic) and so on. What is required to take advantage of the challenges which contemporary physics presents are mathematicians who understand physics, physicists who understand mathematics, and philosophers who understand both. (One might well add the requirement that both of the former groups possess a modicum of philosophical sophistication, especially since to theorize on fundamental matters is, ipso facto, to philosophize, but in view of the fact that these groups have done more to stir up philosophy recently than vice versa we shall pass over that requirement.) Recently people of each variety have begun appearing with increasing frequency.

The purpose of the conference, whose proceedings are presented herein, was to bring together physicists, mathematicians and philosophers who were able to take advantage of the exciting possibilities for mutual interaction that the rise of modern quantum theory presents. The invited participants and observers made up three mathematicians, seven physicists and four philosophers. Conference participants showed about the same distribution by professions. These people all, however, had deep interests in the other disciplinary areas.

The conference was a *working* conference. Between the formally scheduled sessions were held many discussion sessions, scheduled as the desire and/or need arose and determined by the expertise of those attending. It is regrettable that these informal sessions could not be represented in the volume since some of the most valuable 'cross-fertilization' came during these periods.

No attempt was made to cover the entire spectrum of contemporary physics, nor, for that matter, even of quantum theory. Rather a narrower range of topics was concentrated on; roughly, these were concentrated around issues to do with characteristics of the mathematical foundations of quantum theory that are of importance to physics and philosophy. In this way we hoped to make clear to ourselves, and now to you, the reader, the unique significance of that theory.

Especially in respect of its treatment of the logico-algebraic approach to quantum theory, this volume does make an attempt to cover the material within its scope sufficiently comprehensively for the senior student beginning on these problems to use it as a reference text. We therefore ask the indulgence of the experienced researcher if some passages seem to him to be presenting elementary material. The failure to comprehend clearly the foundations of a subject is usually the source of later confusions that arise.

Quantum theory has spawned many controversies and many philosophical interpretations. Its own nature is evidently such that physicists and philosophers find difficulty relating it directly to the world without introducing all manner of philosophically and scientifically controversial doctrines. Such doctrines range from the resignation of Scientific Realism, through postulations of a unique action of mind on matter, or the view that microscopic reality is radically different from macroscopic reality, to the view that the logical form of the world is non-classical.

Of course all those who put forward doctrines designed to render quantum theory intellectually transparent took cognisance of the fundamental mathematical structure of that theory – it was precisely to accommodate that peculiar structure that the doctrines were created. But even though quite profound discussions of the mathematical structure of quantum theory have made their appearance from almost the outset of the theory it has not been until relatively recently that those discussions have been presented in a fashion that is designed to shed clear light on its conceptual structure. It was the hope of the conference to illuminate that fruitful union thus helping to create a new rapport among the three anciently united disciplines.

One of the clearest cases of a philosophically important mathematical study of quantum theory is the study of the algebraic structure of that theory culminating in what is now known colloquially as 'quantum logic'. This latter doctrine is a philosophic development stemming from the early work of von Newmann on the "working logic" of quantum mechanics and later developed by Mackey.[1] It is based deep in the algebraic structure of quantum theory and has proven an exciting development to all three disciplines: (i) it has stimulated the development of lattice theory and more generally, of partially ordered sets,[2] (ii) it has led to a penetrating way in which to uncover the essential structure of a physical theory and stimulated various mathematical generalizations of existing theory,[3] (iii) it has led to a broadened conception of logics and to a new philosophic appreciation of the role of logic in a physical theory.[4] Some of the central papers in this volume are devoted to an exposition of this approach (the papers by Professors Bub, van Fraassen, Gudder and Greechie, with some further remarks found in Hooker) and together they form, I believe, one of the finest critical presentations of the subject now available.

Professors Gudder and Greechie concentrate on the algebraic aspects though developing the structures with an eye to their physical and philosophical interpretation. They discuss critically the various approaches to quantum logic and the non-existence 'proofs' for 'hidden variables' (proving in the process that a hidden variable theory of a certain kind always exists!, it is related to van Fraassen's Anti-Copenhagen interpretation).

The question of whether quantum logic admits a conditional as an element of the logic is answered negatively. Finally they go on to explore the challenging field of combinatorial quantum logics.

Professor van Fraassen develops an important relationship between modal logics and the probability calculus to show the former may be exploited in the interpretation of quantum theory. He is able to develop a 'Copenhagen' interpretation and an 'Anti-Copenhagen' interpretation for quantum theory.

Professor Bub applies the approach to the problem of the 'completeness' of quantum theory as a description of the world, examining critically the various 'proofs' taken to exclude 'hidden variables' (which are all essentially based on the algebraic-logical structure of the theory) and developing, in a rigorous mathematical context, the philosophically exciting idea that the structure (logical form) of the world may be non-boolean.

My own remarks in this context are essentially limited to pointing out the perhaps hitherto unappreciated variety of propositional structures in physical theories and the ways in which these relate to the mathematical and ontological structures of those theories.

Professor Greechie has shown us, in an important theorem, that we may view the quantum algebraic (logical) structure as a collection of boolean (classical) sub-algebras (sub-logics) 'pasted' together in a certain way. Clearly how that pasting is done is important – it is what distinguishes quantum from classical theory. Traditionally a certain pasting procedure has been determined by the rule that quantum states and observables statistically indistinguishable from one another should be identified. Gudder and Greechie adopt this rule (see their Axiom 2), Bub discusses it critically (cf. especially Section VI of his paper). The breaking of it is what permits van Fraassen to develop his controversial 'Anti-Copenhagen' alternative.

Though it is traditional to discuss logical structures whose propositions have the form "The value of observable A lies in borel subset S of the real line", one also finds discussions whose propositions have the form "The value of observable A in statistical state W lies in borel subset S of the real line, with probability r". The connection between the two propositional structures lies in the fact that (essentially by Gleason's theorem)[5] when $r = 1$ the two sorts of propositions can be made to correspond uniquely to exactly the same subspaces of the Hilbert space, if W is a projection operator. This common referent which these special cases of the two propositions share is what permits many writers to treat the two equivalently though, as Bub points out, the distinction between them is actually of quite crucial importance to understanding the logical structure of quantum theory – truth is not to be identified with probability 1. In the usual discussions the structure of the generalized probability statements is not pursued further, the attention essentially being directed to the special sub-case which coincides with the non-probabilistic structure. Indeed, we seem to know relatively little about the general structure and it is to be hoped that we shall see more studies in this direction in the future.[6]

The 'quantum logic' approach, by taking the logical component of a theory's structure seriously, raises the important question of the existence of 'logic-free' formulations of theory. Just as a theory had a specific geometry build into its formulation prior to the development of the general theory of Riemannian manifolds, so also had theories a specific logical structure build into them. The former development raised the possibility of 'metric-free' or 'co-ordinate-free' formulation of physical laws, i.e. formulations of laws such that they could be represented in any continuous Riemannian manifold. The developments in quantum logic raise the analogous question: Can we formulate theories so that they can be represented in various logical structures (e.g. classical-boolean or quantum)? The formulation of the C^*-algebra of observables is already a step in this direction, but to date the idea has received little or no attention – again it is to be hoped that it will in the future.

Whatever the parallels between the 'logical' statuses of geometry and logic, it is a hard fact that the space-time geometry of non-relativistic quantum theory is Euclidean and the working logic of General Relativity boolean. It is not an easy thing to bring the two theories together. Though insight into the geometrico-logical situation may help considerably in this task, it is be noted that the quantum logical approach has yet to deal in a thorough-going fashion with the dynamics of a theory, which bring together the propositional structure of the theory (to do with the algebra of observables) and the space-time symmetry structure of the theory. (This is one of the future challenges facing the quantum logic approach.)[7] Even within the usual tradition of mathematical physics the formulation of a mathematically acceptable special relativistic quantum theory is lacking as yet. (Quantum field theory suffers from the well-known divergences and no non-trivial solution has yet been shown to exist.) Professor Komar's reaction is to attempt to bring the two together by quantising the gravitational field (quantising General Relativity). In the present paper he discusses the problems and the progress that has been made with respect to this particular programme. In keeping with the spirit of these papers, he begins the discussion with a penetrating analysis of what a 'canonical quantization programme' amounts to for a theory, arriving at a general characterization in terms of the algebraic structure of the theory. This characterization emphasizes the important role of the conceptual apparatus of a physical theory in formulating this programme clearly. Every theory has such a descriptive apparatus and it is tied intimately to the mathematical structure of the theory. (For example, the concept of a signal, absolutely central to the standard exposition of boolean General Relativity, is not well defined in non-boolean quantum theory because of the non-commutativity of the position and momentum operators.) Thus the quantization programme is intimately tied, through the structure, to the conceptual structure of the theory to be quantized. Komar makes this connection clear. The quantization programme for General Relativity is then analysed in this context. Not unexpectedly, the peculiar difficulty of this quantization programme lies in the new algebraic structure brought to the theory by the space-time symmetry group of General Relativity (the Einstein group) and this is connected intimately with the non-linearity of the gravitational field equations. Despite its incomplete state, however, the analysis of the problem itself throws a penetrating light on the significance of the structure of physical theories.

Complementing Komar's paper is that by Professor F. D. Peat. Commencing from a frank discussion of the philosophical and methodological motivation for establishing a unified physics, Peat discusses the presently known connections between relativity theory and quantum theory, arguing that it now seems reasonable to expect both theories to be significantly modified during the process of unification. He then enters upon a discussion of what is certainly one of the most ambitious and promising programmes of this sort, namely the attempt to develop a unified theory in terms of twistors (generalization of spinors). This programme is philosophically, as well as scientifically, especially interesting because it contains the seeds of the possibility of viewing quantum phenomena as *geometric* features of the space-time manifold, a possibility seriously considered only very recently.

Also deeply concerned with the joining of relativity theory and quantum mechanics is Professor Mendel Sachs who has devoted the past decade to the elaboration of a non-linear general relativistic spinor field theory designed to asymptoticly approximate quantum theory in the appropriate low energy-momentum, non-relativistic limit. In addition Sachs introduces a generalized Mach Principle as a fundamental requirement (i.e. that no part of the universe be physically independent, or have its physical magnitudes determined independently, of the rest of the universe). Sachs' theory is able to entirely renounce the concept of quantization as an intrinsic feature of the physics – discrete spectra arise in his theory only as approximations to continuous *non-linear* field equations. Sachs' theory is also, strikingly, a 'no-photon' theory of electromagnetism, the electromagnetic interaction being interpreted as a circumlocution for a distant interaction of matter with matter (interpreted in accordance with the Mach principle). It is clear that Sachs' theory proceeds in the converse direction to Komar's programme; while Komar attempts to quantize the non-linear gravitational field, Sachs' attempts to show quantization an approximation to a non-linear continuous field theory. Without endorsing either approach (time will judge between them), both theories are of considerable importance, I believe, in showing us something of the route future theoretical developments must take. The spinor formulation is sufficiently general to capture the traditional Dirac and Klein-Gordon equations (in the appropriate limit) and the non-linearity sufficient to at least break down the superposition principle thus opening a way through the difficulties of quantum theory. (Komar himself has suggested this.)[8] Indeed, in a succinct and penetrating history of the subject, Sachs reviews the situation with respect to the theory of light, develops the logical and interpretational difficulties surrounding the traditional quantum mechanical account of light and shows how his own theory leads to a resolution of them. The paper concludes with a discussion of a fully unified description of man and his world.

The papers by Professors Cohen and Gerjuoy concentrate on two formal features of the mathematical structure of quantum theory central to its unique character. Professor Cohen explores rigorously the suggestion that quantum theory permits the definition of joint probability distributions for observables represented by non-commuting operators. His answer is negative. In the process he generalizes upon such past attempts as that of Professor Wigner, generating an infinite variety of distributions whose marginal distributions match the quantum mechanical marginal distributions. Professor Gerjuoy's work concentrates on the superposition principle which stands at the heart of all the 'paradoxes' of quantum mechanics. (It may even be regarded as forming the foundation of Professor Cohen's negative conclusion.) He presents us with the most recent experimental evidence for that principle – quite striking evidence – and carefully analyzes some important earlier experiments (in particular the Wu-Shaknov experiment so important to the Einstein-Podolsky-Rosen argument)[9] to assess the precise degree to which they support the principle. (Even so these experiments are apparently not yet sufficient to bear upon the interesting conjecture by Professor Komar that the superposition principle should break down for sufficiently separated systems.)[10]

My own paper concerns itself with what may be learned from a systematic exploration of the structures – mathematical, conceptual and ontological – of physical theories. Specifically I find that there is an intimate connection between the mathematical structure a theory displays and its conceptual and ontological structures. Examination shows, moreover, that there are two great and highly disparate ontological structures in the tradition of mathematical physics: the atomic and the plenum schemes. The characteristics of these schemes are outlined. An analysis of the mathematical structures of Classical Particle Mechanics and Classical Field theory then illustrates the intimate relationship between mathematical structure and ontology. This then prepares the way for a corresponding analysis of Quantum Theory which reveals the overall mathematical structure of a field theory but the detailed dynamical structure of a particle theory. The hybrid structure, seen in the light of the preceding discussion, offers insight into the interpretational difficulties encountered with quantum theory and suggests avenues of future research.

Professor Wigner's paper, whose location has to do with the first letter of his surname and not at all with its importance, brings to the volume the wisdom of a lifetime's fundamental research in the mathematical and philosophic foundations of physics. With penetrating eye he surveys those major lacunae in our understanding of quantum theory that spring from its deeper structure. His analysis commences with those formal structural features of quantum theory which are at the centre of the discussion concerning the interpretation of the theory (the superposition principle, the measurement process) and at the centre of forming a conceptually coherent relativistic quantum theory (the concepts of simultaneity and succession, particularly with respect to measurement outcomes, and to probabilistic statements concerning temporal sequences). Attention is then directed to the question of the 'natural' epistemology and ontology for quantum theory as is – Wigner concludes that quantum theory functions only as a pure probability theory of observations and offers no description of 'reality' in itself. Finally he analyses critically possible extensions and modifications of the theory that would affect its epistemology. (These include hidden variable theories, quantum logical theories of measurement, his own views on the role of consciousness and so on.)

And so we come full circle. Quantum theory has, besides being a revolutionary advance in theorizing, rendered us an invaluable service; for, like its sister, Relativity Theory, it has stimulated physicists, mathematicians and philosophers to communicate and cooperate anew.

ACKNOWLEDGEMENT

The conference was made possible through the support of the Faculties of Arts and Science and the Departments of Applied Mathematics, Philosophy and Physics at the University of Western Ontario. I especially wish to express my gratitude to the deans of those faculties, respectively Dean J. G. Rowe and A. E. Scott, and to the heads of those departments, respectively Professors J. H. Blackwell, R. E. Butts and J. W. McGowan, for the generosity and enthusiasm with which they supported the project. Finally my unbounded gratitude to the staff of the department of philosophy and in particular to two secretaries without whom the conference and this volume would hardly have been possible: to Mrs. Marie Leung who literally created with her bare hands and a pen a coherent, smoothly run conference and to Mrs. Alice Smith who created with her bare hands and a typewriter the manuscript of this volume from papers, notes and scribblings.

February, 1972   C. A. HOOKER

NOTES

[1] See G. Birkhoff and J. von Neuman, 'The Logic of Quantum Mechanics', *Annals of Mathematics* 37 (1936), 835; G. W. Mackey, *Mathematical Foundations of Quantum Mechanics*, W. A. Benjamin, Inc., New York, 1963.

[2] For a convenient resumé of the origins of the lattice theoretic approach see J. C. Abbott, *Trends in Lattice Theory*, Van Nostrand, New York, 1970.

[3] Thus it forms the basis of Kochen and Specker's powerful demonstration that quantum theory cannot be reconstrued as a classical phase space theory (while preserving the relations among observables) and it has, for example, led to the abstracting of the C^*-algebra of observables common to classical particle mechanics and quantum particle and field theories – see the bibliography in Hooker, this volume, pp. 281–7.

[4] Cf. Bub's paper in this volume, p. 45 f and H. Putnam, 'Is Logic Empirical?', *Boston Studies in the Philosophy of Science*, Vol. V, D. Reidel Publ. Co., Dordrecht, Holland, 1969.

[5] See A. M. Gleason, 'Measures on the Closed Subspaces of a Hilbert Space', *Journal of Mathematics and Mechanics* 6 (1957), No. 6.

[6] Cf. the remarks in the papers of van Fraassen and Hooker.

[7] Mackey, in his pioneering work, has begun the development of this approach. See G. W. Mackey *op. cit.*

[8] A. Komar, 'Indeterminate Character of the Reduction of the Wave Packet in Quantum Theory', *Physical Review* 126 (1962), 135.

[9] This relationship is examined in detail in C. A. Hooker, 'The Nature of Quantum Mechanical Reality: Einstein Versus Bohr', *Paradox and Paradigm*, University of Pittsburgh Press, 1972.

[10] Gerjuoy – private communication.

AUTHOR BIOGRAPHIES

JEFFREY BUB

Professor Jeffrey Bub B.Sc. (with Honors, Physics, University of Cape Town, 1962) Ph.D. (Theoretical Physics, University of London at Birkbeck College, 1966) is currently Associate Professor of Philosophy at the University of Western Ontario, London 72, Ontario, Canada. Professor Bub has published in the foundations of quantum mechanics, earlier especially on the formulation of the problem of hidden variable theories. Later his research into logical structures arising within mathematical physics led him increasingly toward philosophy; he is currently researching problems which arise in the quantum logical interpretation of quantum mechanics.

LEON COHEN

Professor Leon Cohen B.S. (Physics, City College of the City University of New York, 1962) M.S. (Theoretical Physics, Yale University, 1964) Ph.D. (Theoretical Physics, Yale University, 1966) is currently Associate Professor of Physics at Hunter College of the City University of New York, 695 Park Avenue, New York, N.Y. 10021. Professor Cohen has worked on the mathematical foundations of quantum mechanics with particular emphasis on the probabilistic structure of the theory. Recently he has been researching also the statistical mechanics of N point particles interacting via the gravitational force.

BASTIAAN CORNELIS VAN FRAASSEN

Professor Bas van Fraassen B.A. (with Honors, Philosophy, The University of Alberta, 1963) M.A. (Philosophy, The University of Pittsburgh, 1964) Ph.D. (Philosophy, The University of Pittsburgh, 1966) is Professor of Philosophy at the University of Toronto, Toronto, Ontario, Canada. He is widely known for his research work on the nature of space and time and also for work in logic and formal semantics. He has published books as well as articles in both fields. His work on logic led him to investigate logical structures arising in mathematical physics.

EDWARD GERJUOY

Professor Edward Gerjuoy B.S. (Physics, City College of the City University of New York, 1937) M.A. (Physics, The University of California at Berkeley, 1940) Ph.D (Physics, The University of California at Berkeley, 1942) is Professor of Physics at the University of Pittsburgh, Pittsburgh, Pennsylvania 15213, U.S.A. In recent years his unclassified university research has been concerned mainly with the application of scattering theory to atomic and molecular collisions.

RICHARD J. GREECHIE

Professor Dick Greechie B.A. (Mathematics, Boston College, 1962) Ph.D. (Mathematics, the University of Florida, 1966) is Associate Professor in the Department of Mathematics at Kansas State University, Manhattan, Kansas 66502. His research has been devoted to the elaboration of mathematical structures of relevance to modern theoretical physics, especially to those partially ordered sets arising in connection with quantum mechanics.

STANLEY P. GUDDER

Professor Stanley Gudder B.S. (Engineering Physics, Washington University, St. Louis, 1958) M.S. (Mathematics, University of Illinois, 1960) Ph.D. (Physics, University of Illinois, 1964) is currently Associate Professor of Mathematics, The University of Denver, Denver, Colorado 80210, U.S.A. His research interests in pure mathematics lie in the area of functional analysis, probability theory and lattice theory. Correspondingly, Professor Gudder has research interests in physics in the axiomatics and foundations of quantum mechanics and he has published in each of these areas. He is currently conducting research in axiomatic models for quantum mechanics, operational quantum mechanics and concerning the *-algebras of operators on Hilbert space.

CLIFFORD ALAN HOOKER

Professor Cliff Hooker B.Sc. (with Honors, Physics, The University of Sydney, 1964) B.A. (equivalent, with Honors, Philosophy, The University of Sydney, 1968) Ph.D. (Physics, The University of Sydney, 1968) Ph.D. (Philosophy, York University, Toronto, Ontario, 1970) is Associate Professor of Philosophy and Environmental Engineering at the University of Western Ontario, London 72, Ontario, Canada. His earlier research centered on high frequency oscillations in plasmas but interest in the foundations of physical theory drove him toward philosophy. His recent research interests include ontology and epistemology in pure philosophy, the structure and function of science in general and the foundations of modern mathematical physics in particular as well as the systematic structure of environmental problems.

ARTHUR BARAWAY KOMAR

Professor Art Komar A.B. (Physics, Princeton University, 1952) Ph.D. (Theoretical Physics, Princeton University, 1956) is Professor of Physics and Dean of Belfer Graduate School of Science at Yeshiva University, Amsterdam Avenue & 186th Street, New York, N.Y. 10033. Professor Komar's research interests range from the foundations of the general theory of relativity through quantum field theory and the quantum theory of measurement. He is especially interested in the quantization program for general relativity.

FRANCIS DAVID PEAT

Doctor F. David Peat B.Sc., M.Sc., Ph.D. (Physics, University of Liverpool, 1961, 1962, 1965, respectively) joined the National Research Council of Canada in 1966 after spending a year teaching at Queen's University, Ontario. His earlier work centered on the density matrix approach to quantum theory and its particular application to solid state physics. Lately he has turned his attention to relativity and particle physics and to the connection between these and quantum theory. These studies have driven him to fundamental research on the general mathematical and conceptual foundations of physics and the use of language in theorising.

MENDEL SACHS

Professor Mendel Sachs A.B. (Physics, The University of California at Los Angeles, 1949) M.A. (Physics, The University of California at Los Angeles, 1950) Ph.D. (Physics, The University of California at Los Angeles, 1954) is Professor of Physics and Astronomy at the State University of New York at Buffalo, Buffalo, N.Y. 14214. His research interests cover the theoretical and philosophical foundations of modern physical science and the history of science. His research, which includes solid state theory and nuclear theory, currently centers on the theory of general relativity, high energy and elementary particle physics and the relations of these to the foundations of quantum theory. He has also recently published a book in the history and philosophy of science concerning the history of matter theory.

EUGENE PAUL WIGNER

Professor Eugene Wigner Dr. Eng. (Chemical Engineering, Technische Hochschule, Berlin, 1925) is also recipient of numerous honorary degrees from some of the world's leading institutions and is Professor Emeritus at Princeton University, Jadwin Hall, P.O. Box 708, Princeton, New Jersey 08540, U.S.A. His research interests cover a broad spectrum in physics and mathematics, including the rates of chemical reactions, the quantum theory of metallic cohesion and the theory of nuclear reactions, but he has in particular concentrated upon the problems of invariants and conservation laws, particularly as applied to quantum mechanical theories of atomic and molecular spectra and the relativistic equations of particles, and on the epistemological consequences of the quantum theory. Besides many papers, he has published books covering these latter areas.

JEFFREY BUB

ON THE COMPLETENESS OF QUANTUM MECHANICS*

> The Heisenberg-Bohr tranquilizing philosophy – or
> religion? – is so delicately contrived that, for the time
> being, it provides a gentle pillow for the true believer from
> which he cannot very easily be aroused. So let him lie there.
>
> ALBERT EINSTEIN[1]

My purpose in this paper is to formulate a general completeness problem
for statistical theories, and to show that the quantum theory is complete.
It follows as a corollary to the completeness theorem for quantum mechanics that a phase space reconstruction of the quantum statistics is excluded, i.e., there are no 'hidden variables' underlying the quantum statistics.

The following discussion is divided into six sections and an appendix.
The first five sections are concerned with the completeness problem as it
appears in the literature. In Section I, I isolate a particular thesis concerning the completeness of quantum mechanics that characterizes Heisenberg's version of the Copenhagen interpretation, and indicate how Bohr's
interpretation involves the rejection of this thesis. This analysis makes use
of certain logical and algebraic notions, which are developed briefly in the
appendix. Section II is a reconstruction of the Einstein-Podolsky-Rosen
argument[2] as an objection to Heisenberg's completeness thesis. In Section
III, I discuss von Neumann's completeness proof[3] and compare this with
related theorems by Gleason[4], and by Kochen and Specker[5]. Section IV
deals with the proof of Jauch and Piron[6], and Section V is a detailed analysis of J. S. Bell's result[7] concerning so-called 'local' hidden variable theories. Finally, Section VI concludes with a general formulation of the completeness problem as that of demonstrating the isomorphism between two
logical spaces, together with some remarks on the significance of this
problem for the interpretation of quantum mechanics.

The first five sections involve a deliberate ambiguity, which is only resolved in Section VI with the formulation of the completeness problem.
This ambiguity, which is characteristic of the historical controversy con-

*Hooker (ed.), Contemporary Research in the Foundations and Philosophy of Quantum
Theory*, 1–65. *All rights reserved*
Copyright © 1973 *by D. Reidel Publishing Company, Dordrecht-Holland*

cerning the completeness of quantum mechanics, derives from the implicit assumption that logic is *a priori* and Boolean, in the context of the completeness problem. Once this assumption is rejected, the completeness of quantum mechanics is seen to follow from Gleason's theorem, and the measurement problem is resolved. The alternative is to retain the Boolean assumption, in which case quantum mechanics becomes 'irrational', at least in the sense that the measurement problem cannot be avoided in any obvious way. The artificiality of this view is only disguised by confusing the two logical spaces.

I. THE COPENHAGEN INTERPRETATION

To begin with classical mechanics: The physical magnitudes of a classical mechanical system (e.g. energy) are real-valued (Borel) functions of generalized position and momentum coordinates, the phase variables. These variables parametrize a linear vector space, the phase space Z. For motion in one spatial dimension, the phase space is 2-dimensional, with coordinates Q (for position) and P (for momentum), i.e. the point $z \in Z$ is an ordered pair $\langle Q, P \rangle$. The classical mechanical equations of motion – Hamilton's equations: $\dot{Q} = \partial H / \partial P$, $\dot{P} = -\partial H / \partial Q$ – determine a trajectory in phase space, given the initial values of the variables Q and P. The quantity H, the Hamiltonian, is a function of the phase variables and characterizes the particular system involved. Since the physical magnitudes A, B, C, ... are functions of the phase variables, the values of these quantities are defined for every point on the phase trajectory of the system, and are determined for all time via Hamilton's equations by any point on the trajectory i.e. by an assignment of values to the canonically conjugate pair $\langle Q, P \rangle \in Z$. Such an assignment of values is a classical mechanical *state*.

Now, the singleton subsets $\{z\} \in \mathscr{F}$, where $\mathscr{F}$ is the field of Borel subsets of phase space, are *atoms*[8] in $\mathscr{F}$. The Boolean algebra $\mathscr{F}$ is a perfect reduced field of sets, and so every ultrafilter[9] in $\mathscr{F}$ is determined by a point in Z, i.e. there is a one-one correspondence between atoms in $\mathscr{F}$ and ultrafilters in $\mathscr{F}$. An ultrafilter Φ in $\mathscr{F}$ corresponds to a maximal consistent set of propositions, representing a possible world, a maximal totality of situations in logical space. A typical proposition[10] in the set is expressed by the theoretical sentence: val $(A) \in S$. (Read: the value of the physical magnitude A lies in the (Borel) set S of real numbers.)

The sentence val $(A) \in S$ corresponds to the Borel set:

$$A^{-1}(S) = \{z: A(z) \in S\}$$

in the phase space Z.[11]

The representation of the states of classical mechanics as the atoms of a perfect reduced field of subsets of a space Z, which functions in the theory like the Stone space of a Boolean logic[12], shows that the logical space of events is Boolean.[13]

The sense in which the specification of a point in Z is a *state description* is just this: A point in Z corresponds to an ultrafilter in $\mathscr{F}$, which represents a maximal totality of situations in logical space – maximal in the sense that these situations are related by the logical relations of a Boolean ultrafilter.

Quantum mechanics is a statistical theory. For the purposes of this analysis, a statistical theory may be regarded as characterized by a set of physical magnitudes forming an algebraic structure of a certain kind, together with an algorithm for assigning probabilities to ranges of possible values of these magnitudes. That is to say, the theory involves a set of 'statistical states' which assign probabilities to theoretical sentences val$(A) \in S$.

Two physical magnitudes, A and B, are *statistically equivalent* in a statistical theory just in case

$$p_w(\text{val}(A) \in S) = p_w(\text{val}(B) \in S)$$

for every statistical state W[14] and every Borel set S of real numbers. So, if

$$p_w(\text{val}(X) \in S) = p_w(\text{val}(A) \in g^{-1}(S))$$

for every statistical state W and every Borel set S, where $g: R \to R$ is a real-valued Borel function on the real line, then X and $g(A)$ are statistically equivalent.[15] With respect to *this* relation of equivalence, two magnitudes A_1 and A_2 are said to be *compatible* if and only if there exists a third magnitude, B, and Borel functions $g_1: R \to R$ and $g_2: R \to R$ such that

$$A_1 = g_1(B)$$
$$A_2 = g_2(B).\text{[16]}$$

This definition of the compatibility of two magnitudes is due to Kochen and Specker.[17] A linear combination of two compatible magnitudes may be defined via the linear combination of associated functions as:

$$a_1 A_1 + a_2 A_2 = (a_1 g_1 + a_2 g_2)(B),$$

where a_1, a_2 are real numbers, and similarly the product may be defined as:

$$A_1 A_2 = (g_1 g_2)(B).$$

With linear combinations and products of compatible magnitudes defined in this way, the magnitudes of a statistical theory form a *partial algebra*.[18]

Evidently, if all the magnitudes of a statistical theory are pairwise compatible, the partial algebra is a commutative algebra, and the set of idempotent magnitudes is a Boolean algebra. Now, there is a one-one correspondence between the set of idempotent magnitudes and the set of theoretical sentences:[19] each idempotent magnitude corresponds to a theoretical sentence which is true if and only if the value of the magnitude is 1, and false if and only if the value of the magnitude is 0. In this case, the set of sentences to which probabilities are assigned by the algorithm of the theory forms a Boolean algebra, i.e. the 'working logic'[20] of the theory is Boolean.

On the other hand, if the magnitudes are not all pairwise compatible, the idempotents will not form a Boolean algebra, but a *partial Boolean algebra*.[21] Essentially, a partial Boolean algebra is a partially ordered set with a reflexive and symmetric relation of compatibility, such that each maximal compatible subset is a Boolean algebra. (Thus, a partial Boolean algebra may be pictured as 'pasted together' from its maximal Boolean subalgebras.) There is a problem here – the generalization of the probability concept in a non-Boolean context.

Usually, a probability measure is understood as a normed, countably additive, real-valued set function on a σ-field of subsets of a space X, i.e. a map $p: \mathscr{F} \to R$ satisfying the conditions:

(i) $p(0) = 0, \quad p(R) = 1$

(ii) $0 \leqslant p(E) \leqslant 1 \quad \text{for} \quad E \in \mathscr{F}$

(iii) $p\left(\bigcup_{i=1}^{\infty} E_i \right) = \sum_{i=1}^{\infty} p(E_i),$

where E_i is a countable class of disjoint sets in $\mathscr{F}$, i.e. $E_i \cap E_j = 0$, $i \neq j$. The triple $\langle X, \mathscr{F}, p \rangle$ is referred to as a probability space. The set X may be understood as the set of distinct possible outcomes of an experiment, and the set $\mathscr{F}$ as the set of events to which probabilities are assigned. Equivalently, because of the isomorphism between fields of sets and Boolean algebras, the concept of probability may developed formally on a Boolean algebra. In the same way, probabilities may be assigned to the sentences of a Boolean logic, provided that logically equivalent sentences are assigned the same probability.[22] In the case of a partial Boolean algebra there is an obvious generalization: A probability measure in the generalized sense is any assignment of values between 0 and 1 to the elements of a partial Boolean algebra, which satisfies the usual conditions for a probability measure on each maximal compatible subset of the partial Boolean algebra.

Two sentences in the working logic of a statistical theory are compatible if and only if they generate a Boolean sublogic under the logical operations of conjunction, disjunction, and negation. This is to be understood in the following sense: The working logic of a statistical theory is defined via the isomorphism between the partial Boolean algebra of idempotent physical magnitudes and the Lindenbaum-Tarski algebra[23] of the logic. Two sentences are compatible, then, if and only if their associated elements in the Lindenbaum-Tarski algebra generate a Boolean subalgebra under the operations of infimum, supremum, and orthocomplement. This is the case if and only if the corresponding idempotent magnitudes are compatible in the sense of the original definition of compatibility for magnitudes.

The algorithm of quantum mechanics involves the representation of the statistical states of the theory by a certain class of operators in Hilbert space, the statistical operators, and physical magnitudes by hypermaximal Hermitian operators in Hilbert space. Each statistical operator W assigns a probability to the sentence val $(A) \in S$ according to the rule:

$$p_w(\text{val}(A) \in S) = \text{Tr}(WP_A(S)),$$

where $P_A(S)$ is the projection operator associated with the Borel set S by the spectral measure of A.

The peculiarity of quantum mechanics as a statistical theory lies in the fact that the logic of the sentences to which probabilities are assigned is not Boolean. Each sentence val $(A) \in S$ corresponds to a projection operator $P_A(S)$, and hence to a subspace $\mathcal{K}_A(S)$ in the Hilbert space $\mathcal{H}$. The projection operators represent the idempotent magnitudes. A set of sentences corresponding to a set of mutually commuting projection operators, i.e. mutually compatible idempotent magnitudes, corresponds to a set of mutually compatible subspaces, i.e. a set of subspaces generating a Boolean algebra under the operations of infimum, supremum, and orthocomplement. The Boolean algebras of subspaces of a Hilbert space $\mathcal{H}$ may be partially ordered by set inclusion. A maximal Boolean algebra of subspaces is maximal with respect to this ordering. A set of sentences corresponding to a maximal Boolean algebra of subspaces of $\mathcal{H}$ is a Boolean logic. Such a set of sentences corresponds to a maximal compatible set of idempotent magnitudes. The logic of the sentences to which probabilities are assigned by the quantum algorithm is not Boolean, because the set of subspaces of a Hilbert space is not a Boolean algebra (there exist incompatible subspaces) or, equivalently, because the set of physical magnitudes cannot all be expressed as functions of a single magnitude (there exist incompatible magnitudes, i.e. non-commuting operators).

Now, the statement that the working logic of quantum mechanics is non-Boolean, in the sense that the Lindenbaum-Tarski algebra of this logic is isomorphic to the partial Boolean algebra of subspaces of Hilbert space, is hardly more than a reformulation of the more familiar statement that the physical magnitudes of the theory form a non-commutative algebra, as opposed to the physical magnitudes of classical mechanics. The peculiarities of the quantum statistical relations, and the anomalies to which they give rise, are in no sense *explained* by pointing to the non-Boolean character of the logic of the sentences to which probabilities are assigned by the quantum theory. I understand the Copenhagen interpretation of the theory as a proposed explanation for these anomalies. On *Heisenberg's version*[24], it amounts to something like the following:

The measurement procedures by which we determine the truth or falsity of a proposition expressed by a theoretical sentence s disturb the system in such a way that the truth values of propositions incompatible with s are altered. Furthermore, this disturbance is unavoidable and uncontrollable, i.e. 'theoretically opaque'. This means that we cannot assign truth values to a maximal consistent set of propositions in the sense of a Boolean ultrafilter, without implying something like an 'operational contradiction'. Such an assignment would be possible only if we could determine the extent of the disturbance associated with each measurement procedure. That such a determination is excluded is revealed by a precise analysis of measurement procedures at the micro-level: any procedure for measuring position, for example, involves an interference with the system in such a way that the truth value of a momentum proposition is changed in an indeterminate way.

Thus, Heisenberg's proposal is to understand the non-Boolean working logic of quantum mechanics in terms of the theoretical opacity of measurement disturbances. It is the claim that a certain theory of measurement is true that explains the appropriateness of a theory with a non-Boolean working logic for the description and explanation of the behaviour of micro-systems. And *this* claim in turn is supported by arguments of the kind involved in Heisenberg's γ-ray microscope thought experiment. The peculiarities of the quantum description of micro-events are taken to reflect, ultimately, the truth of a theory concerning the observability of micro-events.

The statistical operators of quantum mechanics divide naturally into two sets: those which are idempotent (i.e. satisfy the equation $W^2 = W$) and those which are not. An idempotent statistical operator is a projection operator onto a 1-dimensional subspace, and in fact there is a one-one correspondence between the set of 1-dimensional subspaces of $\mathcal{H}$ and the set of idempotent statistical operators on $\mathcal{H}$.[25] But each 1-dimensional subspace may be associated with a unit vector in $\mathcal{H}$, and so, for idempotent statistical operators, the algorithm for assigning probabilities may be formulated in terms of the associated Hilbert space vector ψ:

$$p_\psi \left(\mathrm{val}\,(A) \in S \right) = \| P_A(S)\,\psi \|^2 = \left(\psi, P_A(S)\,\psi \right).$$

In the case that A has a pure discrete spectrum with non-degenerate eigenvalues a_i, this becomes:

$$p_\psi \left(\mathrm{val}\,(A) \in S \right) = \sum_{a_i \in S} \| P_{a_i} \psi \|^2$$
$$= \sum_{a_i \in S} |(\alpha_i, \psi)|^2\,,$$

where P_{a_i} is the projection operator onto the 1-dimensional subspace spanned by the eigenvector α_i corresponding to a_i.

Now, the idempotent statistical operators are *homogeneous* or *pure*, in the sense that no idempotent statistical operator is expressible as a convex sum of two (or more) different statistical operators, i.e. if W is idempotent and

$$W = p_1 W_1 + p_2 W_2\,(p_1 + p_2 = 1;\quad p_1 > 0, p_2 > 0)$$

then $W = W_1 = W_2$.[26] This means that the probability assigned by W to the sentence s cannot be reduced to a convex sum of probabilities assigned to s by statistical operators different from W, i.e.,

$$p_W(S) \neq p_1 p_{W_1}(S) + p_2 p_{W_2}(S)$$

for any p_1, p_2, W_1, W_2. The non-idempotent statistical operators are all expressible as convex sums of idempotent statistical operators. Putting this another way: the statistical operators form a convex set, with the pure statistical operators as extrema.

The dispersion, $\Delta_W A$, of a physical magnitude A is defined for each statistical operator W by:

$$(\Delta_W A)^2 = \text{Exp}_W (A - \text{Exp}_W (A))^2$$
$$= \text{Exp}_W (A)^2 - (\text{Exp}_W (A))^2,$$

where $\text{Exp}_W(A)$ is the expectation value of A, i.e.

$$\text{Exp}_W (A) = \int_{-\infty}^{\infty} rd \left(\text{Tr} (W P_A(r)) \right)$$
$$= \text{Tr} (WA) \quad {}^{27}$$

In the case of a pure statistical operator corresponding to a vector ψ, I write

$$(\Delta_\psi A)^2 = \text{Exp}_\psi (A)^2 - (\text{Exp}_\psi (A))^2$$

and

$$\text{Exp}_\psi (A) = \int_{-\infty}^{\infty} rd \left(\psi, P_A (r) \psi \right) = (\psi, A\psi).$$

Heisenberg's relation – a reciprocal relation between the dispersions of two incompatible magnitudes A, B satisfying the commutation relation $AB - BA = ihI/2\pi$ – may be derived as a theorem in quantum mechanics:

$$\Delta_\psi A \Delta_\psi B \geqslant h/4\pi, \text{ for every } \psi. {}^{28}$$

Since the statistical operators form a convex set, *a fortiori*:

$$\Delta_W A \Delta_W B \geqslant h/4\pi$$

for every impure statistical operator W. It is an immediate corollary to this theorem that no statistical operator W is dispersion-free for every magnitude A, i.e. there is no statistical operator W satisfying the equation:

$$\Delta_W (A) = 0, \text{ for all } A.$$

Heisenberg's interpretation provides an explanation for this curious feature of the quantum statistics. In particular, the absence of dispersion-free statistical operators in the theory is understood as reflecting the *physical* impossibility of assigning truth values simultaneously to incompatible propositions in any measurement process, because of the unavoidable and uncontrollable disturbance of the values of magnitudes incompatible with A involved in every measurement of A.

Classically, the expectation value of a random variable A on a probability space $\langle X, \mathscr{F}, p \rangle$ is defined by:

$$\mathrm{Exp}\,(A) = \int_X A\,\mathrm{d}p$$

and the dispersion of A by:

$$(\Delta A)^2 = \int_X (A - \mathrm{Exp}\,A)^2\,\mathrm{d}p$$
$$= \mathrm{Exp}\,(A^2) - (\mathrm{Exp}\,(A))^2.$$

A dispersion-free probability measure p satisfies the condition $\Delta A = 0$ for all A. In particular, $\Delta P = 0$ for all idempotent random variables P. An idempotent random variable P satisfies the condition

$$P(x)^2 = P(x), \quad \text{for all} \quad x \in X$$

i.e. $P(x) = 0$ or 1. So, $P: X \to \{0,1\}$ is the characteristic function of some Borel set $E \subseteq X$, and hence

$$\mathrm{Exp}\,(P) = \int_X P\mathrm{d}p$$
$$= p(E)$$

Now,
$$\Delta P = \mathrm{Exp}\,(P^2) - (\mathrm{Exp}\,(P))^2$$
$$= \mathrm{Exp}\,(P) - (\mathrm{Exp}\,(P))^2$$
$$= p(E) - p^2(E)$$

i.e.
$$p(E) = 1 \text{ or } 0.$$

Thus, a dispersion-free probability measure is a 2-valued measure, an assignment of probabilities 1 or 0 to each Borel set in $\mathscr{F}$.
Moreover:

$$p(E') = 1 - p(E), \text{ where } E' = X - E \text{ }[29]$$

and

$$p(E \cap F) = p(E) \, p(F), \text{ for all Borel sets } E, F \subseteq X \text{ }[30]$$

By the isomorphism between $\mathscr{F}$ and $\mathscr{L}$, the Lindenbaum-Tarski algebra of a Boolean logic, a dispersion-free probability measure on $\mathscr{L}$ is a 2-valued measure on $\mathscr{L}$. But a 2-valued measure on $\mathscr{L}$ is a 2-valued homomorphism on $\mathscr{L}$, and so corresponds to an ultrafilter in $\mathscr{L}$, i.e. to a maximal consistent set of propositions.[31] A dispersion-free probability measure on the phase space of classical mechanics corresponds to a point in phase space – a classical mechanical state.

For Heisenberg, there are no dispersion-free statistical operators in quantum mechanics, because the statistical operators of the theory represent those (and only those) probability assignments that are compatible with *our possible knowledge of the micro-level*, in the light of the theoretical opacity of measurement disturbances. That there is no analogue in quantum mechanics of the classical mechanical state – the Boolean ultrafilter, or dispersion-free probability measure – is taken to reflect the (contingent) truth of a theory concerning possible measurements at the micro-level. The pure statistical operators of the theory are compatible with maximal *knowledge*, not maximal *truth*. It is in this sense that the vectors in Hilbert space are the legitimate successors in the quantum description to the classical mechanical states: they represent various possible totalities of situations in logical space that are *maximal with respect to what can be known simultaneously*. These are the state descriptions of quantum mechanics, and the impure statistical operators represent less than maximal knowledge, probability measures over quantum states.

To put this another way: On the basis of a discovery concerning measurement disturbances at the micro-level (i.e. on the assumption that a certain measurement theory is true), a distinction is made between a possible world in the sense of a maximal totality of situations in logical space, and a possible world in the sense of a totality of knowable situations that is maximal with respect to what can be known simultaneously. The state descriptions of classical mechanics represent possible worlds in the first sense; the state descriptions of quantum mechanics represent possible worlds in the second sense. It is (physically) impossible, by any observation procedure whatsoever, to determine simultaneously the truth values of a set of propositions corresponding to a classical state description. The maximal sets of propositions that are simultaneously decidable empirically (at least in principle) are those sets of propositions corresponding to the quantum state descriptions. The non-Boolean working logic of quantum mechanics is the logic of these state descriptions that are less than maximal in logical space – they are maximal only with respect to what can be known simultaneously. In *this* sense, a set of propositions assigned probability 1 by a Hilbert space vector is the quantum logical analogue of the Boolean ultrafilter. One might distinguish, too, between a classical fact ('*c*-fact') and a quantum fact ('*q*-fact'). A *c*-fact is a situation in logical space represented by a true proposition. A *q*-fact is a situation in logical space represented by a proposition that can be established as true by a measurement procedure that does not alter the truth value of any other proposition representing a *q*-fact. (So the set of *q*-facts is a subset of the set of *c*-facts.) If a distinction is made on this interpretation of the quantum theory between classical reality and quantum reality it must be understood relative to *this* distinction between *c*-facts and *q*-facts, and not as a thesis concerning the ontological status of micro-objects.[32]

Thus, Heisenberg's version of the Copenhagen interpretation is characterized by the thesis that quantum mechanics is both statistical and complete, i.e. irreducibly statistical, in the sense that the Hilbert space vectors represent state descriptions that are as close as possible to Boolean ultrafilters (classical states), where 'as close as possible' is understood with reference to the theory that measurement disturbances are theoretically opaque at the micro-level. This is the thesis concerning the completeness of quantum mechanics that I wish to isolate: that the vectors in Hilbert space are to be interpreted as representing '*q*-maximal'[33] totalities of situations in logical space – totalities of knowable situations that are maximal with respect to what can be known simultaneously.

Now, it would be quite wrong to understand Bohr's version of the Copenhagen interpretation as simply another story that explains *why* such-and-such totalities of situations are maximal in this sense, i.e. in terms of the 'wholeness' of instrument and measured object, the ultimacy of classical concepts, etc., as opposed to Heisenberg's story, which involves a theory about the irreducibility and uncontrollability of measurement disturbances. Bohr's position differs radically from Heisenberg's.

For Bohr, what is fundamental is a theory of the applicability of concepts. A concept is applicable under certain conditions, and the applicability conditions of two different concepts may be mutually exclusive. This, Bohr claims, is the case for the space-time and energy-momentum concepts of classical physics, i.e. the conditions for the applicability of space-time concepts exclude the conditions for the applicability of energy-momentum concepts. Quantum mechanics is a rational generalization of classical mechanics in the following sense: Each theoretical sentence val $(A) \in S$ is associated *either* with the group of space-time concepts, *or* with the group of energy-momentum concepts, but not with both. A theoretical sentence expresses a proposition *if and only if the conditions for the applicability of the associated concept are satisfied.*

I shall return to a critical evaluation of Bohr's interpretation in Section VI. At this point, I wish only to indicate that Bohr's view involves the rejection of Heisenberg's completeness thesis. For Bohr, the probability assigned to a theoretical sentence val $(A) \in S$ is the probability that, if the conditions for the applicability of the associated concept were to be satisfied, the corresponding proposition would be true (i.e. represent a fact). Bohr denies that the Hilbert space vectors represent q-maximal totalities of situations in logical space in Heisenberg's sense. There is no c-maximal totality of situations in which the q-maximal totality is embedded, if we define a q-maximal totality as the totality of situations corresponding to a maximal set of theoretical sentences that could simultaneously express propositions (i.e. such that the conditions for the applicability of their associated concepts are simultaneously satisfiable).

II. THE EINSTEIN-PODOLSKY-ROSEN ARGUMENT

I understand the argument of Einstein, Podolsky and Rosen as a refutation of Heisenberg's thesis concerning the completeness of quantum mechanics.

In the first place, they propose a necessary condition of completeness for a physical theory:

Every element of the physical reality must have a counterpart in the physical theory.[34]

I reformulate this condition as follows:

Every fact in logical space must be representable in the theory, i.e. the corresponding proposition must be expressible by a consistent theoretical sentence.

Secondly, they propose a sufficient criterion of physical reality:

If, without in any way disturbing a system, we can predict with certainty (i. e. with probability equal to unity) the value of a physical quantity, then there exists an element of physical reality corresponding to this physical quantity.[35]

This criterion is applied in a particular context – that of two separated systems, S_1 and S_2, with correlations between the values of their magnitudes (established during a previous interaction), so that the measurement of an S_1-magnitude makes it possible to assign a value to the correlated S_2-magnitude, without disturbing S_2 by the S_1-measurement. In quantum mechanics, the composite system $(S_1 + S_2)$ is associated with the Hilbert space $\mathcal{H} = \mathcal{H}_1 \otimes \mathcal{H}_2$. Each Hilbert space, $\mathcal{H}$, $\mathcal{H}_1$, $\mathcal{H}_2$, may be associated with a corresponding working logic, L, L_1, L_2, through the isomorphism between the Lindenbaum- Tarski algebra of the logic and the partial Boolean algebra of subspaces in the corresponding Hilbert space. I introduce the terms 'S_1-situation' and 'S_2-situation' to distinguish situations represented by S_1-propositions (i.e. by propositions expressed by theoretical sentences in L_1) from situations represented by S_2-propositions Similarly, I refer to 'S_1-facts' and 'S_2-facts'. My reformulation of the criterion of reality is this:

In the case of two separated physical systems, S_1 and S_2, an S_i-situation is an S_i-fact if the S_i-proposition representing this situation can be established as true without thereby altering the truth value of any other proposition representing an S_i-fact.

Einstein, Podolsky, and Rosen explicitly propose their criterion as a *sufficient* condition of reality:

Regarded not as a necessary, but merely as a sufficient, condition of reality, this criterion is in agreement with classical as well as quantum-mechanical ideas of reality.[36]

Thus, whether or not the term 'S_i-fact' is understood as referring to a

q-fact (in the sense attributed to Heisenberg in Section I), the criterion is acceptable, because there is no implication that an S_i-situation is *not* an S_i-fact if the condition is not met. (Presumably, a situation is a c-fact if the criterion is not met, but a situation is not a q-fact if the criterion is not met.) In no sense does this criterion of reality beg the question of the ontological status of micro-objects on Heisenberg's version of the Copenhagen interpretation – not even if Heisenberg's position is understood as involving the thesis that a proposition is neither true nor false unless it can be known to be true or false (where the sense of 'can' here refers to possibilities compatible with the theory of measurement disturbances).

The argument proceeds as follows (the terms 'q-maximal' and 'q-fact' are to be understood in the sense attributed to Heisenberg in Section I): If the quantum theory is complete (according to the condition of completeness) and the Hilbert space vectors are complete state descriptions in the sense that they represent q-maximal totalities of situations, then a position-proposition (i.e. a proposition expressed by a theoretical sentence assigning a very small range of values to the magnitude position) and a momentum-proposition cannot both represent q-facts. For, if both propositions represented q-facts, the q-fact represented by the conjunction of these propositions would be expressible by a consistent theoretical sentence. But the theoretical sentence expressing this conjunction of propositions is inconsistent, as can easily be shown. (It corresponds to the null subspace in Hilbert space.)

Now consider a composite system consisting of two substems, S_1 and S_2, which have interacted at some time so that the values of certain of their magnitudes are correlated. Assume that the Hilbert space vector ψ of the composite system $(S_1 + S_2)$ – a vector in the Hilbert space $\mathscr{H} = \mathscr{H}_1 \otimes \mathscr{H}_2$ – is a complete state description in the above sense. In the example considered by Einstein, Podolsky, and Rosen, ψ does not determine a unique state description for S_1, nor does ψ determine a unique state description for S_2. What ψ does determine are the correlations between S_1-situations and S_2-situations. These correlations are such that the assignment of a truth value to an S_1-position-proposition determines the truth value of an S_2-position-proposition, and conversely. Also, the assignment of a truth value to an S_1-momentum-proposition determines the truth value of an S_2-momentum-proposition, and conversely. Applying the criterion of reality, it follows that an S_2-position-proposition

represents an S_2-fact, because it *can* be established as true without altering the truth value of any other proposition representing an S_2-fact.
Similarly, an S_2-momentum-proposition represents an S_2-fact. The truth
in each case can be established by an S_1-measurement, without altering
the truth values of other propositions representing S_2-facts, if it is assumed that the two systems are separated, so that there is no possibility
of any interaction between them.

Thus, the criterion of reality together with the assumption that the
Hilbert space vector is a complete state description in Heisenberg's sense
leads to a contradiction. Specifically: If the quantum theory is complete,
then a position-proposition and a momentum-proposition cannot both
represent q-facts. But, in a specific case, it follows from the conjunction
of the completeness assumption with the criterion of reality that a position-proposition and a momentum-proposition do both represent q-facts.

In the above argument, the term 'S_2-fact' is to be understood in the
Heisenberg sense of 'q-fact'. Thus, the argument does not establish that
quantum mechanics is incomplete (granted the acceptability of the criterion
of reality) in the sense that the Hilbert space vector does not represent a
maximal (i.e. c-maximal) totality of situations in logical space, but rather
only a totality of knowable situations that is maximal with respect to what
can be known simultaneously (i.e. q-maximal totality of situations). *This
is a presupposition of the argument.* Rather, what is established is that the
completeness assumption/of Heisenberg's interpretation is inconsistent
with a criterion of reality that is consistent with Heisenberg's interpretation. In other words, Heisenberg's interpretation turns quantum mechanics into an inconsistent theory.[37] The Hilbert space vector ψ of the composite system specifies correlations between S_1-situations and S_2-situations,
in such a way that we can establish that a certain S_2-situation is an S_2-q-
fact on the basis of an S_1-measurement. The S_1-measurement itself does
not establish that any S_1-situation is an S_1-q-fact. It establishes that an
S_2-situation is an S_2-q-fact only because of the correlations specified by ψ
and the assumption that measurements on S_1 cannot disturb S_2.

The point is that there *are* no S_1-q-facts, nor are there any S_2-q-facts,
on the assumption that ψ is a complete state description in Heisenberg's
sense. There are only $(S_1 + S_2)$-q-facts, and correlations between the truth
values of propositions representing S_1-situations and propositions representing S_2-situations. But the criterion of reality consistent with Heisen-

berg's interpretation *demands* that certain S_2-situations are S_2-q-facts. This is the contradiction.

It remains to point out that the impossibility of simultaneously determining the truth values of an S_1-position-proposition and an S_1-momentum-proposition by any measurement on S_1 in no way affects the conclusion of the argument. This would only be a valid objection if it were supposed that the conjunction of two propositions each representing a *q-fact* need not itself represent a q-fact, or that which S_2-situations are S_2-q-facts depends on what kinds of measurements we choose to make on S_1 (i.e. whether a position-proposition represents an S_2-q-fact or a momentum-proposition represents an S_2-q-fact depends on whether we choose to establish the truth of an S_1-position-proposition or an S_1-momentum-proposition).

Bohr's interpretation is unaffected by the Einstein-Podolsky-Rosen argument. The criterion of reality presupposes that each sentence in L_1 expresses a proposition and each sentence in L_2 expresses a proposition. In fact, on Bohr's view, if the Hilbert space vector of the composite system is ψ, no L_i-sentence expresses a proposition, because the conditions for the applicability of the concepts associated with L_i-sentences are not satisfied (only L-sentences express propositions). Moreover, the appropriate L_i-sentences express an S_1-position-proposition just in case the corresponding L_2-sentences express an S_2-position-proposition, because the conditions for the applicability of the associated space-time concepts are simultaneously satisfiable and exclude the conditions for the applicability of the energy-momentum concepts. Thus, an S_1-position-measurement (which involves satisfying the conditions for the applicability of S_1-space-time concepts) excludes the possibility of satisfying the conditions for the applicability of S_2-momentum-energy concepts, and hence excludes the possibility that any L_2-sentences express an S_2-momentum proposition.

From the fact that the truth values of an S_1-position-proposition and an S_1-momentum-proposition are not simultaneously decidable empirically, it does *not* follow (in the context of the Einstein-Podolsky-Rosen example) that the truth value of an S_2-position-proposition (or S_2-momentum-proposition) cannot be determined without altering the value of some other S_2-proposition representing a q-fact in Heisenberg's sense, i.e. it does *not* follow that an S_2-position-proposition and an S_2-momentum-proposition cannot both represent q-facts. Indeed, they *must* both represent q-facts, on Heisenberg's view. This is the Einstein-Podolsky-Rosen refutation of Heisenberg's version of the Copenhagen interpretation. However, from the fact that the conditions for L_1-sentences to express an S_1-position-proposition exclude the conditions for L_1-sentences to express an S_1-momentum-proposition, and that the conditions for the applicability of S_1-space-time concepts are satisfiable simultaneously with the conditions for the applicability of S_2-space-time concepts (and exclude the conditions for the applicability of energy-momentum concepts), it *does* follow that no set of facts represented by propositions expressed by L_1 or L_2 sentences can include two facts represented respectively by a position-proposition and a momentum-proposition. This is Bohr's reply.

III. VON NEUMANN'S COMPLETENESS PROOF

The Einstein-Podolsky-Rosen argument shows that a certain completeness thesis is inconsistent with quantum mechanics, that the structure of Hilbert space is such that the vectors in Hilbert space cannot be interpreted as representing totalities of knowable situations that are maximal with respect to what can be known simultaneously in the light of a disturbance theory of measurement, because the disturbance theory is inadequate to explain the quantum mechanical correlations between separated systems. This suggests a general problem of completeness for statistical theories, which von Neumann posed as the problem of distinguishing between two alternative interpretations of the statistics:[38]

I.   The individual systems $s_1, ..., s_N$ of our ensemble can be in different states, so that the ensemble $[s_1, ..., s_N]$ is defined by their relative frequencies. The fact that we do not obtain sharp values for the physical quantities in this case is caused by our lack of information: we do not know in which state we are measuring, and therefore we cannot predict the results.

II.   All individual systems $s_1, ..., s_N$ are in the same state, but the laws of nature are not causal. Then the cause of the dispersions is not our lack of information, but is nature itself, which has disregarded the 'principle of sufficient cause'.

This is the question: Under what conditions is it possible to associate either Case I or Case II *uniquely* with a given statistical theory? Von Neumann's descriptions of the two cases are not formulated very carefully, but we may suppose that what is at issue here is the distinction between a purely epistemological 'ignorance' interpretation of the statistics and an interpretation of the theory as complete, or strongly 'irreducibly statistical' in a formal sense (and not merely in the sense, say, of Heisenberg, that the probability assignments generated in the theory are empirically irreducibly statistical).

Von Neumann proposed the existence or non-existence of dispersion-free probability measures on the algebra of physical magnitudes, *whether or not such measures can be generated via the statistical algorithm of the theory*,[39] as the criterion of demarcation between Case I and Case II, i.e. ɐ statistical theory is inconsistent with a Case I interpretation if and only if there does not exist a dispersion-free probability measure on the algebra of physical magnitudes.

Let us examine this criterion. If the logic of the sentences to which probabilities are assigned by the algorithm of a statistical theory is Boolean, then a dispersion-free probability measure on the partial algebra of physical magnitudes is a 2-valued measure in the usual sense on the Boolean working logic of the theory, and such a measure always exists.[40] Thus, on von Neumann's criterion of demarcation between Case I and Case II, no statistical theory with a commutative algebra of physical magnitudes is formally inconsistent with a Case I interpretation. A Case I interpretation could only be excluded on extra-theoretical grounds.

On the other hand, if the working logic of a statistical theory is non-Boolean, then the notion of a dispersion-free probability measure is undefined. If we understand a dispersion-free probability measure as a 2-valued measure in the usual sense, then trivially there are no such measures on the working logic of the theory, simply because this logic is non-Boolean. If we understand a dispersion-free probability measure as an assignment of probabilities satisfying only the condition:

$$\text{Exp}(A^2) = (\text{Exp}(A))^2$$

for all magnitudes A, then equally obviously such measures always exist· The condition requires that

$$\text{Exp}(P^2) = (\text{Exp}(P))^2$$

for every idempotent magnitude, i.e.

$$\text{Exp}(P) = (\text{Exp}(P))^2$$

so that
$$\text{Exp}(P) = 1 \text{ or } 0.$$

In addition,

$$\text{Exp}(1 - P) = 1 - \text{Exp}(P).$$

where 1 is the unit magnitude in the partial Boolean algebra (corresponding to the set of tautologies in the logic). In terms of the associated logic, a dispersion-free probability measure is a bivalent assignment of truth values, 1 or 0, to the sentences, satisfying only the condition:[41]

$$v(\sim s) = 1 - v(s)$$

and such valuations always exist on a partial Boolean logic.

If we understand a probability measure as a generalized probability in the sense defined in Section I, then the notion of a dispersion-free probability measure is well-defined: it amounts to a bivalent assignment of truth values to the sentences of the partial Boolean logic, which is required to satisfy the usual semantic definition of the logical connectives for *compatible* sentences only, i.e. such that

$$v(\sim s) = 1 - v(s)$$
$$v(s \wedge t) = v(s)\, v(t) \text{ if } s \text{ and } t \text{ are compatible},$$

from which it follows that

$$v(s \vee t) = v(s) + v(t) - v(s)\, v(t) \text{ if } s \text{ and } t \text{ are compatible}.$$

I shall call this assignment of truth values a 'partial Boolean valuation'.

Now, the existence or non-existence of a dispersion-free probability measure in the generalized sense on the partial algebra of magnitudes of a statistical theory is a purely mathematical question, which in principle can be answered for any statistical theory. In the case of quantum mechanics, Kochen and Specker[42] have shown that the partial Boolean algebra of idempotent magnitudes is not embeddable in a Boolean algebra, i.e. that partial Boolean valuations do not exist on the logic.

The non-existence of dispersion-free probability measures follows as corollary to Gleason's theorem[43], that every generalized probability measure on the partial Boolean algebra of idempotent magnitudes of the theory (which is isomorphic to the partial Boolean algebra of subspaces of Hilbert space) is expressible in the form

$$p_W(s) = \mathrm{Tr}(WP_s),$$

where W is a statistical operator and s is the theoretical sentence corresponding to the idempotent magnitude P_s. (There are no dispersion-free statistical operators.) Gleason's theorem holds for Hilbert spaces of 3 or more dimensions.

It is not difficult to prove directly that no dispersion-free probability measure in the generalized sense is definable on the partial Boolean algebra of subspaces of a 3-dimensional Hilbert space, i.e. there is no partial Boolean valuation on the logic associated with $\mathscr{H}_3$.[44]

Only 1-dimensional subspaces are relevant, and so it is convenient to regard the valuation as defining a map assigning a 1 or a 0 to every vector in the space, the same number being assigned to vectors in the same 1-dimensional subspace. Since each vector is assigned a 1 or a 0, there must be pairs of arbitrarily close vectors that are assigned different values. I shall prove that if two vectors are assigned different values, then they cannot be arbitrarily close. This is a contradiction, and so the map does not exist.

The conditions on the map require that for any orthogonal triple of vectors, at least one and only one vector is mapped onto 1, the remaining two vectors being mapped onto 0. If $\{\alpha, \alpha', \alpha''\}$ is an orthogonal triple, and $v(\alpha) = 1$, then $v(\alpha^\perp) = 0$. But

$$\begin{aligned}
v(\alpha^\perp) &= v(P_\alpha\perp) = v(P_{\alpha'} \vee P_{\alpha''}) \\
&= v(P_{\alpha'}) + v(P_{\alpha''}) - v(P_{\alpha'})\, v(P_{\alpha''}) \\
&= v(\alpha') + v(\alpha'') - v(\alpha')\, v(\alpha'')
\end{aligned}$$

and so $v(\alpha') = v(\alpha'') = 0$. It follows that if two vectors α and β are orthogonal, and $v(\alpha) = v(\beta) = 0$, then $v(a\alpha + b\beta) = 0$, for all a, b, i.e. any vector in the plane spanned by α, β is assigned the value 0.

Now, consider any pair of vectors, α and β, such that $v(\alpha) = 1$ and $v(\beta) = 0$. We may suppose, without loss of generality, that α is of unit length, and that $\beta = \alpha + p\alpha'$, where α' is a unit vector orthogonal to α. Let α'' be the remaining member of the orthogonal triple of unit vectors $\{\alpha, \alpha', \alpha''\}$.

If it were possible to demonstrate that $v(\alpha + \alpha'') = v(\alpha - \alpha'') = 0$, then it would follow that $v(2\alpha)=0$, since $\alpha+\alpha''$ and $\alpha-\alpha''$ are orthogonal and sum to 2α. But $v(2\alpha)=v(\alpha)$, which was assumed equal to 1. Now, it would follow that $v(\alpha+\alpha'')=0$ if $\alpha+\alpha''=\gamma+\delta$, where γ is a linear combination of β and α'', and δ is a linear combination of α' and α'', and γ and δ are orthogonal. For this to be possible, it is sufficient that

$$\gamma = \beta + q^{-1}p\alpha'' = \alpha + p\alpha' + q^{-1}p\alpha''$$

and

$$\delta = -p\alpha' + qp\alpha'',$$

with

$$p(q + q^{-1}) = 1, \text{ i.e. } p \leqslant \tfrac{1}{2}.$$

Similarly, if $p \leqslant \tfrac{1}{2}$, there are real values of q such that $p(q+q^{-1})=-1$, i.e. such that $\alpha-\alpha''=\gamma+\delta$. Thus a contradiction follows unless $p \rangle \tfrac{1}{2}$, i.e. there cannot be arbitrarily close pairs of vectors wich are assigned different values.

Von Neumann's own proof[45] of this result is rather curious. What von Neumann apparently sets out to prove is (a) that every generalized probability measure on the subspaces of Hilbert space is expressible in terms of a statistical operator by the formula:

$$p_W(\mathscr{X}) = \text{Tr}(WP),$$

where P is the projection operator whose range is the subspace $\mathscr{X}$, so that the expectation value of a magnitude A is expressible as:

$$\text{Exp}_W(A) = \text{Tr}(WA)$$

(assuming that the partial algebra of physical magnitudes of the theory is isomorphic to the partial algebra of hypermaximal Hermitian operators in Hilbert space), and (b) that no statistical operator is dispersion-free, not even the homogeneous or pure statistical operators. Now, a generalized probability measure requires that

$$\text{Exp}(aA) + \text{Exp}(bB) + \cdots = a\,\text{Exp}(A) + b\,\text{Exp}(B) + \cdots$$

for *compatible* magnitudes A, B, ... Von Neumann makes the additional assumption (and this was first pointed out by J. S. Bell)[46] that the above relation holds for *all* magnitudes. It is quite apparent that he regarded this as an assumption for which some justification is required, because there is a weak attempt to provide an argument for replacing the additivity requirement for compatible magnitudes (which is essentially the defining characteristic of a generalized probability measure in the sense defined above) by the unrestricted additivity requirement:[47]

> But the algorithm of quantum mechanics contains still another operation, which goes beyond the one just discussed: namely, the addition of two arbitrary quantities, which are not necessarily simultaneously observable. This operation depends on the fact that for two Hermitian operators, R, S, the sum $R+S$ is also an Hermitian operator, even if the R, S do not commute, while, for example, the product RS is again Hermitian only in the event of commutativity. In each state ϕ the expectation values behave additively:
>
> $$(\phi, R\phi) + (\phi, S\phi) = (\phi, (R+S)\phi).$$
>
> The same holds for several summands. We now incorporate this fact into our general set-up (at this point not yet specialized to quantum mechanics).

This makes nonsense of the proof, which now amounts to the proof of the trivial (i.e. well-known and easily provable) theorem that *no statistical state in the quantum theory is dispersion-free.* In order to prove (a), von Neumann must solve the problem of specifying all possible probability measures in the generalized sense on Hilbert space. He cannot assume *in advance* that every probability assignment is such that the expectation value of a physical magnitude A is expressible in terms of scalar product expressions of the form $(\psi, A\psi),$

where ψ is a vector in Hilbert space. Of course, Gleason's theorem justifies this assumption for Hilbert spaces of 3 or more dimensions, i.e. if we assume additivity for *compatible* magnitudes only, then it *follows* that:

$$\text{Exp}(A) = \text{Tr}(WA)$$

for some W, so that

$$\text{Exp}(A + B) = \text{Tr}(W(A + B)) = \text{Tr}(WA) + \text{Tr}(WB)$$
$$= \text{Exp}(A) + \text{Exp}(B)$$

for *any* pair of magnitudes, compatible or not. But this is only to say that von Neumann's proof of (a) is justified by the theorem he sets out to prove. More charitably, ʌon Neumann's result is correct for Hilbert spaces of 3 or more dimensions, on the basis of Gleason's theorem, which shows that the statistical operators of quantum mechanics exhaust all possible generalized probability assignments to the partial algebra of physical magnitudes of the theory. The 2-dimensional case is in a sense degenerate – a Case I interpretation is *not* inconsistent with the theory of a spin-$\frac{1}{2}$ particle, for example.[48]

The impossibility of a Case I interpretation of quantum mechanics, except in the 2-dimensional case, means that it is impossible to introduce a 'phase space', Z, and represent each physical magnitude by a real-valued Borel function on Z, in such a way that each maximal compatible set of magnitudes is represented by a set of phase space functions which preserve the functional relationships between the magnitudes. That is to say, if the magnitudes $A_1, A_2, \ldots$ are all functions of the magnitude B, i.e. $A_1 = g_1(B)$, $A_2 = g_2(B), \ldots$, then if B is represented by the phase space function f_B, A_1 is represented by the phase space function $f_{g_1(B)} = g_1(f_B)$, A_2 is represented by the phase space function $f_{g_2(B)} = g_2(f_B)$, etc. – this cannot be achieved. Thus, there is no phase space reconstruction of the quantum statistics which preserves the functional relations between compatible sets of physical magnitudes. Of course, the phase space Z need not be a phase space in the sense that it is parametrized by generalized position and momentum coordinates. It is a phase space in the sense that the points of this space define 2-valued probability measures in the generalized sense, i.e. partial Boolean valuations on the working logic of the theory, or assignments of values to the magnitudes satisfying the above functional relationships.[49]

At this point it might be objected that von Neumann's theorem – or better, Gleason's theorem or the theorem of Kochen and Specker – shows only that the mere statistical equivalence of two magnitudes in the theory does not guarantee equivalence with respect to the actual values of these magnitudes.

A phase space reconstruction of the quantum statistics cannot – and, indeed, ought not – do more than reproduce the equivalence between A_1 and A_2 for those probability measures which correspond to statistical states of the quantum theory. In particular, A_1 and A_2 will not be equivalent for a dispersion-free state, i.e. A_1 and A_2 may be assigned different values by the same point in phase space.

To put the point another way: If we assume that equivalence in the partial algebra is not merely statistical, but represents equivalence with respect to the values assigned to magnitudes which are equivalent in the partial algebra, then no phase space reconstruction of the quantum statistics is possible. On the other hand, we might equally well assume that a phase space reconstruction of a statistical theory is always a possibility (or, at least, a possibility in the case of quantum mechanics), in which case it follows that equivalence in the partial algebra is merely statistical, and cannot be extended to the values of the magnitudes.

This is a crucially important point, to which I shall return in Section VI. Here I wish only to emphasize that the issue raised by von Neumann concerns the question of excluding a Case I interpretation of a statistical theory, i.e. a phase space reconstruction of the statistics.

There are other analyses of the completeness problem which I find quite inadequate. In the following sections I shall discuss two rival completeness theorems.

IV. THE JAUCH AND PIRON COMPLETENESS PROOF

There is, firstly, the work of Jauch and Piron.[50] For definiteness, I shall confine my remarks to the discussion in Jauch's book.[51] For various reasons (which I am not criticizing here), Jauch proposes a lattice theoretical generalization of Hilbert space quantum mechanics. That is to say, the theory is developed as a generalized probability calculus on a complete, orthocomplemented, weakly modular, atomic lattice of 'propositions', where a proposition is formally 'a class of physical yes-no experiments, all of which measure the same proposition.'[52] The probability measures on this non-Boolean lattice are generalized probabilities, defined by the following conditions:

(1) $0 \leqslant p(a) \leqslant 1$,   for every $a \in \mathscr{L}$

(2) $p(o) = 0$, $p(I) = 1$, where O and I are the minimum and maximum elements of $\mathscr{L}$

(3)   If $\{a_i\}$ is a sequence of orthogonal elements in $\mathscr{L}$ (i.e. $a_i \leqslant a_k \perp$, $i \neq k$), then $p(\bigvee_i a_i) = \sum_i p(a_i)$

(4)   For any sequence $\{a_i\}$, if $p(a_i) = 1$ for all i then $p(\bigwedge_i a_i) = 1$

(5a)   If $a \neq 0$, then there exists a probability assignment such that $p(a) \neq 0$

(5b)   If $a \neq b$, then there exists a probability assignment such that $p(a) \neq p(b)$.

Now, two elements in the lattice are compatible just in case they generate a Boolean sublattice under the operations of infimum, supremum, and orthcomplementation. Equivalently, a and b are compatible just in case there exist three mutually orthogonal elements, a_1, b_1, c, such that

$$a = a_1 \vee c$$
$$b = b_1 \vee c$$

from which it follows that

$$a_1 = a \wedge c^\perp$$
$$b_1 = b \wedge c^\perp$$
$$c = a \wedge b$$

or

$$a = \left(a \wedge (a \wedge b)^\perp\right) \vee (a \wedge b)$$
$$b = \left(b \wedge (a \wedge b)^\perp\right) \vee (a \wedge b).$$

Thus, if a and b are compatible,

$$a \vee b = (a \wedge b^\perp) \vee (a \wedge b) \vee (b \wedge a^\perp),$$

where $(a \wedge b^\perp)$, $(a \wedge b)$, $(b \wedge a^\perp)$ are mutually orthogonal. By (3), then:

$$p(a \vee b) = p(a \wedge b^\perp) + p(a \wedge b) + p(b \wedge a^\perp)$$

and also:

$$p(a) + p(b) = p(a \vee b) + p(a \wedge b)$$

so that:

$$2p(a \wedge b) = p(a) + p(b) - p(a \wedge b^\perp) - p(b \wedge a^\perp)$$

if a and b are compatible. From this expression it is easy to show that:

$$p_i(a \wedge b) = 0 \quad \text{if} \quad p(a) = 1, \quad p(b) = 0$$
$$p_*(a \wedge b) = 0 \quad \text{if} \quad p(a) = 0, \quad p(b) = 1$$
$$p(a \wedge b) = 0 \quad \text{if} \quad p(a) = 0, \quad p(b) = 0$$

and it follows from (4) that

$$p(a \wedge b) = 1 \quad \text{if} \quad p(a) = 1, \quad p(b) = 1.$$

The dispersion of a probability measure for the proposition a is defined as:

$$\sigma(a) = p(a) - p^2(a)$$

A probability measure is dispersion-free if

$$\sigma = \sup_{a \in \mathscr{L}} \sigma(a) = 0$$

in which case:

$$p(a) = 1 \text{ or } 0$$

Hence, a dispersion-free probability measure defines a map $v: \mathscr{L} \to \{0,1\}$ such that:

$$v(a^\perp) = 1 - v(a)$$
$$v(a \wedge b) = v(a) \wedge (b) \text{ for compatible } a, b$$

with the additional requirement that

$$v(a \wedge b) = 1 \quad \text{if} \quad v(a) = v(b) = 1$$

whether or not a and b are compatible.

Jauch proves the theorem that every generalized probability measure (in Jauch's sense) is expressible as a weighted integral of dispersion-free measures, only if all the elements of $\mathscr{L}$ are pairwise compatible, i.e. only if $\mathscr{L}$ is Boolean. He remarks:[53]

The conclusion of the theorem is seen to be very strong, since it affirms compatibility for *all* pairs of propositions. Thus it suffices to exhibit a single pair of noncompatible propositions to establish that hidden variables are empirically refuted. Now we have seen that the occurrence of non-compatible propositions is the essence of quantum mechanics, since the lattice is Boolean and the system behaves classically if every pair of propositions is compatible. Because of this result we may simply affirm: a quantum system cannot admit hidden variables in the sense in which we have defined them. With this result the quest for hidden variables of this particular kind has found its definitive answer in the negative.

But this theorem is uninteresting as a solution to von Neumann's problem. It is not a lattice theoretical generalization of the Gleason or Kochen and Specker theorem for Hilbert space quantum mechanics. The possibility of expressing every generalized probability measure on a non-Boolean lattice as a weighted integral of dispersion-free measures is immediately excluded by condition (4) *even in the spin-1/2 case, i.e. $\mathscr{L}_2$.* If a and b are incompatible in this case, $a \wedge b = 0$, i.e. $p(a \wedge b) = 0$. If $p(a) = 1$, it follows (by condition (4)) that $p(b) = 0$ for every b incompatible with a, and so, trivially, no statistical state of the theory assigning a finite probability to any proposition *b* incompatible with a can be expressed as a weighted integral of dispersion-free probability measures in Jauch's sense, because *every single one of these dispersion-free measures necessarily (by condition (4)) assigns a probability of O to b.*

V. BELL'S COMPLETENESS PROOF

J. S. Bell was the first to isolate this condition of Jauch and Piron as suspect in the context of an enquiry concerning the possibility of a phase space reconstruction of the quantum statistics.[54] For Bell, the problem at issue is the development of a viable 'hidden variable' theory as an alternative to quantum mechanics, and not primarily the question of interpretation. He argues that a hidden variable theory cannot be excluded on the basis of arguments of the kind proposed by von Neumann, Jauch and Piron, or Gleason, because in each case the assumptions are unreasonably restrictive: The arguments assume, to a greater or lesser extent, an equivalence with respect to the values assigned to statistically equivalent magnitudes by hypothetical dispersion-free probability measures, whereas a hidden variable theorist might very plausibly assume that equivalence in the partial algebra of quantum magnitudes is merely statistical.

It will be urged that these analyses leave the real question untouched. In fact it will be seen that these demonstrations require from the hypothetical dispersion-free states, not only that appropriate ensembles thereof should have all measurable properties of quantum mechanical states, but certain other properties as well. These additional demands appear reasonable when results of measurement are loosely identified with properties of isolated systems. They are seen to be quite unreasonable when one remembers with Bohr 'the impossibility of any sharp distinction between the behaviour of atomic objects and the interaction with the measuring instruments which serve to define the conditions under which the phenomena appear'.[55]

Now, von Neumann's problem was the problem of excluding a Case I interpretation of the quantum theory, i. e. a phase space reconstruction of the quantum statistics. In Section III I pointed out that we could assume either (a) that equivalence in the partial algebra of magnitudes of quantum mechanics is not merely statistical, but represents equivalence with respect to the values assigned to statistically equivalent magnitudes, in which case it follows from Gleason's theorem that no phase space reconstruction of the quantum statistics is possible, or (b) that a phase space reconstruction of the quantum statistics is possible, in which case it follows from

Gleason's theorem that equivalence in the partial algebra is merely statistical, and cannot be extended to the values of the magnitudes.

A phase space reconstruction of the quantum statistics is a hidden variable theory of quantum phenomena, in the sense that the dispersion free probability measures represented by the phase space points are variables which play a hidden role in determining the statistical relations of the quantum theory (cf. classical statistical mechanics). By a hidden variable theory one might also understand a classical (i. e. non-quantum) theory which is empirically adequate with respect to the domain of experimental phenomena currently explained by quantum mechanics, but which contradicts quantum mechanics for a class of experiments which have not yet been performed (and might perhaps be technologically unfeasable at the present time). The Gleason or Kochen and Specker theorem says nothing at all about the possibility or impossibility of such a theory. Indeed, there could be no *a priori* grounds for excluding a hidden variable theory in this sense, but the hypothesis that the quantum theory is false is quite uninteresting in this context and of no relevance to the problem of interpretation considered by von Neumann.

For Bell, the possibility of hidden variables underlying the quantum statistics can be excluded only if it can be shown that a physically reasonable condition on the hidden variable theory conflicts with the ability of the theory to recapture the statistical relations of quantum mechanics. Bell proposes a proof of the impossibility of reproducing the statistical relations of quantum mechanics *precisely or arbitrarily closely* by a *local* hidden variable theory. This result is interpreted as showing that no phase space theory satisfying a *locality assumption* can reproduce the quantum statistics, whether or not the functional relationships between compatible sets of physical magnitudes are preserved, i. e. whether or not the equality $f_{g(A)} = g(f_A)$ holds in general (where f_A, $f_{g(A)}$ are the phase space functions representing the magnitudes A, $g(A)$, respectively). And this is regarded as a significant improvement over previous proofs, because the rejection of hidden variable theories is based on "a very small (and very rational) restriction on the nature of hidden variables."[56] Presumably the onus is now on the hidden variable theorist to show either the superiority of a non-local hidden variable theory over quantum mechanics, or that the quantum theory is in fact false with respect to the statistical relations referred to in Bell's proof. Clauser *et al.*[57] have proposed an experiment which in principle will decide whether the second alternative is open or not.

The condition Bell imposes on a hidden variable theory is that in the case of an experiment involving two separated systems, S_1 and S_2, as in the example considered by Einstein, Podolsky, and Rosen, the *result* of a measurement on S_1 cannot be affected by the *kind* of measurement performed on S_2 (e.g. whether it is a position measurement or a momentum measurement). That is to say, the value of an S_1-magnitude A_1 depends only on the values of the hidden variables and the experimental arrangement at S_1, and not at all on the experimental arrangement at S_2, if S_1 and S_2 are sufficiently separated in space. Bell demonstrates, quite correctly and ingeniously, that no hidden variable theory satisfying this condition can recover the quantum statistics, not even arbitrarily closely.

Now, to parody Bell, it seems to me that

these additional demands appear reasonable when results of measurement are loosely identified with properties of isolated systems. They are seen to be quite unreasonable when one remembers with Bohr 'the impossibility of any sharp distinction between the behaviour of atomic objects and the interaction with the measuring instruments which serve to define the conditions under which the phenomena appear.'

Why should the hidden variable theorist assume that the value assigned to a magnitude represented by an operator expressible in the form $A_1 \otimes I_2$ in the case of a Hilbert space $\mathscr{H}_1 \otimes \mathscr{H}_2$ (where A_1 is an operator in $\mathscr{H}_1$ and I_2 is the unit operator in $\mathscr{H}_2$) is independent of the value assigned to a magnitude represented by an operator $A_2 \otimes I_1$? Clearly, this assumption depends on *some* interpretation of the magnitudes represented by Hilbert space operators, or rather, the claim that this assumption is reasonable involves an interpretation of the magnitudes.

On might interpret Bell's result as a refutation of a Bohrian[58] interpretation of quantum mechanics. I have presented the Einstein-Podolsky-Rosen argument as a refutation of Heisenberg's interpretation of the theory, according to which a measurement on a microsystem establishes the value of a physical magnitude associated with the system alone, i. e. establishes that the system possesses a certain property, or that a certain proposition is true of the system. The non-existence of dispersion-free statistical states is explained by a theory of irreducible measurement disturbances, i.e. the geometrical relations of Hilbert space are interpreted as

a generalization of the disturbance relations exhibited in the γ-ray microscope thought experiment.

Now, one might suppose that Bohr's interpretation differs from Heisenberg's in the following respect: The magnitudes represented by Hilbert
space operators are interpreted as *relations* holding between macroscopic
measuring instruments and micro-systems, or, more precisely, a measurement on a micro-system establishes that a certain relation holds between
the measuring instrument *qua* macro-environment and the micro-object
measured. On this view, incompatible magnitudes represent incompatible
relations holding between a micro-system and different macro-environments. The Einstein-Podolsky-Rosen argument does not exclude this interpretation of quantum mechanics: From the possibility of establishing
a relation of a certain kind between S_1 and an appropriate macroscopic
measuring instrument without influencing the macro-instrument at S_2, and
from the fact that a particular relation of this kind is associated uniquely
with a particular relation of the same kind between S_2 and an appropriate
macro-instrument, in the case that the macro-environment at S_2 constitutes such a measuring instrument, it does not follow that two incompatible relations hold simultaneously between S_2 and its environment. By
extending the Einstein-Podolsky-Rosen argument to the statistical correlations between values of S_1-magnitudes and values of S_2-magnitudes, it
seems that Bell has shown that this interpretation is consistent with the
quantum statistics only in the case of measurements of the same kind at
S_1 and S_2. Thus, the Bohrian only apparently escapes the Einstein-Podolsky-Rosen objection.

Bell considers a composite system, $S_1 + S_2$, consisting of a pair of spin-
1/2 particles formed in the singlet spin state and moving freely in opposite
directions. He denotes the spin of S_i by σ_i, and the spin component in a
particular direction by $\sigma_i \cdot \mathbf{a}$, where $\mathbf{a}$ is a unit vector. The spin component
$\sigma_i \cdot \mathbf{a}$ is measured by a macro-instrument at S_i, say a Stern-Gerlach magnet. On the relational theory of quantum magnitudes, $\sigma_i \cdot \mathbf{a}$ represents a
relation of a certain kind between S_i and an appropriate macro-environment, a Stern-Gerlach magnet oriented along the direction $\mathbf{a}$. The value
of $\sigma_i \cdot \mathbf{a}$ in a particular case, i. e. ± 1, represents a particular relation of
this kind. Thus, the quantum statistical relations refer to a statistical en-
semble, the individuals of which are pairs of system-apparatus wholes,
i.e. composites of the form $S_1 \cdot M_1 + S_2 \cdot M_2$, where M_i represents the Stern-

Gerlach magnet at S_i. The value of $\sigma_i \cdot \mathbf{a}$ is determined by two parameters: $\mathbf{a}$ and a parameter ranging over individuals in the ensemble, say λ. Bell writes the value of $\sigma_1 \cdot \mathbf{a}$ as A and the value of $\sigma_2 \cdot \mathbf{a}$ as B, so that:

$$A(\mathbf{a}, \lambda) = \pm 1$$
$$B(\mathbf{b}, \lambda) = \pm 1.$$

The locality assumption is simply that A is independent of $\mathbf{b}$ and B is independent of $\mathbf{a}$.

The expectation value of the product of $\sigma_1 \cdot \mathbf{a}$ and $\sigma_2 \cdot \mathbf{b}$ is

$$P(\mathbf{a}, \mathbf{b}) = \int A(\mathbf{a}, \lambda) B(\mathbf{b}, \lambda) p(\lambda) \, d\lambda,$$

where $p(\lambda)$ is the probability distribution over λ. The Einstein-Podolsky-Rosen argument applied to this system would exploit only the quantum mechanical relations between the value of $\sigma_1 \cdot \mathbf{a}$ and the value of $\sigma_2 \cdot \mathbf{a}$: if $\mathrm{val}(\sigma_1 \cdot \mathbf{a}) = 1$, then $\mathrm{val}(\sigma_2 \cdot \mathbf{a}) = -1$, and conversely. Bell shows that the above expression for $P(\mathbf{a}, \mathbf{b})$ is quite consistent with the relation

$$P(\mathbf{a}, \mathbf{a}) = - P(\mathbf{a}, -\mathbf{a}) = -1$$

and

$$P(\mathbf{a}, \mathbf{b}) = 0 \quad \text{if} \quad \mathbf{a} \cdot \mathbf{b} = 0$$

but inconsistent with the quantum statistics in the general case, where $\mathbf{a}$ and $\mathbf{b}$ are not collinear and not perpendicular, i. e. no distribution over λ – no distribution over individuals $S_1 \cdot M_1 + S_2 \cdot M_2$ – can reproduce the quantum mechanical expectation value:

$$P(\mathbf{a} \cdot \mathbf{b}) = - \mathbf{a} \cdot \mathbf{b}$$

Commenting on Gleason's theorem, Bell remarks:[59]

It was tacitly assumed that measurement of an observable must yield the same value independently of what other measurements may be made simultaneously. ...The result of an observation may be reasonably depend not only on the state of the system (including hidden variables) but also on the complete disposition of the apparatus; see again the quotation from Bohr.

On this interpretation of Bell's result, the position excluded by Bell's extension of the Einstein-Podolsky-Rosen argument is the view here attributed to Bohr: a relational interpretation of the quantum magnitudes

Now, one might argue that the inconsistency of a relational interpretation of quantum magnitudes with the theory poses, in the first place, a difficulty for the interpretation of quantum mechanics, and only derivatively an obstacle for the hidden variable theorist. It is just *because* a relational interpretation is ruled out that no hidden variable theory explicitly committed to such an interpretation can recover the quantum statistics.

Implicit in Bell's analysis of the problem of hidden variables (and the related work of Wigner, Clauser *et al.*) is the suggestion of a hierarchy of proofs of the impossibility of hidden variable theories, each involving less restrictive assumptions than the preceding proof. Thus, von Neumann's proof is followed by the proofs of Gleason and Kochen and Specker, which avoid the additivity assumption but retain the Hilbert space. Jauch and Piron formulate their proof in the context of a generalized lattice theoretical version of quantum mechanics. All these proofs involve restrictions on the values assigned to quantum magnitudes that go beyond the purely statistical relations of quantum mechanics, and a hidden variable theorist might reject these restrictions. Finally, Bell's proof depends only on a physically plausible locality assumption that a hidden variable theorist could not reasonably reject.

This hierarchy completely distorts the problem and the contributions of the various authors. Fundamentally, the problem is the interpretation of quantum mechanics and not the problem of developing a rival hidden variable theory. The proof of Kochen and Specker is an independent proof of a corollary to Gleason's theorem, a result that apparently solves a specific sub-problem posed by von Neumann – the problem of excluding a Case I interpretation of quantum mechanics. The theorem of Jauch and Piron is a muddled and abortive attempt to improve von Neumann's theorem. It seems that Bell's result is not a contribution to the debate on von Neumann's problem after all. Rather, it appears to be an extension of the Einstein-Podolsky-Rosen argument in such a way as to exclude a relational interpretation of quantum mechanics, and hence relational hidden variable theories. The characterization of these theories as 'local' hidden variable theories is inappropriate without qualification. The first characteristic of such a theory is that the value of a magnitude represented by a Hermitian operator in Hilbert space represents a *relation* between a microsystem and its macro-environment, so that the value of this magnitude is determined by two parameters: the 'state' of the micro-system and the 'state' of the macro-environment. This is a 2-level theory – a macro-system is not reducible to a complex of interacting microsystems. The second characteristic of the theory is that the macro-environment of a micro-system need not include the whole universe. This is the locality assumption. It seems that the class of hidden variable theories excluded by Bell's result is the class of non-degenerate 2-level relational theories, in which quantum magnitudes are represented as relations between a microsystem and its macro-environment, and two distinct micro-systems can have different macro-environments.

I say *it seems* that this is so, because I do not believe that this is the correct interpretation of Bell's result, in spite of the fact that the relational interpretation is strongly suggested by Bell's own remarks. For this interpretation completely trivializes the result. In the singlet spin state, the systems S_1 and S_2 are *mirror images* of each other. If the value of the spin component $\sigma_1 \cdot \mathbf{a}$ of S_1 is $+1$, the value of the spin component $\sigma_2 \cdot \mathbf{a}$ of S_2 is -1. Moreover, the probability that the value of the spin component $\sigma_2 \cdot \mathbf{b}$ is $+1$, *given that the value of $\sigma_1 \cdot a$ is $+1$*, is $1/2 \sin^2 1/2\, \theta ab$, where θab is the angle between the directions $\mathbf{a}$ and $\mathbf{b}$. That is, the probability is equal to the product of the probability that the value of $\sigma_2 \cdot \mathbf{a}$ is -1 with the probability that the value of $\sigma_2 \cdot \mathbf{b}$ is $+1$, *given that the value of $\sigma_2 \cdot a$ is -1*, or

$$p\left(b_1^2 \mid a_1^1\right) = P\left(a_2^2\right) P\left(b_1^2 \mid a_2^2\right) = 1/2 \sin^2 1/2\theta_{ab},$$

where a_1^1 is the theoretical sentence val $(\sigma_1 \cdot \mathbf{a}) = +1$, a_2^2 is the sentence val $(\sigma_2 \cdot \mathbf{a}) = -1$, and b_1^2 is the sentence val $(\sigma_2 \cdot \mathbf{b}) = +1$.

Now, it is obviously impossible to reproduce mirror-image correlations of this sort on the basis of a theory in which the value of $\sigma_1 \cdot \mathbf{a}$ is determined by two parameters (λ and $\mathbf{a}$) and the value of $\sigma_1 \cdot \mathbf{b}$ is determined by two parameters (λ and $\mathbf{b}$), *with the requirement that a and b are independently variable* – this is the assumption that A, the value of $\sigma_1 \cdot \mathbf{a}$ is independent of $\mathbf{b}$, and B, the value of $\sigma_2 \cdot \mathbf{b}$, is independent of $\mathbf{a}$. Clearly, if B is a function of a parameter $\mathbf{b}$ which is *required* to be independent of the parameter $\mathbf{a}$, of which A is a function, the probability distribution of B cannot depend on $\mathbf{a}$. But, according to quantum mechanics, the probability distribution of B does depend on $\mathbf{a}$. Hence, trivially, a (non-degenerate) relational theory cannot reproduce the mirror image statistical correlations of quantum mechanics.

Bell remarks[60] that a relational hidden variable theory can reproduce the quantum mechanical correlations in the special case considered by Einstein, Podolsky, and Rosen, i.e. when $\mathbf{b}=\mathbf{a}$, or $\mathbf{b}=-\mathbf{a}$, or $\mathbf{a}\cdot\mathbf{b}=0$. Of course! If the measurement or environment parameters are essentially the *same* at S_1 and S_2, a relational theory can reproduce mirror-image correlations. Similarly, a degenerate relational theory (and only a *degenerate* relational theory) can reproduce the mirror-image correlations in general.

Bell concludes:

In a theory in which parameters are added to quantum mechanics to determine the results of individual measurements, without changing the statistical predictions, there must be a mechanism whereby the setting of one measuring device can influence the reading of another instrument, however remote. Moreover, the signal involved must propagate instantaneously, so that such a theory could not be Lorentz invariant.

This echoes his comment on Bohm's hidden variable theory[61], that

in this theory an explicit causal mechanism exists whereby the disposition of one piece of apparatus affects the results obtained with a distant piece. In fact, the Einstein-Podolsky-Rosen paradox is resolved in the way which Einstein would have liked least.

But this conclusion is utterly trivial if the hidden variable theory considered is a relational theory, for it amounts to the claim that only a degenerate relational theory – a 'non-local' hidden variable theory – can reproduce the mirror image correlations of the quantum statistics in the Einstein-Podolsky-Rosen experiment.

I now want to propose an alternative interpretation of Bell's result, making use of Wigner's elegant reformulation of the argument[62]. Wigner begins by pointing out that a hidden variable theory which introduces a different hidden variable for each quantum mechanical magnitude, and also for each distinct sequence of magnitudes, can reproduce the quantum statistics. Obviously, an absurdly *ad hoc* theory of this sort can reproduce any statistics at all. Recipe: Associate different domains of values of the hidden variable λ_A (corresponding to the magnitude A) with the various possible values of A. Treat a sequence of magnitudes as a new magnitude, with a distinct eigenvalue for each possible sequence of values of the component magnitudes. Let each hidden variable form a dimension of the 'phase space'. Introduce a probability measure on this space for each quantum mechanical statistical state as the product measure of factors of the form $\mu_{\psi A}$, where $\mu_{\psi A}$ is a probability measure defined on the 1-dimensional subspace parametrized by λ_A. Postulate that the value of the magnitude A is determined by the variable λ_A, irrespective of the values of any other hidden variables, and that the probability of a particular value of A is the measure, defined by $\mu_{\psi A}$, of the domain associated with that value.

This is, in effect, what I have called the degenerate relational theory, because the value of any magnitude is determined by two parameters – the position in the 'phase space', and the projection of this position vector onto a particular axis corresponding to the magnitude measured. Two magnitudes associated with two separate subsystems are treated as a distinct third magnitude: it is in this sense that the theory is a *degenerate* relational theory. Now Wigner remarks that

it is very surprising that an apparently very small (and very natural) restriction on the nature of the hidden variables renders it impossible to define a distribution which gives for certain measurements (actually nine measurements) the same probabilities as follow from quantum mechanical theory.

The nine measurements considered are measurements of the spin in three directions, **a**, **b**, **c**, on two spin-1/2 particles. Let the magnitudes involved be A^1, B^1, C^1; A^2, B^2, C^2 (corresponding to Bell's $\sigma_1 \cdot \mathbf{a}$, $\sigma_1 \cdot \mathbf{b}$, $\sigma_1 \cdot \mathbf{c}$; $\sigma_2 \cdot \mathbf{a}$, $\sigma_2 \cdot \mathbf{b}$, $\sigma_2 \cdot \mathbf{c}$). Each sequence of magnitudes ($A^1 \cdot A^2$, $A^1 \cdot B^2$, $A^1 \cdot C^2$, $B^1 \cdot A^1$, etc.) is treated as a distinct magnitude, with four possible values, corresponding to spin measurements $++$, $+-$, $-+$, $--$. The degenerate relational theory introduces a different hidden variable for each such sequence of magnitude, i. e. nine hidden variables, or nine dimensions of 'phase space'. Each dimension is divided into four domains, corresponding to the four possible values of the magnitude. Thus, there are 4^9 domains in the 9-dimensional subspace of 'phase space' corresponding to these measurements.

Bell's locality assumption has the consequence that instead of the 4^9 different domains of 'phase space' corresponding to these measurements, there are only 2^6 different domains. Instead of treating each sequence of magnitudes $A^1 \cdot A^2$, $A^1 \cdot B^2$, ..., as a distinct magnitude requiring a new dimension of 'phase space', only six hidden variables are introduced, i. e. six 'phase space' dimensions corresponding to the six magnitudes A^1, B^1, C^1, A^2, B^2, C^2, because now the value of the spin of S_1 is regarded as independent of the value of the spin of S_2. There are 2^6 domains, because each dimension is divided into two domains, corresponding to the two possible values of the magnitude.

A brief argument[63] now shows that in the case of the singlet spin state, i.e. mirror image correlations, no probability measure over the 6-dimensional subspace of 'phase space' can reproduce the quantum mechanical statistics. Let the possible values of A, B, C be $a_1, a_2; b_1, b_2; c_1, c_2$; where a_1, b_1, c_1 are the spin eigenvalues for positive spin in the $\mathbf{a}$, $\mathbf{b}$, $\mathbf{c}$ directions, respectively, and a_2, b_2, c_2 are the spin eigenvalues for negative spin. Let $(a_i^1,\ b_j^1,\ c_k^1;\ a_1^2,\ b_m^2,\ c_n^2)$ denote the *probability* (defined by the probability measure over 'phase space') of the domains corresponding to these values of the variables, where the superscript refers to the subsystem (S_1 or S_2).

In the case of the singlet spin state, only eight of the 2^6 regions are assigned non-zero probabilities. (The probability $(a_i^1,\ b_j^1,\ c_k^1;\ a_1^2,\ b_m^2,\ c_n^2)$ $=0$ if $i=1$ or $j=m$ or $k=n$.) The probability that the value of A^1 is a_1^1, and C^2 is c_1^2 is:

$$p(a_1^1 \cdot c_1^2) = \sum_{jklm} (a_1^1, b_j^1, c_k^1; a_1^2, b_m^2, c_1^2)$$
$$= (a_1^1, b_1^1, c_2^1; a_2^2, b_2^2, c_1^2) + (a_1^1, b_2^1, c_2^1; a_2^2, b_1, c_1)$$
$$= w + z, \text{ say}$$

but

$$p(b_1^1 \cdot c_1^2) = (a_1^1, b_1^1, c_2^1; a_2^2, b_2^2, c_1^2) + (a_2^1, b_1^1, c_2^1; a_1^2, b_2^2, c_1^2)$$
$$= w + x, \text{ say}$$

and

$$p(a_1^1 \cdot b_1^2) = (a_1^1, b_2^1, c_1^1; a_2^2, b_1^2, c_2^2) + (a_1^1, b_2^1, c_2^1; a_2^2, b_1^2, c_1^2)$$
$$= y + z, \text{ say}$$

according to quantum mechanics

$$p(a_1^1 \cdot c_1^2) = \tfrac{1}{2} \sin^2 \tfrac{1}{2}\theta_{ac}, \text{ where } \theta_{ac} \text{ is the angle between } \mathbf{a} \text{ and } \mathbf{c}$$
$$p(b_1^1 \cdot c_1^2) = \tfrac{1}{2} \sin^2 \tfrac{1}{2}\theta_{bc}$$
$$p(a_1^1 \cdot b_1^2) = \tfrac{1}{2} \sin^2 \tfrac{1}{2}\theta_{ab}$$

It follows that

$$p(a_1^1 \cdot c_1^2) \leqslant p(b_1^1 \cdot c_1^2) + p(a_1^1 \cdot b_1^2),$$

i. e.

$$\sin^2 \tfrac{1}{2}\theta_{ac} \leqslant \sin^2 \tfrac{1}{2}\theta_{ab} + \sin^2 \tfrac{1}{2}\theta_{bc}.$$

But, as Wigner shows, this condition is violated whenever $\mathbf{b}$ bisects the angle between $\mathbf{a}$ and $\mathbf{c}$ (in fact, whenever $\mathbf{a}$, $\mathbf{b}$, and $\mathbf{c}$ are coplanar).

At first sight, Wigner's argument appears to be no more than a rather

neat way of deriving Bell's conclusion. The important difference is this:
Wigner's analysis evidently applies equally well to *any* phase space reconstruction of the quantum statistics, and not merely to a relational theory
in which the domains of the 'phase space' correspond to hidden variables
associated with the quantum mechanical magnitudes. In any phase space
theory there will be different domains corresponding to the various possible values of the magnitudes. We can denote the probability of the phase
space domain corresponding to the values a_i^1, b_j^1, c_k^1; a_1^2, b_m^2, c_n^2 by
the symbol $(a_i^1, b_j^1, c_k^1; a_1^2, b_m^2, c_n^2)$, irrespective of the way in which
the space is parametrized. Thus, Wigner has apparently shown that a
phase space reconstruction of the quantum statistics in excluded, without
a restriction corresponding to Kochen and Specker's assumption: $f_{g(A)}$
$= g(f_A)$. This appears to vindicate Bell, in particular Bell's evaluation of
his result relative to the theorems of Gleason and Kochen and Specker.
The achievement is all the more surprising, because the argument would
appear to apply just as well to a single spin-1/2 particle, and according to
Kochen and Specker a phase space reconstruction of the quantum statistics *is* possible in this case. Indeed, they explicitly produce such a reconstruction.[64]

Consider the three magnitudes *A*, *B*, *C* for a single spin-1/2 particle.
There are eight regions of phase space corresponding to the possible results a_i, b_j, c_k of these magnitudes, i.e. eight non-zero probabilities. Let
$p(c_1 \mid a_2)$ denote the *conditional* probability that the value of *C* is c_1, given
that the value of *A* is a_2. Then the probability that a measurement of *A*
yields the result a_2 and a subsequent measurement of *C* yields the result
c_1 ought to be given by:

$$p(a_2)\, p(c_1 \mid a_2) = (a_2, b_1, c_1) + (a_2, b_2, c_1).$$

Also:

$$p(b_2)\, p(c_1 \mid b_2) = (a_1, b_2, c_1) + (a_2, b_2, c_1)$$

and

$$p(a_2)\, p(b_1 \mid a_2) = (a_2, b_1, c_1) + (a_2, b_1, c_2).$$

It follows that:

$$p(a_2)\, p(c_1 \mid a_2) \leqslant p(b_2)\, p(c_1 \mid b_2) + p(a_2)\, p(b_1 \mid a_2),$$

i.e.

$$p(c_1 \mid a_2) \leqslant p(b_2)/p(a_2)\, p(c_1 \mid b_2) + p(b_1 \mid a_2)$$

and so

$$p(c_1 \mid a_2) \leqslant p(c_1 \mid b_2) + p(b_1 \mid a_2)$$

if

$$p(b_2) \leqslant p(a_2).$$

Hence, again:

$$\sin^2 \tfrac{1}{2}\theta_{ac} \leqslant \sin^2 \tfrac{1}{2}\theta_{ab} + \sin^2 \tfrac{1}{2}\theta_{bc}$$

In their demonstration that a phase space reconstruction of the quantum statistics is possible for $\mathscr{H}_2$, Kochen and Specker are concerned to show only that the probabilities assigned by any statistical state of quantum mechanics to the theoretical sentences may be computed, in the case of $\mathscr{H}_2$, by a probability measure over phase space, where a point in phase space assigns a value to every magnitude, i.e. a truth value to every sentence. Nothing at all is said about *conditional* probabilities. This is the clue to the relation between Bell's result and the theorem of Kochen and Specker.

Let $p(s \cdot t)$ denote the probability that a first measurement satisfies the sentence s (i.e. s is found to be true in the measurement) and a subsequent measurement satisfies t. I use the notation $p(s \mid t)$ for the conditional probability of s given t, and $p(s \wedge t)$ for the joint probability of s and t, i.e. the probability that the conjunction, $s \wedge t$, is true. Generally, in quantum mechanics

$$p(s \cdot t) \neq p(t \cdot s)$$

unless s and t are compatible. For example,

$$p_\psi(a_1 \cdot b_1) = |(\alpha_1, \psi)|^2 \, |(\beta_1, \alpha_1)|^2$$
$$p_\psi(b_1 \cdot a_1) = |(\beta_1, \psi)|^2 \, |(\alpha_1, \beta_1)|^2 . \text{ [65]}$$

Since

$$|(\beta_1, \alpha_1)|^2 = |(\alpha_1, \beta_1)|^2$$

and

$$|(\alpha_1, \psi)|^2 \neq |(\beta_1, \psi)|^2$$

it follows that

$$p_\psi(a_1 \cdot b_1) \neq p_\psi(b_1 \cdot a_1).$$

In a phase space theory, however,

$$p(s) = \mu(\Phi_s),$$

where Φ_s is the set of phase space points satisfying s. And

$$p(s \mid t) = \mu(\Phi_s \cap \Phi_t)/\mu(\Phi_t)$$

i.e. the probability of s given t is the measure of the set of phase space points satisfying s in the set Φ_t, with respect to a renormalized measure assigning a probability of 1 to the set Φ_t. But then

$$p(s \mid t) = \frac{p(s \wedge t)}{p(t)}$$

and so

$$p(t \cdot s) = p(t)\, p(s \mid t) = p(s \wedge t) = p(s \cdot t).$$

It follows that if we take $p(t \cdot s)$ as equal to $p(t)\, p(s \mid t)$, the phase space rule for conditional probabilities must be violated in a phase space reconstruction of the quantum statistics. Instead of computing $p_W(s \mid t)$ according to the rule:

$$p_W(s \mid t) = \mu_W(\Phi_s \cap \Phi_t)/\mu_W(\Phi_t) = \mu'_W(\Phi_s)$$

we must use the rule:

$$p_W(s \mid t) = \mu_{W'}(\Phi_s) = \mu_{W'}(\Phi_s \cap \Phi_t),$$

where $\mu_{W'}$ is the phase space measure associated with the statistical state W' and the transition $W \to W'$ is given by the projection postulate. For example,

$$p_\psi(b_1 \mid a_1) = \mu_{a_1}(\Phi_{b_1}) = \mu_{a_1}(\Phi_{b_1} \cap \Phi_{a_1}).$$

The difference between μ'_W and $\mu_{W'}$ is this: μ'_W is the *original* measure renormalized to the set Φ_t, whereas $\mu_{W'}$ is a *uniform* probability measure over the set Φ_t. Both measures assign probability 1 to the set Φ_t, i.e.

$$\mu'_W(\Phi_t) = \mu_{W'}(\Phi_t) = 1$$

but the relative probability of subsets in Φ is different. According to the straight phase space rule, the relative probability of subsets in Φ_t is unchanged by the additional information that t is true. But in a phase space reconstruction of the quantum statistics, we must assume that any initial information concerning the relative probability of subsets in Φ_t is somehow invalidated by the additional information that t is true (or false). The fact that an initial probability measure is reduced or 'collapses' to the set Φ_t is not problematic here. What is problematic is that this reduction is accompanied by a *randomization* process, i. e. the reduced probability measure becomes uniform over Φ_t

It is only if we use the straight phase space rule for conditional probabilities that the analogue of Bell's result can be derived in the case of a single spin-1/2 particle. Here the computation is perhaps too transparent to seriously embarrass a hidden variable theorist. It is immediately obvious that there must be something wrong in taking $p(t \cdot s)$ as equal to $\mu(\Phi_s \cap \Phi_t)$, because $p(t \cdot s) \neq p(s \cdot t)$ in general if s and t are incompatible. What the computation amounts to is this:

$$p(a_2 \cdot c_1) = p(a_2) p(c_1 \mid a_2) = \mu(\Phi_{a_2} \cap \Phi_{c_1})$$
$$p(b_2 \cdot c_1) = p(b_2) p(c_1 \mid b_2) = \mu(\Phi_{b_2} \cap \Phi_{c_1})$$
$$p(a_2 \cdot b_1) = p(a_2) p(b_1 \mid a_2) = \mu(\Phi_{a_2} \cap \Phi_{b_1}).$$

Since

$$\mu(\Phi_{a_2} \cap \Phi_{c_1}) \leqslant \mu(\Phi_{b_2} \cap \Phi_{c_1}) + \mu(\Phi_{a_2} \cap \Phi_{b_1}) \text{ [66]}$$

it follows that

$$p(a_1) p(c_1 \mid a_2) \leqslant p(b_2) p(c_1 \mid b_2) + p(a_2) p(b_1 \mid a_2)$$

and hence

$$\sin^2 \tfrac{1}{2}\theta_{ac} \leqslant \sin^2 \tfrac{1}{2}\theta_{bc} + \sin^2 \tfrac{1}{2}\theta_{ab}$$

if

$$p(b_2) \leqslant p(a_2).$$

In the mirror-image 2-particle case, however, the sequential probability $p(a_1^1 \cdot c_1^2) = p(a_1^1) p(c_1^2 \mid a_1^1)$ is equal to the joint probability $p(a_1^1 \wedge c_1^2)$, because a_1^1 and c_1^2 are compatible sentences, and also equal to $p(a_2^2) \cdot p(c_1^2 \mid a_2^2)$ – this is the sense in which S_2 is the mirror image of S_1. The analogous computation is:

$$p(a_1^1 \cdot c_1^2) = p(a_1^1 \wedge c_1^2) = \mu(\Phi_{a_1 1} \cap \Phi_{c_1 2})$$
$$p(b_1^1 \cdot c_1^2) = p(b_1^1 \wedge c_1^2) = \mu(\Phi_{b_1 1} \cap \Phi_{c_1 2})$$
$$p(a_1^1 \cdot b_1^2) = p(a_1^1 \wedge b_1^2) = \mu(\Phi_{a_1 1} \cap \Phi_{b_1 2})$$

and so

i. e.

$$\mu(\Phi_{a_1^1} \cap \Phi_{c_1^2}) \leqslant \mu(\Phi_{b_1^1} \cap \Phi_{c_1^2}) + \mu(\Phi_{a_1^1} \cap \Phi_{b_1^2})\ ^{67}$$

$$p(a_1^1 \cdot c_1^2) \leqslant p(b_1^1 \cdot c_1^2) + p(a_1^1 \cdot b_1^2)$$
$$p(a_1^1)\, p(c_1^2 \mid a_1^1) \leqslant p(b_1^1)\, p(c_1^2 \mid b_1^1) + p(a_1^1)\, p(b_1^2 \mid a_1^1)$$
$$p(a_2^2)\, p(c_1^2 \mid a_2^2) \leqslant p(b_2^2)\, p(c_1^2 \mid b_2^2) + p(a_2^2)\, p(b_1^2 \mid a_2^2)$$
$$p(a_2^2 \cdot c_1^2) \leqslant p(b_2^2 \cdot c_1^2) + p(a_2^2 \cdot b_1^2)$$
$$\tfrac{1}{2}\sin^2\theta_{ac} \leqslant \tfrac{1}{2}\sin^2\theta_{bc} + \tfrac{1}{2}\sin^2\theta_{ab}.$$

It appears that the computation cannot be rejected here, because the straight phase space rule for conditional probabilities does not apply. The Bell-Wigner argument is that since S_2 is the mirror image of S_1, the sequential probability $p(a_2^2 \cdot c_1^2)$ *must* be computable as:

$$p(a_2^2 \cdot c_1^2) = \mu(\Phi_{a_1^1} \cap \Phi_{c_1^2})$$

and the significance of the inequality which cannot be satisfied in general is that this expression for $p(a_2^2 \cdot c_1^2)$ cannot be equal to the sequential probability computed according to the quantum mechanical rule for conditional probabilities. Thus, the Bell-Wigner argument is that the straight phase space rule for conditional probabilities cannot be replaced by the quantum mechanical rule *if there exist composite systems, $S_1 + S_2$, where S_2 is the mirror image of S_1.*

The quantum mechanical rule for conditional probabilities is essentially the projection postulate. The problem of conditional probabilities – the problem of giving an explanation for this rule, which appears puzzling – is well-known (if not well-named) as the *measurement problem*. Classically,

$$p(s) = p(t)\, p(s \mid t) + p(\sim t)\, p(s \mid \sim t)$$

but this is not the case in general in quantum mechanics. For example:

$$p_\psi(a_1) = |(\alpha_1, \psi)|^2$$
$$p_\psi(b_1) = |(\beta_1, \psi)|^2$$
$$p_\psi(b_2) = |(\beta_2, \psi)|^2$$
$$p_\psi(a_1 \mid b_1) = |(\alpha_1, \beta_1)|^2$$
$$p_\psi(a_1 \mid b_2) = |(\alpha_1, \beta_2)|^2.$$

Now,

$$\psi = (\beta_1, \psi)\, \beta_1 + (\beta_2, \psi)\, \beta_2$$

so

$$(\alpha_1, \psi) = (\beta_1, \psi)(\alpha_1, \beta_1) + (\beta_2, \psi)(\alpha_1, \beta_2)$$

i.e.

$$|(\alpha_1, \psi)|^2 = p_\psi(a_1) = p_\psi(b_1)\,p_\psi(a_1 \mid b_1) +$$
$$+ p_\psi(b_2)\,p_\psi(a_1 \mid b_2) + \text{'interference terms'}.$$

Usually one says that the system has been disturbed by the measurement of B, which transforms the pure state, ψ, into a mixture of pure states, β_1 and β_2, with weights $|(\beta_1, \psi)|^2$ and $|(\beta_2, \psi)|^2$, respectively. The probability of the value a_1 of A in the pure ensemble specified by ψ is different from the probability in the mixed ensemble.

In a phase space theory, the measurement problem is transformed into the problem of providing an explanation for the assumption that an initial probability measure is randomized as well as reduced (or re-normalized) with additional information. But the fact that the problem is merely transformed and not resolved by a phase space reconstruction of the statistics in no way shows the impossibility of such a reconstruction. There is no inconsistency in replacing the straight phase space rule for conditional probabilities by the quantum mechanical rule. What is immediately obvious is that in the case of *incompatible* sentences, s and t, the joint probability distribution

$$p_W(s \wedge t) = \mu_W(\Phi_s \cap \Phi_t)$$

bears no direct relation to the conditional probability:

$$p_W(s \mid t) = \mu_{W'}(s).$$

One might suppose that in case of *compatible* sentences

$$p_W(s \mid t) = \frac{p_W(s \wedge t)}{p_W(t)} = \frac{\mu_W(\Phi_s \cap \Phi_t)}{\mu_W(\Phi_t)}.$$

What the Bell-Wigner calculation shows (and this is *all* that it shows) is that *this relation does not hold in general even for compatible sentences.* (It fails for compatible sentences referring to two different subsystems in the mirror-image case.)

Now, the phase space theorist is perfectly at liberty to define joint probabilities in terms of sequential probabilities in a phase space reconstruction of the quantum statistics, *even for compatible sentences*. The sequential probability is defined by:

$$p_W(s \cdot t) = p_W(s) \, p_W(t \mid s)$$
$$= \mu_W(\Phi_s) \, \mu_{W'}(\Phi_t),$$

where the transition $W \to W'$ is given by the projection postulate. The joint probability, $p_W(s \wedge t)$, is undefined, unless

$$p_W(s \cdot t) = p_W(t \cdot s)$$

in which case

$$p_W(s \wedge t) = p_W(s \cdot t) = p_W(t \cdot s).$$

Thus, the measure $\mu_W(\Phi_s \cap \Phi_t)$ will have no direct relation to the joint probability $p_W(s \wedge t)$ in general.

This is sufficient to block the Bell-Wigner argument. For now in the case of the two particles in the singlet spin state

$$\psi = \frac{1}{\sqrt{2}} \alpha_1^1 \alpha_2^2 - \frac{1}{\sqrt{2}} \alpha_2^1 \alpha_1^2$$
$$p_\psi(a_1^1 \cdot c_1^2) = p_\psi(a_1^1 \wedge c_1^2) \neq \mu_\psi(\Phi_{a_1^1} \cap \Phi_{c_1^2}).$$

Rather:
$$p_\psi(a_1^1 \cdot c_1^2) = p_\psi(a_1^1) \, p_\psi(c_1^2 \mid a_1^1) = \mu_\psi(\Phi_{a_1}) \, \mu_{a_1^1}(\Phi_{c_1^2})$$
$$= \mu_\psi(\Phi_{a_1}) \, \mu_{a_2^2}(\Phi_{c_1^2}) = \tfrac{1}{2} \sin^2 \tfrac{1}{2}\theta_{ac}$$

and the inequality cannot be derived.

It cannot be objected that the rule for the calculation of probabilities is mysterious here, tantamount to a violation of the locality condition, because the calculation simply reflects the consistent application of the quantum mechanical rule for conditional probabilities as opposed to the straight phase space rule. Since the straight phase space rule is evidently inapplicable in a phase space reconstruction of the quantum statistics, it is entirely natural to define $p_W(s \cdot t)$ as $\mu_W(\Phi_s) \, \mu_{W'}(\Phi_t)$, so that the equality

$$\mu_W(\Phi_s) \, \mu_{W'}(\Phi_t) = \mu_W(\Phi_s \cap \Phi_t)$$

which holds for some s and t, is without theoretical significance. At any rate, there is no *more* reason to reject this calculation in a phase space theory than there is to reject the randomization assumption. If the response is that the straight phase space rule *must* be applied in a phase space theory, then *this* is the condition on the basis of which hidden variable theories are excluded, and not the locality assumption. And I cannot see that this condition is in any way more plausible than the Kochen and Specker assumption: $f_g(_A) = g(f_A)$. Rather, this condition (which has nothing whatsoever to do with a locality problem) excludes hidden variables as immediately and as trivially as the condition of Jauch and Piron rejected by Bell.

One final comment: The Clauser-Horne-Shimony-Holt experiment [68] now seems pointless. What it amounts to is a crucial test between quantum mechanics and a phase space theory in which sequential probabilities are calculated according to *both* rules [69]

(i) $$p_W(s \cdot t) = p_W(\Phi_s \cap \Phi_t)$$

and

(ii) $$p_W(s \cdot t) = p_W(\Phi_s) \cdot p_{W'}(\Phi_t)$$

in the case of compatible sentences. But these rules are inconsistent, at least if the theory is to describe the quantum statistical correlations between mirror-image pairs. If the experimental results were to contradict quantum mechanics, this would throw doubt on the existence (or at least the stability) of mirror-image pairs at the microlevel, and not on the appropriateness of replacing the straight phase space rule by the quantum mechanical rule in a hidden variable theory. And the confirmation of quantum mechanics in this experiment could only exclude the assumption that in a phase space theory

$$p(s \wedge t) = \mu(\Phi_s \cap \Phi_t)$$

in the case of compatible sentences (the equality would have to fail for compatible sentences referring to different subsystems in mirror-image pairs). Certainly, the experiment could not possibly exclude any significant class of hidden variable theories.

VI. THE PROBLEM OF COMPLETENESS

In Section I, I characterized a statistical theory as involving a set of physical magnitudes forming an algebraic structure of a certain kind, together with an algorithm for assigning probabilities to ranges of possible values of these magnitudes. I shall refer to the given algebraic structure of the idempotent magnitudes as the *logical space* $\mathscr{L}_1$ of a statistical theory.

The relation of statistical equivalence (as defined in Section I) is an equivalence relation on $\mathscr{L}_1$. With sums and products of magnitudes defined for compatible magnitudes, in terms of the compatibility relation defined via this equivalence relation, the magnitudes of the theory form a partial algebra, and the idempotent magnitudes form a partial Boolean algebra. I shall refer to the partial Boolean algebra of idempotents defined in this way as the *logical space* $\mathscr{L}_2$ of a statistical theory.

Evidently, the two algebraic structures, $\mathscr{L}_1$ and $\mathscr{L}_2$, are different. *I shall say that a statistical theory is complete if and only if $\mathscr{L}_1$ and $\mathscr{L}_2$ are isomorphic.* For this is the case if and only if the statistical states of the theory generate all possible probability measures on the logical space $\mathscr{L}_1$. I shall refer to the problem of demonstrating isomorphism as *the completeness problem for a statistical theory.*

There are two possible interpretations of the role of the Hilbert space in quantum mechanics. The first interpretation takes the partial Boolean algebra of subspaces of Hilbert space as the logical space $\mathscr{L}_1$ of microevents. A vector in Hilbert space (or rather, a 1-dimensional subspace) then represents an elementary event, not a statistical state. The problem of specifying all possible probability measures in the generalized sense on such an event structure has been solved by Gleason. Gleason's theorem states that in a Hilbert space of 3 or more dimensions, all possible probability measures in the generalized sense on the partial Boolean algebra of subspaces may be generated by the statistical operators, W, according to the algorithm:

$$p_W(\mathscr{X}) = \mathrm{Tr}\,(WP),$$

where P is the projection operator onto the subspace $\mathscr{X}$. It follows immediately that the logical spaces $\mathscr{L}_1$ and $\mathscr{L}_2$ are isomorphic: Gleason's theorem solves the completeness problem for quantum mechanics. Kochen and Specker pointed out a further corollary to this theorem: Because there are no dispersion-free statistical operators, no dispersion-free probability measure in the generalized sense is definable on $\mathscr{L}_1$, except in the case of a 2-dimensional Hilbert space. It follows that there are no partial Boolean valuations on the logical space $\mathscr{L}_1$, that it is impossible to embed $\mathscr{L}_1$ in a Boolean algebra, and hence that there is no phase space reconstruction of the quantum statistics which preserves the functional relations between compatible sets of physical magnitudes.

The second interpretation takes the Hilbert space as the space of statistical states of a statistical theory, with each unit vector ψ representing a statistical state for the algorithm

$$p_\psi\left(\mathrm{val}\,(A)\in S\right) = \|P_A\,(S)\,\psi\|^2$$

'Mixtures' of such 'pure' states may be defined as more general statistical states, specifying probability assignments representable as weighted sums of probability assignments generated by pure ψ-states. It follows that there is a many-one correspondence between mixtures and statistical operators (so that the decomposition of a 'mixture' specified by a statistical operator into its constituent pure states is not unique). On the second interpretation, then, the Hilbert space is in effect taken as the logical space $\mathscr{L}_2$ of a statistical theory.

Now on this interpretation of the Hilbert space the completeness problem is undefined. And whether there are hidden variables underlying the statistics captured by the quantum algorithm depends on what we take as the logical space $\mathscr{L}_1$. I shall refer to these two interpretations as the *logical interpretation* of the Hilbert space and the *statistical interpretation* of the Hilbert space respectively.

There can be no motive for the statistical interpretation other than the prejudice for a Boolean logical space $\mathscr{L}_1$, i.e. the implicit assumption is that the Boolean character of logic is *a priori*. Then quantum mechanics is incomplete, because $\mathscr{L}_1$ and $\mathscr{L}_2$ are not isomorphic. It seems to me that this assumption is the basis of the Einstein-Podolsky-Rosen argument. In 1950, Einstein wrote to Schrodinger:[70]

If one wants to consider the quantum theory as final (in principle), then one must believe that a more complete description would be useless because there would be no laws for it. If that were so then physics could only claim the interest of shopkeepers and engineers; the whole thing would be a wretched bungle.

The parenthetical qualification 'in principle' is sufficient to show that Einstein was not at all concerned with the pseudo-problem of whether the quantum theory is complete in the sense of being the *last word*, the ultimate rock-bottom theory, the final story about the building-blocks of the universe. This is an absurd notion, a question that does not deserve serious consideration. Rather, Einstein saw clearly that a statistical theory excluding dispersion-free probability measures cannot possibly be complete if the logical space $\mathscr{L}_1$ is Boolean, because dispersion-free probability measures do exist on a Boolean algebra. The Copenhagen interpretation of quantum mechanics involves the statistical interpretation of Hilbert space, together with the claim that the theory is complete. The realist version of the Copenhagen interpretation is what I have reconstructed as Heisenberg's version in Section I.[71] On this view, the logical space of micro-events is Boolean, but only ensembles generated by the statistical states of quantum mechanics are *in fact* constructable. More precisely defined ensembles corresponding, say, to dispersion-free probability measures, would involve an over-specification of physical parameters, in the sense that the additional parameters required for such a description would be random, i.e. not subject to physical law. But if this were so, the universe would be 'irrational' – quantum physics would reduce to a collection of rules-of-thumb for describing the 'theoretically opaque' statistical behaviour of micro-systems. This is a broad philosophical objection to Heisenberg's interpretation. The Einstein-Podolsky-Rosen argument is a more precise objection to Heisenberg's measurement theory, the proposed explanation for the non-existence of dispersion-free ensembles.

Bohr's version of the Copenhagen interpretation is a Kantian position. For a Kantian, there is no logical space $\mathscr{L}_1$ of events – there is only the logical space $\mathscr{L}_2$, if the theory is a truly fundamental theory of mechanics. The structure of this space is *a priori*, representing a necessary presupposition of all experience. The transition from classical to quantum mechanics represents, ultimately, a discovery about ourselves, about our conceptual framework for the description and communication of experience. What we have discovered is that the conditions for the applicability of space-time and dynamical concepts are not simultaneously satisfiable. Thus we see that the structure of logical space *ought* to be non-Boolean. Clearly, the completeness problem as I have formulated it does not exist for a Kantian. There is no inconsistency in this position, but it seems to me uninteresting. Recall Einstein's remark about shopkeepers and engineers.

Von Neumann seems at first to distinguish (at least implicitly) between the logical spaces $\mathscr{L}_1$ and $\mathscr{L}_2$, but his formulation of the *general* problem as that of demonstrating the non-existence of dispersion-free probability measures shows his prejudice for a Boolean logical space $\mathscr{L}_1$. His proof amounts to the demonstration that the logical space $\mathscr{L}_2$ is not embeddable in a Boolean algebra. To claim that this is a proof of the impossibility of a phase space reconstruction of the quantum statistics, the impossibility of hidden variables underlying the quantum statistics, without showing the isomorphism between $\mathscr{L}_1$ and $\mathscr{L}_2$, is to beg the question at issue, unless (as I suspect) von Neumann regards the Boolean structure of $\mathscr{L}_1$ as *a priori*, in which case his position is very close to Heisenberg's.

Feyerabend comments on von Neumann's proof as follows:[72]

> This proof allows for the application, to individual systems, of probabilities in the sense of relative frequencies. Hence, any operation with an *ensemble* which leads from a statistical operator W to another statistical operator, W', can also be interpreted as an operation with an *individual system* (which may, or may not be completely known), leading from the state W to the state W' and vice versa, a procedure which would not be possible in the classical case.

I take this as an expression of Heisenberg's position. What it amounts to is the claim that ensembles more precisely definable than those generated by the statistical states of quantum mechanics do not in fact exist, because there is an irreducible, 'irrational' randomness in the variables required for their specification. This is the sense in which the behaviour of individual systems is inherently probabilistic.

The theorems of Jauch and Piron, Bell, and Wigner are not contributions to the completeness problem. The Jauch and Piron proof is no more than a demonstration that a particular $\mathscr{L}_2$ logical space is non-Boolean. One might say that Jauch and Piron generalize von Neumann's mistake. The Bell-Wigner result, on the other hand, is relevant to the *measurement problem*.

At the end of Section III, I formulated two alternative interpretations of the Gleason and Kochen and Specker solution to von Neumann's problem as follows: If we assume that equivalence in the working logic of quantum mechanics (the partial Boolean algebra of idempotent magnitudes) represents logical equivalence and not merely statistical equivalence, then no phase space reconstruction of the quantum statistics is possible (in the case of Hilbert spaces of three or more dimensions); on the other hand, if we assume that a phase space reconstruction of the quantum statistics is possible, then equivalence in the working logic represents statistical equivalence of the corresponding idempotent magnitudes, but not logical equivalence, i.e. two equivalent sentences in the partial Boolean algebra may be assigned different truth values.

In terms of the logical spaces $\mathscr{L}_1$ and $\mathscr{L}_2$, the *assumption* of the first alternative is just the logical interpretation of Hilbert space, from which it follows (by completeness) that statistical equivalence (i.e. equivalence in $\mathscr{L}_2$) is logical equivalence (i.e. equivalence in $\mathscr{L}_1$). The second alternative is the statistical interpretation of Hilbert space, with a Boolean structure postulated for the logical space $\mathscr{L}_1$. On the statistical interpretation quantum mechanics is incomplete, and so the proposal is empty without a blue-print for a new Boolean mechanics. Moreover, any such mechanics, simply because $\mathscr{L}_1$ is Boolean and $\mathscr{L}_2$ is non-Boolean, will involve the measurement problem. The measurement problem is characteristic of the statistical interpretation. In Section V, I pointed out that in a Boolean or phase space reconstruction of the quantum statistics, we must assume that any initial information concerning the relative probability of subsets in Φ_t – the set of ultrafilters containing the sentence t, or the set of phase space points satisfying t – is somehow invalidated by the additional information that t is true (or false). What is problematic is that the renormalization of an initial probability measure over the set Φ_t is accompanied by a randomization process, i.e. the renormalized probability measure becomes uniform over Φ_t. Thus, in a phase space theory, the measurement problem becomes the problem of providing an explanation for the assumption that an initial probability measure is randomized as well as renormalized with additional information.

For this reason, I do not regard the statistical interpretation of Hilbert space as providing the basis for an interpretation of quantum mechanics on a par with the logical interpretation. It does not lead to an interpretation of quantum mechanics, but demands a new physics. I am not suggesting that this problem is insoluble – this was the mistake of Bell and Wigner. But certainly the story that explains how the Boolean logical space $\mathscr{L}_1$ is related to the non-Boolean logical space $\mathscr{L}_2$, reflecting the probability structure of quantum mechanics, will be a very different kind of explanation from anything hitherto considered. For in actual fact, not a single hidden variable theory has been proposed in which the slightest attempt has been made to solve the measurement problem.

On the statistical interpretation of the Hilbert space, quantum mechanics is incomplete, because the (necessarily) Boolean logical space $\mathscr{L}_1$ of micro-events is not isomorphic to the logical space $\mathscr{L}_2$ of the theory. It follows that the idempotent magnitudes of the theory cannot directly represent statistical properties of micro-systems, and that the statistical states of the theory refer to a certain set of statistical ensembles, each of which satisfies the uncertainty principle for the magnitudes of the theory, but not for the fundamental magnitudes which are determined by states defined as Boolean ultrafilters, or points in phase space. On this view, a physical magnitude represented by the function $g(A)$ in quantum mechanics is defined by the condition:

$$p_W\left(\mathrm{val}\left(g(A)\right)\in S\right) = p_W\left(\mathrm{val}(A)\in g^{-1}(S)\right)$$

for every statistical operator W in Hilbert space and every Borel set S, and it need not be the case that the sentences associated with the idempotent magnitudes $P_{g(A)}(S)$ and $P_A(g^{-1}(S))$ are logically equivalent in L_1, i.e. assigned the same truth value by every Boolean valuation.

For example, in $\mathscr{H}_3$ consider the magnitudes:

$$A = a_1 P_1 + a_2 P_2 + a_3 P_3$$
$$B = P_1$$
$$C = c_1 P_1 + c_2 Q_2 + c_3 Q_3,$$

where P_1, P_2, P_3 are three idempotent magnitudes represented by mutually orthogonal projection operators, and P_1, Q_2, Q_3 is another such orthogonal triple, with Q_2 and Q_3 incompatible with P_2 and P_3. The idempotent magnitude B is a function of both A and C in the statistical sense, i.e. B is compatible with A and C, but A is incompatible with C. Although the sentences a_1, b_1, c_1 associated with the idempotent magnitudes:

$$P_{a_1} = P_A(\{a_1\}) = P_A(g_1^{-1}\{1\})$$
$$P_{b_1} = P_B(\{1\})$$
$$P_{c_1} = P_C(\{c_1\}) = P_C(g_2^{-1}\{1\})$$

are equivalent in the logic L_2, i.e. $P_{a_1} = P_{b_1} = P_{c_1}$ in $\mathscr{L}_2$, they are not assigned the same truth value by every Boolean valuation. That there is no Boolean valuation (or even partial Boolean valuation) on $\mathscr{L}_2$ is simply taken as a confirmation of the incompleteness of the theory on this view. The complete theory will represent $g_1(A)$ and $g_2(C)$ (both statistically equivalent to the magnitude B in quantum mechanics) as different idempotent magnitudes in $\mathscr{L}_1$, i.e. different phase space functions, $f_{g_1(A)}$ and $f_{g_2(C)}$, while reproducing, in terms of phase space measures, the statistical equivalence:

$$p_W\left(\text{val}\left(g_1(A)\right)\in S\right) = \mu_W\left(f_{g_1(A)}^{-1}(S)\right) = \mu_W\left(f_{g_2(C)}^{-1}(S)\right)$$
$$= p_W\left(\text{val}\left(g_2(C)\right)\in S\right)$$

for every Borel set S and every phase space measure *corresponding to a statistical state in quantum mechanics*.

The various hidden variable theories all amount to different schemes for 'completing' quantum mechanics in this sense. They are 'classical' precisely because they involve the rejection of the logical interpretation of Hilbert space, and the stipulation that the structure of the logical space $\mathscr{L}_1$ is Boolean.

In his original hidden variable theory,[73] Bohm pointed out the irrelevance of von Neumann's proof

since in our interpretation of measurements of the type that can now be carried out, the distribution of hidden parameters varies in accordance with the different mutually exclusive experimental arrangements of matter that must be used in making different kinds of measurements.

Now, tailoring the phase space probability measure to the relevant maximal Boolean sub-algebra in the logical space $\mathscr{L}_2$ is formally equivalent to introducing a fixed measure for each quantum mechanical state, and representing a single quantum magnitude (which is a function of various incompatible maximal magnitudes) by different phase space functions, i.e. it is formally equivalent to replacing each theoretical sentence in the logic L_2 by a family of sentences (one for each maximal Boolean sub-algebra) in the Boolean logic L_1. The distinction between Bohm's original theory and the more recent Bohm-Bub theory[74] is just this: The Bohm-Bub theory introduces a fixed phase space measure for each quantum statistical state, and *a different map associating phase space points with the truth values of sentences for each maximal Boolean sub-algebra*. Evidently, relativizing the bivalent valuations to a maximal Boolean sub-algebra is formally equivalent to introducing a different sentence for each maximal Boolean sub-algebra.

The maps are generated by the equation of motion, which involves the phase point (represented by the pair $(\langle\psi, \xi\rangle)$ and a set of operators which specify the maximal Boolean sub-algebra. Given the phase point and a maximal Boolean sub-algebra, the equation describes the transition to a new phase point which assigns particular values to the relevant set of compatible magnitudes, i.e. particular truth values to the sentences val$(A) \in S$ associated with the maximal Boolean sub-algebra. So, the equation of motion plays the role of an algorithm for assigning truth values to the sentences associated with any maximal Boolean sub-algebra in $\mathscr{L}_1$, for any point in phase space. It is a simple matter to introduce a probability measure on this phase space which generates the quantum statistics for each quantum statistical state.

The logical interpretation of the Hilbert space leads to an immediate solution to the measurement problem. On this interpretation, the significance of the transition from classical to quantum mechanics is understood as the proposal – on empirical grounds – that the logical space of events (micro-events) is non-Boolean. Just as the significance of the transition from classical to relativistic mechanics lies in the proposal that geometry can play the role of an explanatory principle in physics, that the geometry of events is not *a priori*, and that it makes sense to ask whether the world geometry is Euclidean or non-Euclidean, so the significance of the quantum revolution lies in the proposal that logic can play the role of an explanatory principle, that logic is similarly not *a priori*.[75]

I shall say that a logic characterizes a 'Boolean logical space' if and only if the Lindenbaum-Tarski algebra is embeddable in a Boolean algebra, and in general that two (or more) logics characterize the same logical space if and only if their Lindenbaum-Tarski algebras are embeddable in the algebraic (or topological) structure characteristic of that space. The embeddability criterion for sameness of logical space means essentially that two logics characterize the same logical space if and only if they generate the same set of tautologies. Thus, the logical space will be the same for two logics which differ purely syntactically, or merely with respect to the interpretation of the connectives (say, a standard versus a non-standard interpretation).

It is clear, then, that the structure of logical space is not parasitic on the syntactic properties of a formalized language. To say that the logical space of events is Boolean or non-Boolean is to specify a very basic structural feature of events or states of affairs, a certain fundamental symmetry of a very specific kind. The tautologies characterize the logical space because they are in a sense the invariants for the set of possible logical transformations. It is important to see that the structure of this space is not conventional. Nor is the logical space *a priori* in the sense that the laws of logic characterize necessary features of any linguistic framework suitable for the description and communication of experience. Ultimately, logic is about the world, not about language. The interpretation of quantum mechanics which takes the Hilbert space as specifying a non-classical logical space is not the proposal that we change our usual patterns of inference in order to avoid anomalies in our description of micro-world. It is not the claim that the language of quantum mechanics involves the adoption of a new set of logical conventions, which are more appropriate for communication about micro-events than the standard logical conventions of ordinary discourse. Rather, the thesis is that quantum mechanics posits a non-Boolean logical structure to events – that the theory is non-classical in the very radical sense of involving a non-Boolean logical space.

The interpretation of quantum mechanics based on the logical interpretation of Hilbert space resolves the measurement problem, because it can be shown that the quantum mechanical rule for conditional probabilities – the projection postulate – is the appropriate rule for the non-Boolean logical space $\mathscr{L}_1$, and not the Boolean or phase space rule.[76] The logical interpretation of Hilbert space was initially proposed by Putnam[77], whose article should be read in conjunction with the mathematical investigations of Kochen and Specker.[78] The work on 'quantum logic' by mathematicians and physicists such as Birkhoff and von Neumann[79], Jauch and Piron[80], Finkelstein[81], Mackey[82], Varadarajan[83], Gudder[84], etc., with the primary aim of providing new axiomatic foundations for quantum mechanics, while important as a contribution to mathematical physics, confuses the logical spaces $\mathscr{L}_1$ and $\mathscr{L}_2$ and so contributes nothing to the interpretation problem, or the traditional conceptual problems of quantum mechanics. Philosophers such as Suppes[85], von Fraassen[86], Fine[87], etc., have also failed to see the distinction between $\mathscr{L}_1$ and $\mathscr{L}_2$. The title of Suppes' article, 'The Probabilistic Argument for a Non-Classical Logic of Quantum Mechanics' gives the game away. Suppes adopts the statistical interpretation of Hilbert space, and his demonstration that the 'working logic' of quantum mechanics is non-Boolean shows only that the logical space $\mathscr{L}_2$ is non-Boolean. Thus, Suppes' 'quantum logical interpretation' is the very antithesis of Putnam's interpretation. (Ironically,

Suppes remarks in an apparently critical aside that Kochen and Specker "deliberately exclude all probability questions in their consideration of the logic of quantum mechanics"[88].) Similarly, the so-called 'modal interpretation' of van Fraassen is no more than a philosopher's formulation of the statistical interpretation. It is in no sense a rival quantum logical interpretation.

To sum up: In this section I have formulated a general completeness problem for statistical theories, in terms of which the controversy concerning the completeness of quantum mechanics can be seen to hinge on the implicit assumption that logic is *a priori* and Boolean. Once the *a priori* status of logic is rejected, once it is recognized that logic is as empirical as geometry, it becomes clear that a *realist* interpretation of quantum mechanics as a *complete* theory demands the logical interpretation of Hilbert space. This analysis shows that the original motivation for the hidden variable program – the argument that quantum mechanics cannot be a complete theory – depends ultimately on a certain philosophy of logic, according to which the structure of the logical space L_1 is necessarily Boolean. Since this is so obviously a metaphysical prejudice of the worst kind, the development of a Boolean physics is seen to be without theoretical justification. Only the empirical inadequacy of the quantum theory could now justify the search for a Boolean physics – and even so the onus is on the hidden variable theorist to show that the failure of quantum mechanics to account satisfactorily for the behaviour of micro-systems is symptomatic of its non-Boolean structure, and not of some other feature of the theory.

APPENDIX

A *Boolean algebra* is a non-empty set on which two binary operations, meet and join, and one unary operation, complementation, are defined which have properties analogous to the operations of intersection, union, and complementation on the subsets of a fixed set or space. More precisely, *a Boolean algebra is a complemented distributive lattice.*

A lattice is a partially ordered set X with a greatest lower bound (infimum) and a least upper bound (supremum) defined for every pair of elements in X. The infimum of x and y is denoted by $x \wedge y$ and is defined as the (unique) element $z \in X$ such that $z \leqslant x$ and $z \leqslant y$, and if $w \leqslant x$ and $w \leqslant y$

then $w \leqslant z$. Similarly, the supremum of x and y is denoted by $x \vee y$ and is defined as the (unique) element z such that $x \leqslant z$ and $y \leqslant z$, and if $x \leqslant w$ and $y \leqslant w$, then $z \leqslant w$.

The lower and upper bounds (maximum and minimum elements) of a lattice are unique if they exist, denoted by 0 and I. The *complement* of a lattice element x is defined as an element x' such that

$$x \wedge x' = 0$$
$$x \vee x' = I.$$

The complement x' of x is not necessarily unique. A complemented lattice is a lattice in which every element has a complement. An *orthocomplemented lattice* is a lattice with an operation, denoted by $\perp$ (the 'orthogonal complement' or orthocomplement), satisfying the conditions:

$$(x^{\perp})^{\perp} = x$$
$$x \leqslant y \text{ if and only if } y^{\perp} \leqslant x^{\perp}$$
$$x \wedge x^{\perp} = 0$$
$$x \vee x^{\perp} = 1$$

The orthocomplement $x^{\perp}$ of x is unique if it exists.

A lattice is *distributive* if and only if

$$x \wedge (y \vee z) = (x \wedge y) \vee (x \wedge z)$$
$$x \vee (y \wedge z) = (x \vee y) \wedge (x \vee z)$$

for every $x, y, z \in z$. Complementation is unique in a distributive lattice and has the properties or orthocomplementation.

A complemented distributive lattice is a Boolean algebra. *Every Boolean algebra is isomorphic to a perfect reduced field of sets.* This is *Stone's representation theorem* for Boolean algebras. I shall denote a Boolean algebra by the symbol $\mathscr{B}$, with or without subscripts. To distinguish the elements of $\mathscr{B}$ from the members of the representative set, X, I shall denote elements of $\mathscr{B}$ by the letters $a, b, c, \ldots$ (from the beginning of the alphabet) and members of X by the letters $\ldots x, y, z$ (from the end of the alphabet). The proof of Stone's theorem involves the concept of an ultrafilter in a Boolean algebra.

A *filter* in a Boolean algebra is a non-empty subset Φ of $\mathscr{B}$, satisfying the conditions:

(i)   if $a, b \in \Phi$, then $a \wedge b \in \Phi$
(ii)   if $a \in \Phi$ and $a \leqslant b$, then $b \in \Phi$.

Equivalently, a filter may be defined as a non-empty subset Φ of $\mathcal{B}$ such that $a \wedge b \in \Phi$ if and only if $a \in \Phi$ and $b \in \Phi$. (The dual notion is that of an *ideal*.)

A *proper* filter is a proper subset of $\mathcal{B}$. The *principal* filter *generated* by the element $a \in \mathcal{B}$ is the set $\{b \in \mathcal{B} : a \leqslant b\}$. The set of all filters in $\mathcal{B}$ is partially ordered with respect to the relation of set inclusion. An *ultra-filter* (maximal filter, prime filter) is a proper filter that is maximal with respect to this ordering, i. e. it is not a proper subset of a proper filter in $\mathcal{B}$.

It can be shown that a necessary and sufficient condition for a proper filter Φ in $\mathcal{B}$ to be an ultrafilter is that for every $a \in \mathcal{B}$, either $a \in \Phi$ or $a' \in \Phi$, but not both.

An atom in a Boolean algebra $\mathcal{B}$ is a minimal element of $\mathcal{B}$ that is not equal to the element 0, i. e. an element $a \in \mathcal{B}$ such that there is no element of $\mathcal{B}$ between 0 and a. More precisely, a is an atom if and only if $a \neq 0$ and $b \leqslant a$ implies $b = 0$ or $b = a$. It follows that a is an atom if and only if the principal filter generated by a is an ultrafilter. Equivalently, a is an atom if and only if there is only one ultrafilter containing a.

A map $h: \mathcal{B}_1 \rightarrow \mathcal{B}_2$ is a *homomorphism* if it preserves the algebraic operations, i. e. if for all $a, b \in \mathcal{B}$:

$$h(a \wedge b) = h(a) \wedge h(b)$$
$$h(a \vee b) = h(a) \vee h(b)$$
$$h(a') = h(a)'.$$

It follows that h maps the maximum and minimum elements of $\mathcal{B}_1$ onto the maximum and minimum elements of $\mathcal{B}_2$, and that if $a \leqslant b$ in $\mathcal{B}_1$, then $h(a) \leqslant h(b)$ in $\mathcal{B}_2$. $\mathcal{B}_1$ and $\mathcal{B}_2$ are *isomorphic* if h is one-one and on-to, i. e. if $h(\mathcal{B}_1) = \mathcal{B}_2$, where $h(\mathcal{B}_1)$ is the image of $\mathcal{B}_1$ under the map, the set of elements in $\mathcal{B}_2$ onto which some element of $\mathcal{B}_1$ is mapped by h. A *2-valued homomorphism* on $\mathcal{B}$ is a homomorphism from $\mathcal{B}$ onto a 2-element Boolean algebra, **2**. The set of elements in $\mathcal{B}$ mapped onto $I \in \mathbf{2}$ by a 2-valued homomorphism, i. e. $\{a \in \mathcal{B} : h(a) = I\}$, is an ultrafilter. Conversely, if Φ is an ultrafilter in $\mathscr{B}$, the map $h: \mathscr{B} \to 2$, such that $h(a) = I$ if $a \in \Phi$ and $h(a) = 0$ if $a \notin \Phi$, is a 2-valued homomorphism. Hence, there is a one-one correspondence between the set of ultrafilters in $\mathscr{B}$ and the set of 2-valued homomorphisms on $\mathscr{B}$.

It is a theorem that for every proper filter Φ in a Boolean algebra there exists an ultrafilter that includes Φ, i. e. that is an extension of Φ. (This theorem depends on the Axiom of Choice.) It follows that each element $a \neq 0$ is contained in some ultrafilter, and that if a and b are distinct elements of $\mathscr{B}$, there is an ultrafilter containing a but not b.

A *field* of sets, $\mathscr{F}$, is a non-empty set of subsets of a fixed set, X, closed with respect to finite unions, intersections, and complements. Obviously, a field of sets is a Boolean algebra. A field $\mathscr{F}$ of subsets of a set X is said to be *reduced* if for every distinct pair $x, y \in X$ there exists a set in $\mathscr{F}$ containing x but not containing y. $\mathscr{F}$ is *perfect* if every ultrafilter in $\mathscr{F}$ is determined by a member of X. Stone's theorem may be formulated as follows: If X is the set of ultrafilters in a Boolean algebra $\mathscr{B}$, and $h(a)$ is the subset of ultrafilters in X containing the element $a \in \mathscr{B}$, then the set of all such subsets, $\mathscr{F} = \{h(a):a \in \mathscr{B}\}$, is a perfect reduced field of subsets of X, and h is an isomorphism from $\mathscr{B}$ onto $\mathscr{F}$. The core of this theorem is the lemma that every element $a \neq 0$ of a Boolean algebra is contained in an ultrafilter, i. e. it is the *existence* of this ultrafilter, guaranteed by the Axiom of Choice, that is crucial for the existence of the isomorphism between the Boolean algebra and the field of sets of ultrafilters.

The *Lindenbaum-Tarski algebra* $\mathscr{L}$ of a logic L is generated by first defining an equivalence relation on the set of sentences in L:

$$s \equiv t \quad \text{if and only if} \quad \vdash s \supset t \quad \text{and} \quad \vdash t \supset s,$$

and then defining a relation (transitive, reflexive, and antisymmetric) on the set of equivalence classes of formulas $\{|s| : s \in L\}$, where

$$|s| = \{t \in L : s \equiv t\} :$$
$$|s| \leqslant |t| \quad \text{if and only if} \quad \vdash s \supset t.$$

The partially ordered set of equivalence classes of formulas is the Lindenbaum-Tarski algebra of the logic.

In the case of the classical sentential calculus L_C, for example, $\mathscr{L}$ is a Boolean algebra. The formulas of the sentential calculus are generated from a countable set of atomic formulas or sentences $\{p_i\}$ by the logical operations of negation (denoted by the symbol $\sim$) and conjunction (denoted by the symbol $\wedge$). Other logical operations or connectives may be defined in terms of negation and conjunction. For example, the disjunction $s \vee t$ is an abbreviation for $\sim (\sim s \wedge \sim t)$, and the conditional $s \supset t$ is an abbreviation for $\sim (s \wedge \sim t)$, i.e. $\sim s \vee t$. The use of the symbols $\wedge$ and $\vee$ for conjunction and disjunction in L_c as well as for the infinum and supremum in $\mathscr{L}_c$ is justified because it can be shown that:

$$|s \wedge t| = |s| \wedge |t|$$

and

$$|s \vee t| = |s| \vee |t|.$$

and it is obvious from the context whether the symbols denote lattice operations or logical operations.

That $\mathscr{L}_c$ is a Boolean in this case is shown by the syntactic or proof theoretic structure of L_c. The semantics concerns the interpretation of the sentences as truth functions of the atomic sentences, i. e. the truth or falssity of a sentence defined by logical operations on a set of atomic sentences is completely determined by the truth values of the atomic sentences. An interpretation is then a particular assignment of the truth values true and false to the atomic sentences or, equivalently, a map from the set of atomic sentences onto the maximum and minimum elements of a 2-element Boolean algebra, with 0 corresponding to false and I corresponding to true. Formally, an interpretation or *realization* of the classical sentential calculus may be defined as a bivalent valuation, v, on the set of atomic sentences, extended to the set of all sentences by the recursive definition:

$$v(s \wedge t) = v(s) \wedge v(t)$$
$$v(\sim s) = v(s)'$$

so that:

$$v(s \vee t) = v(s) \vee v(t).$$

It follows that a 2-valued homomorphism on $\mathscr{L}_c$ defines a realization of L_c, i.e. a bivalent valuation v on the atomic sentences:

$$v(p_i) = h(|p_i|)$$

that may be extended to the set of all sentences:

$$v(s) = h(|s|)$$

because h is a homomorphism (and so if $v(s) = h(|s|)$ and $v(t) = h(|t|)$, for s and t, then $v(s \wedge t) = v(s) \wedge v(t)$ and $v(\sim s) = v(s)'$). Conversely, every realization on L_C corresponds to a 2-valued homomorphism on $\mathcal{L}_C$. Thus there is also a one-one correspondence between realizations on L_C and, ultrafilters in $\mathcal{L}_C$: An ultrafilter in $\mathcal{L}_C$ corresponds to a *maximal consistent set* of sentences in L_C.

A sentence s is *satisfiable* if and only if it is assigned the value I under some realization; s is *valid* if and only if it is satisfied by all realizations; and s is *inconsistent* if and only if it belongs to the equivalence class $|s| = 0$, i. e. if and only if a sentence and its negation are both provable from s on the basis of the axioms and inference rules of L_C. Hence, s is consistent if and only if $|s| \neq 0$, or $\sim s$ is not provable. The *completeness theorem* states that *s is provable if and only if s is valid*. This is equivalent to the statement that *s is consistent if and only if s is satisfiable*. The completeness theorem is the logical analogue of Stone's representation theorem for Boolean algebras. The Boolean algebra $\mathcal{L}_C$ is isomorphic to the field $\mathcal{F} = \{h(|s|): |s| \in \mathcal{L}_C\}$ of subsets of X, where X is the set X of ultrafilters in $\mathcal{L}_C$ and $h(|s|)$ is the subset of ultrafilters containing $|s|$. Now, if $|s| \neq 0$, then $|s|$ is mapped onto a non-empty subset of $h(|s|)$ in $\mathcal{F}$. Every member of the subset $h(|s|)$ is an ultrafilter containing $|s|$ and so corresponds to a 2-valued homomorphism that maps $|s|$ onto I, i. e. to a realization that satisfies s. Conversely, if s is satisfiable, i. e. if there exists a 2-valued homomorphism assigning the value I to $|s|$, then $|s| \neq 0$.

ACKNOWLEDGEMENTS

Many of the ideas in this paper, especially those concerned with quantum logic, arose in discussions and correspondence with William Demopoulos (University of New Brunswick). A joint paper on quantum logic will be published in the *British Journal for the Philosophy of Science* in 1972. I also gratefully acknowledge stimulating discussions with Donald Hockney and Clifford Hooker (University of Western Ontario), and Bas van Fraassen (University of Toronto). More obviously, I am indebted to David Bohm (although we no longer agree on fundamentals), and to Hilary Putnam, whose article 'Is Logic Empirical?'[89] first aroused me from my dogmatic slumbers.

The University of Western Ontario

* Research supported by a grant from the National Science Foundation, U. S. A.

NOTES

[1] From a letter to Schrödinger, dated May 31, 1928, in *Letters on Wave Mechanics* (ed. by K. Przibram, transl. by M. J. Klein), Philosophical Library, New York, 1967, p. 31.

[2] A. Einstein, B. Podolsky, and N. Rosen, *Physical Review* **46** (1935), 777.

[3] J. von Neumann, *Mathematical Foundations of Quantum Mechanics*, Princeton University Press, 1955, Chapter IV, §§1,2.

[4] A. M. Gleason, *Journal of Mathematics and Mechanics* **6** (1957), 885.

[5] S. Kochen and E. P. Specker, *Journal of Mathematics and Mechanics* **17** (1967), 59.

[6] J. M. Jauch and C. Piron, *Helvetia Physica Acta* **36** (1963), 827. Also J. M. Jauch, *Foundations of Quantum Mechanics*, Addison-Wesley, 1968, Chapter 7.

[7] J. S. Bell, *Physics* **1** (1964), 195. E. P. Wigner, *American Journal of Physics* **38** (1970), 1005.

[8] See Appendix.

[9] See Appendix.

[10] The proposition may be identified with the set of possible worlds in which the sentence is true, for those with nominalistic prejudices.

[11] So, if $Y \subseteq Z$ is a Borel set in the ultrafilter ϕ, then the corresponding sentence is:

$$\text{val}(E) = 1, \text{ i. e. val}(E) \in \{1\},$$

where E is the characteristic function of the set Y, i. e.

$$E(z) = 1 \quad \text{if} \quad z \in Y$$
$$E(z) = 0 \quad \text{if} \quad z \notin Y$$

[12] See Appendix.

[13] See Section VI.

[14] I. e., every statistical state *in the theory*.

[15] The magnitude $g(A)$ satisfies the condition

$$p(\text{val}(g(A)) \in S) = p(\text{val}(A) \in g^{-1}(S))$$

for every possible probability assignment, not only those defined by the statistical states of the theory.

[16] I. e. such that

$$p_w(\text{val}(A_1) \in S) = p_w(\text{val}(B) \in g_1^{-1}(S))$$
$$p_w(\text{val}(A_2) \in S) = p_w(\text{val}(B) \in g_2^{-1}(S))$$

for every statistical state W and every Borel set S.

[17] S. Kochen and E. P. Specker, *op. cit.* They use the term 'commeasurability', which I wish to avoid, because it suggests a particular – and, I think, misleading – interpretation of the relation.

[18] The term is due to Kochen and Specker, *op. cit.*, who have investigated the properties of these systems.

[19] Or rather, between the set of idempotent magnitudes and the set of equivalence classes of theoretical sentences.

[20] This term is due to Suppes. See P. Suppes, 'The Probabilistic Argument for a Non-Classical Logic of Quantum Mechanics', *Philosophy of Science* 33 (1966), 14. My use of Suppes' terminology in no way implies endorsement of his position. In fact, I regard Suppes' article as exhibiting, with more than usual clarity, the fundamental confusion that has characterized the treatment of the completeness problem in the literature. I shall comment on Suppes' argument in Section VI.

[21] S. Kochen and E. P. Specker, *op. cit.*

[22] Evidently, a probability assignment to the sentences of a Boolean logic is to be understood more precisely as a probability measure on the Lindenbaum-Tarski algebra of the logic. The condition of countable additivity in this case is the requirement that for every countable set of sentences $\{s_i\}$ such that

$$|s_i| \wedge |s_j| = |s_i \wedge s_j| = 0, \qquad i \neq j$$

$$p\left(\bigwedge_{i=1}^{\infty} s_i\right) = \sum_{i=1}^{\infty} p(s_i).$$

The condition that $|s_i \wedge s_j| = 0$ is equivalent to the conditions that $\vdash \sim (s_i \wedge s_j)$, or $s_i \vdash \sim s_j$, i. e. that s_i and s_j are inconsistent.

[23] See Appendix.

[24] I mean the view commonly attributed to Heisenberg. I am not proposing a historical thesis. The Copenhagen interpretation is a chameleon-like creature, and Heisenberg says different things at different times.

[25] If $W^2 = W$, W is a projection operator because W is Hermitian. Since $\text{Tr}(W) = 1$, W is a projection operator onto a 1-dimensional subspace.

[26] I reproduce the simple proof of London and Bauer. (F. London and E. Bauer, *La Théorie de l'Observation en Mécanique Quantique*, Hermann et Cie, Paris, 1939, pp.31, 32.)

If $W = p_1 W_1 + p_2 W_2$ is idempotent, $W - W^2 = 0$, and

$$\begin{aligned} W^2 &= p_1{}^2 W_1{}^2 + p_2{}^2 W_2{}^2 + p_1 p_2 (W_1 W_2 + W_2 W_1) \\ &= p_1{}^2 W_1{}^2 + p_2{}^2 W_2{}^2 + p_1 p_2 (W_1{}^2 + W_2{}^2 - (W_1 - W_2)^2) \\ &= p_1 W_1{}^2 + p_2 W_2{}^2 - p_1 p_2 (W_1 - W_2)^2. \end{aligned}$$

So $W - W^2 = p_1(W_1 - W_1{}^2) + p_2(W_2 - W_2{}^2) + p_1 p_2 (W_1 - W_2)^2 = 0$.

But, $(W_1 - W_1{}^2)$, $(W_2 - W_2{}^2)$, and $(W_1 - W_2)^2$ are all semi-definite, from which it follows that

$$W_1 - W_1{}^2 = 0$$
$$W_2 - W_2{}^2 = 0$$

and, in particular:

$$(W_1 - W_2)^2 = 0$$

i. e.

$$W_1 - W_2 = 0$$

since $(W_1 - W_2)$ is Hermitian. Thus, $W_1 = W_2$, and from $W = p_1 W_1 + p_2 W_2$ we get $W = W_1 = W_2$.

[27] $P_A(r)$ is defined as $P_A(-\infty, r]$. Recall that

$$A = \int_{-\infty}^{\infty} r dP_A(r).$$

[28] The following proof is standard in the literature, and I reproduce it for completeness:
If

$$A' = A - (\psi, A\psi)I$$
$$B' = B - (\psi, B\psi)I$$

then A', B' satisfy the same commutation relation as A, B:

$$A'B' - B'A' = (ih/2\pi)I.$$

Assuming $\|\psi\| = 1$, $(\psi, (A'B' - B'A')\psi) = (ih/2\pi)(\psi, \psi) = ih/2\pi$.
Now,

$$
\begin{aligned}
(\psi, (A'B' - B'A')\psi) &= (\psi, A'B'\psi) - (\psi, B'A'\psi) \\
&= (A'\psi, B'\psi) - (B'\psi, A'\psi) \\
&= 2i\,\mathrm{Im}(A'\psi, B'\psi) \\
&\leqslant 2i|(A'\psi, B'\psi)| \\
&\leqslant 2i\|A'\psi\|\,\|B'\psi\|.
\end{aligned}
$$

Since

$$
\begin{aligned}
(\Delta_\psi A)^2 &= (\psi, (A - (\psi A\psi)I)^2\psi) \\
&= (\psi, (A')^2\psi) \\
&= (A'\psi, A'\psi) \\
&= \|A'\psi\|^2
\end{aligned}
$$

the inequality becomes

$$\Delta_\psi A \Delta_\psi B \geqslant h/4\pi.$$

[29] Because $p(E \cup E') = p(E) + p(E') = p(X) = 1$.

[30] Because (i) $E \cap F \subseteq E$ and $E \cap F \subseteq F$, and so $p(E \cap F) \leqslant p(F)$, from which it follows that $p(E \cap F) = 0$ if $p(E) = 0$ or $p(F) = 0$; and (ii) if $p(E) = p(F) = 1$, then $p(E' \cup F') = 0$, because $p(E' \cup F') \leqslant p(E') + p(F')$ from which it follows that $p(E \cap F) = 1 - p(E' \cup F') = 1$.

[31] See Appendix.

[32] I ignore the vulgarized variant of Heisenberg's interpretation that constitutes the world of q-facts only, i. e. that says: nothing is a fact unless it can be known to be a fact by an observational procedure that does not disturb any other (known) facts. This thesis is simply irrelevant to the question of completeness, and does not affect the conclusion of the Einstein-Podolsky-Rosen argument at all.

[33] By a 'q-maximal totality of situations in logical space', I mean a totality of knowable situations that is maximal with respect to what can be known simultaneously.

[34] A. Einstein, B. Podolsky, and N. Rosen op. cit., p. 777.

[35] Ibid., p. 777.

[36] Ibid., p. 777.

[37] Thus, Heisenberg's measurement theory and his completeness thesis is false if quantum mechanics is true, and the conjunction of this thesis with quantum mechanics is inconsistent.

[38] J. von Neumann, op. cit., p. 302. I am not suggesting a direct historical connection between the Einstein-Podolsky-Rosen paper and von Neumann's distinction between Case I and Case II. This would be absurd. Rather, I am attempting a logical reconstruction of the evolution of the completeness problem. Von Neumann and Einstein, Podolsky, and Rosen made independent contributions to this problem. My concern is to evaluate these contributions logically, not historically.

[39] I.e. whether or not such measures correspond to statistical states in the theory.

[40] It corresponds to a 2-valued homomorphism, which always exists on a Boolean algebra.

[41] If the idempotent magnitude P corresponds to the sentence s, then for every statistical state W:

$$p_w(s) = \text{Exp}_w(P_S).$$

[42] S. Kochen and E. P. Specker, *op. cit.*

[43] A. M. Gleason, *op. cit.*

[44] The following proof is an adaptation of a similar proof by J. S. Bell in *Reviews of Modern Physics* **38** (1966), 447. Bell's purpose is to demonstrate that in a Hilbert space of three or more dimensions, von Neumann's additivity requirement (see below) cannot be satisfied by dispersion free states, not even for the expectation values of commuting operators.

[45] I mean the proof in J. von Neumann, *op. cit.*

[46] J. S. Bell, *Reviews of Modern Physics* **38** (1966), 447.

[47] J. von Neumann, *op. cit.*, p. 309.

[48] Kochen and Specker produce a phase space reconstruction for the quantum statistics of a spin-1/2 particle. See S. Kochen and E. P. Specker, *op. cit.*, pp. 75-80.

[49] The existence of (sufficient) partial Boolean valuations on the working logic of quantum mechanics is equivalent to the possibility of introducing a phase space, where the points of this space define 2-valued measures in the generalized sense, i.e. assignments of values to magnitudes satisfying the quantum mechanical functional relationships. See S. Kochen and E. P. Specker, *op. cit.*

[50] J. M. Jauch and C. Piron, *Helvetia Physica Acta* **36** (1963), 827.

[51] J. M. Jauch, *Foundations of Quantum Mechanics*, Addison-Wesley, 1968.

[52] J. M. Jauch, *op. cit.*, p. 74.

[53] J. M. Jauch, *op. cit.* p. 118.

[54] J. S. Bell, *Reviews of Modern Physics* **38** (1966), 447.

[55] *Ibid.*, p. 447.

[56] E. P. Wigner in *American Journal of Physics* **38** (1970), 1005, 1006.

[57] J. F. Clauser, M. A. Horne, A. Shimony, and R. A. Holt, *Physical Review Letters* **23** (1969), 880.

[58] I refer to the following as a 'Bohrian' interpretation' and not 'Bohr's interpretation', because I do not think that it characterizes Bohr's view. Nevertheless, it is a view which is often attributed to Bohr.

[59] J. S. Bell, *Reviews of Modern Physics* **38** (1966), 447-451.

[60] J. S. Bell, *Physics* **1** (1964), 195-199.

[61] J. S. Bell, *Reviews of Modern Physics* **38** (1966), 447-452.

[62] E. P. Wigner, *op. cit.*

[63] This is a trivial transcription of Wigner's argument into my own notation, for the purposes of subsequent criticism.

[64] S. Kochen and E. P. Specker, *op. cit.*, pp. 75-80.

[65] Here α_1 and β_1 are the eigenvectors corresponding to the eigenvalues a_1 and b_1 of the incompatible magnitudes A and B. I use the symbol a_1 for the eigenvalue of A and also for the sentence $\text{val}(A) = a_1$, i.e. $\text{val}(A) \in \{a_1\}$. Thus, the symbol $p_\psi(a_1 \cdot b_1)$ denotes the probability, defined by the statistical state ψ, that a first measurement satisfies the sentence $\text{val}(A) = a_1$ and a subsequent measurement satisfies the sentence $\text{val}(B) = b_1$

[66] For any sets Φ_1, Φ_2, Φ'_2 in phase space (where Φ'_2 is the set theoretical complement of Φ_2), and any measure μ:

$$\mu(\Phi_1) = \mu(\Phi_1 \cap \Phi_2) + \mu(\Phi_1 \cap \Phi'_2)$$

Hence:

$$\mu(\Phi_s\cap \Phi_t) = \mu(\Phi_s\cap \Phi_t\cap \Phi_u) + \cap(\Phi_s\cap \Phi_t\cap \Phi'_u)$$
$$\mu(\Phi_s\cap \Phi_u) = \mu(\Phi_s\cap \Phi_t\cap \Phi_u) + \mu(\Phi_s\cap \Phi'_t\cap \Phi_u)$$
$$\mu(\Phi_t\cap \Phi'_u) = \mu(\Phi_s\cap \Phi_t\cap \Phi'_u) + \mu(\Phi'_s\cap \Phi_t\cap \Phi'_u).$$

So:

$$\mu(\Phi_s\cap \Phi_t) \leqslant \mu(\Phi_s\cap \Phi_u) + \mu(\Phi_t\cap \Phi'_u).$$

Taking s as a_2, t as c_1, and u as b_1, we have:

$$\mu(\Phi_{a_2}\cap \Phi_{c_1}) \leqslant \mu(\Phi_{a_2}\cap \Phi_{b_1}) + \mu(\Phi_{c_1}\cap \Phi'_{b_1}).$$

Since the set of phase space points satisfying b_2 is the complement of the set of points satisfying b_1, i.e. $\Phi_{b_2} = \Phi'_{b_1}$, this becomes

$$\mu(\Phi_{a_2}\cap \Phi_{c_1}) \leqslant \mu(\Phi_{a_2}\cap \Phi_{b_1}) + \mu(\Phi_{b_2}\cap \Phi_{c_1}).$$

[67] Consider the inequality:

$$\mu(\Phi_s\cap \Phi_t) \leqslant \mu(\Phi_s\cap \Phi_u) + \mu(\Phi_t\cap \Phi'_u)$$

Let Φ_r be a set such that

$$\mu(\Phi_u\cap \Phi_r) = \mu(\Phi'_u\cap \Phi'_r) = 0$$

then

$$\mu(\Phi_t\cap \Phi'_u) = \mu(\Phi_t\cap \Phi_r).$$

Proof:

$$\mu(\Phi_t\cap \Phi'_u) = \mu(\Phi_t\cap \Phi'_u\cap \Phi_r) + \mu(\Phi_t\cap \Phi'_u\cap \Phi'_r)$$
$$\mu(\Phi_t\cap \Phi_r) = \mu(\Phi_t\cap \Phi_r\cap \Phi_u) + \mu(\Phi_t\cap \Phi_r\cap \Phi'_u)$$

and

$$\mu(\Phi_t\cap \Phi'_u\cap \Phi'_r) = \mu(\Phi_t\cap \Phi_r\cap \Phi_u) = 0.$$

Taking s as $c_1{}^2$, t as $a_1{}^1$, and u as $b_1{}^1$, we have:

$$\mu(\Phi_{a1}{}^1\cap \Phi_{c1}{}^2) \leqslant \mu(\Phi_{b1}{}^1\cap \Phi_{c1}{}^2) + \mu(\Phi_{a1}{}^1\cap \Phi_{b1}{}^1).$$

Now, the quantum statistical mirror-image correlations require that

$$\mu(\Phi_{b1}{}^1\cap \Phi_{b1}{}^2) = \mu(\Phi_{b2}{}^1\cap \Phi_{b2}{}^2) = 0$$

i.e.

$$\mu(\Phi_{b1}{}^1\cap \Phi_{b1}{}^2) = \mu(\Phi'_{b1}{}^1\cap \Phi'_{b1}{}^2) = 0.$$

Hence, taking r as $b_1{}^2$, the above inequality becomes

$$\mu(\Phi_{a1}{}^1\cap \Phi_{c1}{}^2) \leqslant \mu(\Phi_{b1}{}^1\cap \Phi_{c1}{}^2) + \mu(\Phi_{a1}{}^1\cap \Phi_{b1}{}^2).$$

[68] J. F. Clauser, M. A. Horne, A. Shimony, and R. A. Holt, *op. cit.*

[69] Bell's inequality derives from the application of rule (i). The mirror-image correlations are derived from the application of rule (ii).

[70] *Letters on Wave Mechanics, op. cit.*, p. 39.

[71] Bohr was consistently a Kantian. Heisenberg sometimes adopts a realist position.

[72] P. K. Feyerabend, 'On the Quantum Theory of Measurement', in, *Observation and Interpretation in the Philosophy of Physics* (Proceedings of the Ninth Symposium of the Colston Research Society held in the University of Bristol April 1st–April 4th, 1957) (ed. by S. Korner), Dover Publications, Inc., 1962. Quote from p. 122.

[73] D. Bohm, *Physical Review* **85** (1952), 166, 180.

[74] D. Bohm and J. Bub, *Reviews of Modern Physics* **38** (1966), 453. See also J. Bub, *International Journal of Theoretical Physics* 2 (1969), 101; and D. Bohm 'On the Role of Hidden Variables in the Fundamental Structure of Physics', in *Quantum Theory and Beyond* (ed. by Ted Bastin), Cambridge University Press, 1971. Evidently Bohm's view

on this issue, as expressed, for example, in the latter article, is now different from my own.

[75] This analogy was first pointed out by Hilary Putnam in his 'Is Logic Empirical', in *Boston Studies in the Philosophy of Science*, Vol. V (ed. by R. S. Cohen and M. Wartofsky), D. Reidel Publ. Co., 1969.

[76] See my article 'Towards the Interpretation of Quantum Mechanics', *British Journal for the Philosophy of Science*, 1972. Also my article to be published in the proceedings of International Colloquium: *The Meaning and Function of Science in Comtemporary Society*, Pennsylvania State University, September, 1971, (ed. by J. J. Kockelmans).

[77] Hilary Putnam, *op. cit.*

[78] S. Kochen and E. P. Specker, 'The Problem of Hidden Variables Quantum Mechanics', *op. cit.*

'Logical Structures Arising in Quantum Theory', in *Theory of Models*, Proceedings of the 1963 International Symposium at Berkeley, North-Holland Publ. Co., 1965.

[79] G. Birkhoff and J. von Neumann, 'The Logic of Quantum Mechanics', *Annals of Mathematics* 37 (1936), 823.

[80] J. M. Jauch and C. Piron, *Helvetia Physica Acta* 36 (1963), 827. Article in *Quanta* (ed. by P. G. O. Freund, C. J. Goebel, and Y. Nambu), University of Chicago Press, 1970. C. Piron, *Helvetia Physica Acta* 37 (1964), 439. J. M. Jauch, *Foundations of Quantum Mechanics*, Addison-Wesley, 1968.

[81] D. Finkelstein, 'Logic of Quantum Physics', *Transactions of the New York Academy of Science* 25 (1963), 621; 'Matter, Space and Logic', *Boston Studies in the Philosophy of Science* Vol. V (ed. by R. S. Cohen and M. Wartofsky), D. Reidel Publ. Co., 1969.

[82] G. Mackey, *The Mathematical Foundations of Quantum Mechanics*, Benjamin, 1963.

[83] V. Varadarajan, *Geometry of Quantum Mechanics*, Vol. I, Van Nostrand, 1968; Vol. II, 1970.

[84] S. Gudder, 'Axiomatic Quantum Mechanics and Generalized Probability Theory', in *Probabilistic Methods in Applied Mathematics* Vol. 2, Academic Press, 1970. This article contains a bibliography which lists other work in this field by Gudder.

[85] P. Suppes, *Philosophy of Science* 33 (1966), 14. Also: 'Logics Appropriate to Empirical Theories', in *Symposium on the Theory of Models*, North-Holland Publ. Co., 1965.

[86] See van Fraassen's contribution to this volume for references.

[87] A. Fine, 'Some Conceptual Problems of Quantum Theory', in *Paradigms and Paradoxes: The Philosophical Challenge of the Quantum Domain* (ed. by R. G. Colodny), Pittsburgh University Press, 1972.

[88] P. Suppes, *Philosophy of Science* 33 (1966), pp. 14, and 17.

[89] Hilary Putnam, *op. cit.*

LEON COHEN

JOINT PROBABILITY DISTRIBUTIONS IN QUANTUM MECHANICS

I. INTRODUCTION

The notion and methods of calculating probabilities in quantum mechanics is strikingly different from classical mechanics and standard probability theory. Quantum mechanics manipulates mysteriously, from a classical probabilistic point of view, complex functions and operators to nevertheless end up with observable quantities which are interpreted as if they were derived from a standard type of probability theory. Whether the strange methods of obtaining the predictions of the theory are essential, in the sense that given the type of results quantum mechanics predicts implies a quantum mechanical type of mathematical formalism, is not totally clear. Certainly many of the paradoxes seem to be based on the mode of calculating probabilities. For example, the paradox of the double slit experiment arises from the fact that the probability of the particle being at a certain point on the screen is not the same probability that it went through one slit or the other. Since standard probability theory arose out of simplistic considerations it may be that it is restricted and that its future development will generalize it to a degree where quantum mechanics will fit in naturally. Weiner (1963) made some attempts along these lines by showing some examples from ordinary probability theory that yield results similar in mathematical form as in quantum mechanics. Another approach to the understanding of the probabilistic methods is to attempt to formulate quantum mechanics as a traditional probability theory. We shall describe this attempt, and although we shall show that this is strictly speaking impossible, the impossibility will illuminate certain aspects of the theory. In particular we will emphasize the question of joint distributions of non-commuting variables and show the one-to-one relationship between so-called correspondence rules and the distribution functions. As we shall see, the question as to whether there can exist a consistent rule for obtaining quantum mechanical operators (a correspondence rule) from their classical counterpart and the question

Hooker (ed.), Contemporary Research in the Foundations and Philosophy of Quantum Theory, 66–79. All rights reserved

whether there exists a legitimate joint distribution function are one and the same.

For the sake of continuity we contrast the basic ideas and relevant manipulative methods of both traditional probability theory (TPT) and quantum mechanics.

1. Distribution function

In TPT there is a density or distribution function $P(x_1, x_2...x_n)$ of the random variables $x_1, x_2...x_n$, which is the probability per unit phase space volume of obtaining the random variables. The density function, being a probability, is never negative.

In quantum mechanics, instead of a distribution function we have a state function ψ which is generally complex. The density function of these variables for the representation in which the state function is ψ is then $|\psi(\mathbf{x})|^2$. The symbol $\mathbf{x}$ signifies the set of n variables $x_1, x_2...x_n$.

2. Expectation Values

The expectation value of any function $A(\mathbf{x})$ representing an observable is given by

$$\langle A(\mathbf{x}) \rangle = \int A(\mathbf{x}) P(\mathbf{x}) \, d\mathbf{x} \tag{1}$$

in TPT.

In quantum mechanics, observables are associated with operators and the expectation value is obtained by calculating

$$\langle \bar{A} \rangle = \int \psi^*(\mathbf{x}) \bar{A} \psi(\mathbf{x}) \, d\mathbf{x}, \tag{2}$$

where $\bar{A}$ is the operator representing the observable.

3. Distribution of Other Variables

Suppose we wish to find the density function of a set of random variables $\mathbf{y}$ which are functions of $\mathbf{x}$. Classically one calculates

$$M(\theta) = \langle e^{i\theta \cdot \mathbf{y}} \rangle \tag{3}$$

which is known as the characteristic or moment generating function of the variables **y**. If we knew the probability distribution of the **y**, $P(\mathbf{y})$ say, $M(\boldsymbol{\theta})$ would equal

$$M(\boldsymbol{\theta}) = \int e^{i\boldsymbol{\theta}\cdot\mathbf{y}} P(\mathbf{y})\, d\mathbf{y} \qquad (4)$$

and by Fourier inversion

$$P(\mathbf{y}) = \frac{1}{(2\pi)^n} \int e^{-i\boldsymbol{\theta}\cdot\mathbf{y}} M(\boldsymbol{\theta})\, d\boldsymbol{\theta}. \qquad (5)$$

But $M(\boldsymbol{\theta})$ can be found from a knowledge of $P(\mathbf{x})$ since **y** are known functions of **x**,

$$M(\boldsymbol{\theta}) = \int e^{i\boldsymbol{\theta}\cdot\mathbf{y}(\mathbf{x})} P(\mathbf{x})\, d\mathbf{x}. \qquad (6)$$

Hence the density function of the y is

$$P(\mathbf{y}) = \frac{1}{(2\pi)^n} \int e^{-i\boldsymbol{\theta}\cdot\mathbf{y}} \left\{ \int e^{i\boldsymbol{\theta}\cdot\mathbf{y}(\mathbf{x})} P(\mathbf{x})\, d\mathbf{x} \right\} d\boldsymbol{\theta} \qquad (7)$$

which can be written symbolically as

$$\int \delta(\mathbf{y} - \mathbf{y}(\mathbf{x}))\, P(\mathbf{x})\, d\mathbf{x}. \qquad (8)$$

The number of y variables may be less then the number of x variables, in which case then the inner integration of (7) is performed over all $n\, x$ variables, but the number of integrals in the outer integration equals the number of y variables.

In quantum mechanics we have the Born rule which is at the heart of the theory. The eigenvalues α_n and eigenfunctions u_n of the operator $\tilde{A}$ are found from

$$\tilde{A} u_n = \alpha_n u_n. \qquad (9)$$

As one of the basic assumptions is that the operator yields a complete set of eigenfunctions, the state function ψ can thus be expanded in terms of the u's

$$\psi = \sum_n c_n u_n. \qquad (10)$$

Then, according to the Born rule, the numerical values which can be observed are the α_n's and the probability density of observing a particular α_n is $|c_n|^2$.

Once the method (1) of calculating expectation in TPT value is accepted the result given by (7) follows. Similarly, the Born rule is not independent of the method of calculating expectation values via (2) but can actually be proven. This is not usually pointed out in most textbooks where the Born rule is presented more or less ad hoc. It may be of some interest to show how the Born rule can be derived from (2) using characteristic functions. This is done in the Appendix.

4. *Dynamics*

In TPT the equation governing the evolution of the probability density depends of course on the particular physical problem. Examples of such equations are the Liouville, Boltzmann, Fokker-Planck equations. They are first order in time and have the property that the density remains normalized to one for all time if it was so for a particular time.

The quantum mechanical equation of evolution, the Schrödinger equation, evolves ψ rather than the density $|\psi|^2$, but $|\psi|^2$ does remain normalized for all time.

III. THE WIGNER DISTRIBUTION

To formulate Quantum Mechanics in terms of classical probabilistic concepts it is essential to find a distribution function which can be used to calculate expectation values in the traditional way. Wigner (1932) was the first to study such a function. It is

$$P_w(q, p) = \frac{1}{2\pi} \int \psi^*\left(q - \tfrac{1}{2}\tau\hbar\right) e^{i\tau p} \psi\left(q + \tfrac{1}{2}\tau\hbar\right) d\tau. \tag{11}$$

Its basic property is that it yields the quantum mechanical probability density functions of position and momentum. That is

$$\int P_w(q, p)\, dp = |\psi(q)|^2 \tag{12}$$

$$\int P_w(q, p)\, dq = |\phi(p)|^2, \tag{13}$$

where ψ and ϕ are the position and momentum wave functions. The Wigner distribution has been used extensively as a calculational tool. Since it is 'classical' in form, one can use the methods of classical mechanics, using P_w as the distribution function, and obtain results reflecting quantum mechanics.

Although the Wigner distribution gives the correct quantum mechanical marginal densities and therefore gives the same results for expectation values of functions of either position or momentum as does quantum mechanics, it cannot be properly considered a legitimate density function. For one, it can become negative; and furthermore, if it is used to calculate distributions of other variables (as described in Section II) or expectation values of functions of position and momentum, one does not always get the quantum mechanical results. Although Wigner clearly pointed out that (11) is not unique and cannot be thought of as a true joint distribution many authors in later years have been under the impression that P_w is a result forced upon us by quantum mechanics. This is probably due to the paper by Moyal (1949) where he seems to derive the Wigner distribution from first principles. Moyal sets out to find the joint distribution of q and p through the characteristic function method. What has to be found is the quantum mechanical counterpart to (3)

$$M(\theta, \tau) = \langle e^{i\theta q + i\tau p} \rangle, \tag{14}$$

where we have let $\theta = \theta_1$, and $\tau = \theta_2$. Moyal argues that M should be calculated through the 'usual' quantum mechanical procedure

$$M(\theta, \tau) = \langle e^{i\theta \bar{q} + i\tau \bar{p}} \rangle = \langle e^{1/2 i\theta\tau\hbar} e^{i\theta \bar{q}} e^{i\tau \bar{p}} \rangle \tag{15}$$

$$= \int \psi^*(q) \, e^{1/2 i\tau\theta\hbar} e^{i\theta q} e^{i\tau \bar{p}} \psi(q) \, dq$$
$$= \int \psi^*(q - \tfrac{1}{2}\tau\hbar) \, e^{i\theta q} \psi(q + \tfrac{1}{2}\tau\hbar) \, dq. \tag{16}$$

Fourier inversion of (16) readily yields the Wigner distribution.

Moyal does not consider the possibility that the derivation requires a premiss which is outside quantum mechanics. If the derivation was correct and quantum mechanics therefore predicted a probability distribution which could become negative, the whole foundation of the theory would be in jeopardy. Indeed, some authors have claimed that this shows that there must be something wrong with quantum mechanics or at least with the probabilistic interpretation. Others have argued that, since quantum mechanics yields strange probabilistic results we should formulate it in terms of non-classical logics. Not as extreme but still unwarranted, are conclusions relating to the question of joint measurability of position and momentum. Arguments have been given that the Wigner distribution should be used as a criterion for when q and p are jointly measureable. When P_w is positive we can then presumably measure q and p simultaneously and not otherwise. This can not be the case, for even when P_w is always positive it still does not always give the correct quantum mechanical results for observables which are functions of both q and p. Furthermore, as we shall discuss, there are an infinity of distribution functions satisfying (12) and (13) which are not necessarily positive for the same range of q and p.

The derivation given above of the Wigner distribution is 'incorrect' in assuming that the expected value of the classical quantity $e^{i\theta q + i\tau p}$ can be calculated, if at all, by replacing q and p by the operator counterparts in the expression. There is nothing in quantum mechanics which requires this procedure and it is quite clear that by choosing different correspondences for $e^{i\theta q + i\tau p}$ we will obtain different distribution functions. We therefore turn our attention to the question of correspondence rules between classical variables and their quantum mechanical operator counterparts.

IV. CORRESPONDENCE RULES

Since the beginning of quantum mechanics a central question has been the finding of the operator which represents a given physical observable. Classical mechanics was the main guideline and it was found that in the coordinate representation, position is to be represented by $\tilde{q}$ and momentum by $(\hbar/i)(\partial/\partial q)$. This works fine as long as the classical functions are either functions of position or momentum or the sum of such functions. But when we have mixed expressions such as qp, simply replacing q by $\tilde{q}$ and p by $\tilde{p}$ will not work because the resulting expression will not even be Hermitian. Thus there has been a search to find a rule which would enable us to write quantum mechanical operators from their classical counterpart. Such rules are called correspondence rules or rules of association.

That such a procedure should exist seems so ingrained that when a particular rule leads to inconsistent results it is sometimes taken to be a fault with quantum mechanics itself.

What we shall do in this section is list the rules most widely discussed and then derive an expression which allows the writing of any number of such rules (Cohen, 1966). After obtaining the set of all possible correspondence rules we will show there doesn't exist any rule which can consistently be used to obtain operators from their classical counterparts. Since each rule yields a different distribution function we will have at our disposal the set of all possible distribution functions. The impossibility of having a consistent rule of association bears directly on whether there exists a distribution function which can be used to obtain expectation values of quantum mechanical observables in the classical manner.

The rules most commonly cited are:

1. *Von Neumann's Rule*

If $A \leftrightarrow \tilde{A}$

then $F(A) \leftrightarrow F(\tilde{A})$,

where F is any function, and if

$$B \leftrightarrow \tilde{B} \tag{17}$$

then $A + B \leftrightarrow \tilde{A} + \tilde{B}$,

where A and B are the classical functions and $\tilde{A}$ and $\tilde{B}$ are operators. The arrow signifies the correspondence. It is rather remarkable that such a simple rule of association is self contradictory. It can be shown (Groenewold, 1946; Shewell, 1959) that von Neumann's rule implies both

$$p^2 q^2 \leftrightarrow \tilde{q}^2 \tilde{p}^2 - 2i\hbar \tilde{q} \tilde{p} - \tfrac{1}{4}\hbar^2 \tag{18}$$

and

$$p^2 q^2 \leftrightarrow \tilde{q}^2 \tilde{p}^2 - 2i\hbar \tilde{q} \tilde{p} - \hbar^2 . \tag{19}$$

This lack of uniqueness lead to a rather amusing exchange between Temple (1935), Fröhlich and Guth (1935) and Peierls (1935). The argument was that since the only way the above two expressions can be made equal is by taking $\hbar = 0$ – doesn't this invalidate quantum mechanics! Of course a somewhat less drastic idea is to reject the rule.

2. Dirac's *Rule of Associating Commutators with Poisson brackets*

$$\{A, B\} \leftrightarrow -\frac{i}{\hbar} [\tilde{A}, \tilde{B}] \tag{20}$$

This rule is also self contradictory (Groenewold, 1946; Shewell, 1959).

3. *Weyl Rule (Weyl,* 1931*)*

$$e^{i\theta q + i\tau p} \leftrightarrow e^{i\theta \tilde{q} + i\tau \tilde{p}} = e^{1/2i\theta\tau\hbar} e^{i\theta\tilde{q}} e^{i\tau\tilde{p}} \tag{21}$$

or equivalently as shown by McCoy (1932)

$$q^n p^m \leftrightarrow \tfrac{1}{2}n \sum_{l=0}^{n} \binom{n}{l} \tilde{q}^{n-l} \tilde{p}^n \tilde{q}^l \tag{22}$$

4. *Symmetrization Rule*

$$e^{i\theta q + i\tau p} \leftrightarrow \tfrac{1}{2} \{ e^{i\theta\tilde{q}} e^{i\tau\tilde{p}} + e^{i\tau\tilde{p}} e^{i\theta\tilde{q}} \} = e^{1/2i\theta\tau\hbar} \cos \tfrac{1}{2}\theta\tau\hbar e^{i\theta\tilde{q}} e^{i\tau\tilde{p}} \tag{23}$$

or

$$q^n p^m \leftrightarrow \tfrac{1}{2} \{ \tilde{q}^n \tilde{p}^m + \tilde{p}^m \tilde{q}^n \} \tag{24}$$

5. *Rule of Born and Jordan* (1925)

$$e^{i\theta q + i\tau p} \leftrightarrow e^{1/2i\theta\tau\hbar} \frac{\sin \tfrac{1}{2}\theta\tau\hbar}{\tfrac{1}{2}\theta h} e^{i\theta\tilde{q}} e^{i\tau\tilde{p}} \tag{25}$$

or

$$q^n p^m \leftrightarrow \frac{1}{m+1} \sum_{l=0}^{m} \tilde{p}^{m-l} \tilde{q}^n \tilde{p}^l . \tag{26}$$

If the only requirement of a correspondence rule is that it reduce to the correspondences

$$F(q) \leftrightarrow F(\tilde{q}) \\ G(p) \leftrightarrow G(\tilde{p}) \tag{27}$$

for functions of either position or momentum then is it clear that the most general correspondence for $e^{i\theta q + i\tau\alpha}$ is

$$f(\theta, \tau) e^{i\theta\tilde{q} + i\tau\tilde{p}} = f(\theta, \tau) e^{1/2i\theta\tau\hbar} e^{i\theta\tilde{q}} e^{i\tau\tilde{p}}, \tag{28}$$

where $f(\theta, \tau)$ is any function which satisfies

$$f(0, \tau) = f(\theta, 0) = 1 . \tag{29}$$

To find (Cohen, 1966) the operator for a general classical function $A(q, p)$ we first express it in terms of its Fourier expansion

$$A(q, p) = \int \int \gamma(\theta, t) \, e^{i\theta q + i\tau p} \, dq \, dp, \tag{30}$$

where

$$\gamma(\theta, \tau) = \frac{1}{4\pi^2} \int \int A(q, p) \, e^{-i\theta q - i\tau p} \, dq \, dp. \tag{31}$$

The operator $\tilde{A}$ which corresponds to $A(q, p)$ is then

$$\tilde{A} = \int \int \gamma(\theta, \tau) \, f(\theta, \tau) \, e^{i\theta \tilde{q} + i\tau \tilde{p}} \, dq \, dp. \tag{32}$$

If we also require hermiticity then f must satisfy

$$f(\theta, \tau) = f^*(-\theta, -\tau).$$

The rules of Weyl, Symmetrization and of Born and Jordan follow from (28) or (32) by taking $f = 1$, $\cos \frac{1}{2} \theta \tau h$, $\sin \frac{1}{2} \theta \tau h / \frac{1}{2} \theta \tau h$ respectively.

Now the question is whether there is any correspondence rule out of the infinite number given by (32), which can consistently be used to obtain operators from the classical function. By consistently we mean that if we have a rule giving the operator $\tilde{A}$ for $A(q, p)$, the same rule should also give the correct quantum mechanical operator for any function of $A(q, p)$. It has been shown that the rules given by (21), (23), and (25) do not. It is straight forward, although lengthy, to also show that none of the other rules generated by (32) do either.

The fact that no rule will always yield the proper operators from their classical counterparts should not be too surprising. We should not expect classical mechanics to be anything more than a guide. After all, there is no set procedure in classical mechanics for obtaining the important functions of the theory such as the Hamiltonian. What one did in classical mechanics is examine the experimental results for particular situations and then intelligently guess what the functions must be to yield the experimental data. A lack of a formal constructive procedure of particular observables is no more a drawback in quantum mechanics then it is in classical mechanics. Furthermore, there are observables in quantum mechanics which have no classical counterpart.

V. DISTRIBUTION FUNCTIONS

If we suppose for the moment that the correspondence given by (28) is valid for some particular f then the characteristic function becomes

$$M(\theta, \tau) = f(\theta, \tau) \int \psi^*(u - \tfrac{1}{2}\tau\hbar) e^{i\theta u} \psi(u + \tfrac{1}{2}\tau\hbar) du \qquad (33)$$

and Fourier inversion gives

$$P(q, p) = \frac{1}{4\pi^2} \int \int \int e^{-i\theta q - i\tau p + i\theta u} f(\theta, \tau) \psi^*(u - \tfrac{1}{2}\tau\hbar) \times$$
$$\times \psi(u - \tfrac{1}{2}\tau\hbar) du. \qquad (34)$$

It is readily verified that (35) satisfies (12) and (13) for any f constrained by (29). The choice of $f = 1$ produces the Wigner distribution and if we choose $f = \cos \tfrac{1}{2}\tau\hbar$ we obtain

$$P(q, p) = \frac{1}{4\pi} \text{ Real Part} \left\{ \psi(q) \int e^{-i\tau p} \psi^*(q - \tau\hbar) d\tau \right\} \qquad (35)$$

which is derived by Margenau and Hill (1961). The set of distributions given by (35) is the totality of all possible distribution functions yielding the quantum mechanical marginal distributions.

Most choices of f yield distributions which do become negative for an arbitrary state ψ but they do not necessarily become negative at the same regions of q and p. Are there any distributions which are always positive? In Wigner's (1932) paper he states that no positive definite distribution exists which is bilinear in ψ. He has recently given a proof of this (Wigner, 1971). In our notation that means no positive definite distribution exists for those f's which are 'pure' functions of q and p, that is, they are not functionals of ψ. That there are positive distributions when f is a functional of ψ is clear, for the choice of

$$f(\theta, \tau) = \frac{|\psi(q)|^2 |\phi(p)|^2 e^{i\theta q + i\tau p} dq \, dp}{\psi^*(u - \tfrac{1}{2}\tau\hbar) e^{i\theta u} \psi(u + \tfrac{1}{2}\tau\hbar) du} \qquad (36)$$

yields a $P(q, p)$ which equals

$$|\psi(q)|^2 |\phi(p)|^2,$$

a perfectly legitimate distribution function which if taken literally would imply no correlations. I do not know whether there are any other positive distributions but as we will see the question is not very important to the central issue.

Leaving aside the question of positiveness we may ask to what extent one can calculate expectation values in the classical manner. Since the marginal distributions are correctly given it is clear that for any function of the form $A(q) + B(p)$, the expectation value calculated via phase space integration will yield the same answer as the operator formalism. But what if we want to calculate the expectation value of a mixed function of q and p, $A(q, p)$. Simple algebra shows that if we use the correspondence given by (32) between $A(q, p)$ and $\bar{A}$ for any particular f and use the distribution function given by (35) using the same f then

$$\int \int A(q, p) P(q, p) \, dq \, dp = \int \psi^* \bar{A} \psi. \tag{37}$$

Now suppose we want to calculate the expectation value of a function of $A(q, p)$, $F(A(q, p))$. Again using (31) and letting $G(q, p)$ be the classical function corresponding to $F(\bar{A})$ we have

$$\int \int G(q, p) P(q, p) \, dq \, dp = \int \psi^* F(\bar{A}) \psi. \tag{38}$$

But $G(q, p)$ will in general not equal $F(A(q, p))$!

Thus irrespective of the existence of a positive distribution, if we want to formulate quantum mechanics as a normal probability theory we would have to violate one of its basic assumptions, namely that the expected value of a function of an observable is calculated by integrating that function with the probability density.

VI. CONCLUSION

The uncertainty principle is most often given as evidence for the impossibility of the existence of a true joint distribution. If anything, the statistical statement made by the uncertainty relation would lead one to believe that position and momentum do have something to do with each other. An uncertainty relationship by itself does not preclude the possibility of a joint distribution. It is certainly conceivable, for example, that nature could have been constructed according to the probability distribution given by $P(q, p) = |\psi(q)|^2 |\phi(p)|^2$ and that the observables of the theory are expectation values calculated in the classical manner. Many of the results of quantum mechanics would remain unchanged and some, of course, would be different, but the theory would be consistent and complete. The question of joint distribution and as to whether q and p can be measured simultaneously would never arise for it is already implied by the acceptance of $P(q, p)$. But the uncertainty principle would still hold since

$$\Delta q \Delta p = \left[\int (q - \langle q \rangle)^2 P(q, p) \, dq \, dp \right]^{1/2} \times$$

$$\times \left[\int (p - \langle p \rangle)^2 P(q, p) \, dq \, dp \right]^{1/2} =$$

$$= \left[\int |\psi(q)|^2 (q - \langle q \rangle)^2 \, dq \right]^{1/2} \times$$

$$\times \left[\int (p - \langle p \rangle)^2 |\phi(p)|^2 \right]^{1/2}$$

which is the starting point in the usual derivation.

Therefore, what is there in quantum mechanics which necessitates the strange form of probability theory? The standard version of probability theory cannot readily accommodate quantization of observables and dispersionless states. It is possible to have quantization in a phase space formulation but only at the expense of violating other predictions of quantum mechanics. If an observable $A(q, p)$ is in a dispersionless state, then the only classical type of distribution which will accommodate that fact is

$$\delta(a - A(q, p)).$$

This distribution will almost never yield the other quantum-like results.

The formalism developed in the previous sections clarifies to a degree the question of simultaneous measurability of noncommuting observables. As Park and Margenau (1968) have discussed, the usual reasons given for the impossibility of joint measurement are not legitimate. Naturally, it is not totally clear how to formulate our intuitive notion of joint measurement in the context of quantum mechanics. If we mean the existence of a true distribution function with all that it implies in the classical sense, then as the previous discussion has shown, the impossibility is certain. But if by simultaneous measurement we mean the possible existence of an operator which when applied to a state function gives results satisfying our intuitive ideas, then there is nothing in quantum mechanics to contradict that possibility. But we could further argue that such an operator cannot exist, for an ability to make a simultaneous measurement would allow us to construct a joint distribution (since a knowledge of $q^n p^m$ for all n and m yields the characteristic function) which we have already shown not to be possible in the classical sense. Depending on one's commitments there are many ways out of the argument. The obvious is to say that the joint distribution thus found should not be used in finding expectation values of other operators. Distributions of other sets of non-commuting operators should then be derived by finding the measurement operator rather than from the classical functional relationships. Furthermore, it is not apparent that the idea of simultaneous measurement necessarily implies the ability to measure and interpret the classical type quantities $q^n p^m$, which go into the making of the characteristic function. Naturally, the best thing to do for those who believe in simultaneous measurements is to discover an operator which no one could reasonably deny as a proper operator for the joint measurement.

APPENDIX

To prove the Born rule from the assumption that the expected value of the operator $\tilde{A}$ is

$$\int \psi^* \tilde{A} \psi \tag{A1}$$

we proceed by considering the expected value of

$$e^{i\theta \tilde{A}}.$$

The probability density of the observable a is then

$$P(a) = \frac{1}{2\pi} \int e^{-i\theta a} \langle e^{i\theta \tilde{A}} \rangle \, d\theta \tag{A2}$$

and we wish to show that $P(a)$ equals $|C_n|^2$ where C_n is obtained from

$$\psi = \sum C_n U_n; \quad \tilde{A} U_n = \alpha_n U_n$$

and that the only possible values for a are the α_n's.
Now

$$
\begin{aligned}
P(a) &= \frac{1}{2\pi} \int e^{-i\theta a} \int \psi^*(q) \, e^{i\theta \tilde{A}} \psi(q) \, dq \\
&= \frac{1}{2\pi} \int e^{-i\theta a} \sum_{m,n} C_m^* U_m(q) \, e^{i\theta \alpha_n} C_n U_n(q) \, dq \\
&= \frac{1}{2\pi} \int e^{-i\theta a} \sum_n e^{i\theta \alpha_n} |C_n|^2 \, dq \\
&= \sum_n \delta(a - \alpha_n) |C_n|^2 .
\end{aligned}
$$

Hence the probability is zero unless a equals one of the α_n and in which case the probability is $|C_n|^2$.

ACKNOWLEDGEMENT

This work was done during tenure of a grant by the National Science Foundation for a different project.

*Hunter College of the City University
of New York*

BIBLIOGRAPHY

Born, M. and Jordan, P., *Zeitschrift für Physik* **34** (1925), 873.
Cohen, L., *Journal of Mathematical Physics* **7** (1966), 781.
Fröhlich, H. and Guth, E., *Nature* **136** (1935), 129.
Groenewold, H. J., *Physics* **12** (1946), 405.
Margenau, H. and Hill, R. N., *Progress in Theoretical Physics (Kyoto)* **26** (1961), 722.
McCoy, N. H., *Proceedings of the National Academy of Sciences (U.S.A.)* **18** (1932), 674.
Moyal, J. E., *Proceedings of the Cambridge Philosophical Society* **45** (1949), 99.
Park, J. L. and Margenau, H., *International Journal of Theoretical Physics* **1** (1968), **211**.
Peierls, R., *Nature* **136** (1935), 395.
Shewell, J. R., *American Journal of Physics* **27** (1959), 16.
Temple, G. *Nature* **135** (1935), 937.
Weiner, N., *Nonlinear Problems in Random Theory*, The M.I.T. Press, Massachusetts, 1963.
Weyl, H., *The Theory of Groups and Quantum Mechanics*, E. P. Dutton and Co., New York, 1931, p. 275.
Wigner, E. P., *Physical Review* **40** (1932), 749.
Wigner, E. P., *Perspectives in Quantum Theory*, The M.I.T. Press, Massachusetts, 1971.

BAS C. VAN FRAASSEN

SEMANTIC ANALYSIS OF
QUANTUM LOGIC*

This paper has a beginning, a middle, and an end. If these parts are to follow the dramatic unities, they will lead from suffering through recognition to reversal; but of this ideal they may fall short.

To begin I shall sketch the general semantic framework that has been developed in the study of modal logics, and show how quantum logic falls naturally within it. In the middle part I shall sketch developments in quantum logic with special attention to their relations to formal logic generally, and counterfactual implication in particular. To end I shall discuss logical aspects of mixed states, a modal interpretation of quantum logic suggested by this, and the relations between this and hidden variables theories.

For the convenience of the reader, the special symbols used and the sections in which they are introduced are listed in a note.[1]

I. SEMANTIC FRAMEWORK

1. *Generalized Modal Logic*

In the semantic analysis of modal (and related) logics over the past two decades, a general picture has emerged. I do not personally endorse the idea that this picture is adequate to all purposes to which a philosopher may find semantic analysis relevant. But my reservations on this point bear no relation to the present subject, and I shall say no more about them. In my opinion, quantum logic finds a rightful place among modal logics and their ilk, and this I shall now attempt to show.

A language has both a syntax and a semantics[2]. The former specifies the well-formed expressions, and the latter the admissible interpretations. In the picture I am now presenting, the following holds: The well-formed expressions are given by syntactic functions ϕ defined on a basic vocabulary, and each such expression $E = \phi(e_1, ..., e_n)$ is given in each interpretation a *value* $|E|$. This value is a function of the values of its components,

Hooker (ed.), *Contemporary Research in the Foundations and Philosophy of Quantum Theory*, 80–113. *All rights reserved* so that there is a function Φ such that

(1) $|\phi(e_1, ..., e_n)| = \Phi(|e_1|, ..., |e_n|)$.

The assignment of values has furthermore the following specific form: each interpretation is defined with reference to a model structure. Without loss of generality, we can take each *model structure* to be a couple $M = \langle K, R \rangle$, where K is a non-empty set, and R a set of relations and functions defined with reference to K (and, possibly, to standard mathematical objects, such as the real number continuum). The members of K are called the *possible worlds* of M. In Equation (1), the function Φ is taken from R, or defined in terms of it.

Mathematically speaking, the set K acts as an index set. Hence there is something to be said for referring to its elements as indices or points of reference, or by some other name that does not carry the metaphysical overtones of 'possible worlds'. But just as metaphysicians of old found it convenient and satisfying to use spatial imagery in the description of Being, so today we find it convenient and satisfying to use metaphysical imagery in the description of language. No harm can come of it, if the pictorial character of our language-in-use is not forgotten.

2. *Connectives and Operations*

The description of our framework of semantic analysis has so far been quite abstract and I shall now give some examples. These examples are all of the sort that will be relevant to our discussion of quantum logic, and indeed, we will have occasion later to refer to these specific examples.

Well-formed expressions are divided into *nouns, sentences,* and *functors.* A functor is an expression whose concatenation with certain other expressions in a certain way yields a noun or sentence. For example, 'John runs' could be regarded as formed from 'John', a noun, and 'runs', a functor whose role it is to form a sentence when it is fed a noun. Functors which can thus be used to yield sentences from sentences are called 'connectors', or, more frequently, 'connectives'. So 'It is not the case that', which is used to form the denial of a given sentence through concatenation with that sentence, is a connective.

We shall now list several connectors and their interpretation in a model structure $M = \langle K, R \rangle$. If E is a sentence, the value $|E|$ given to E is a set of possible worlds and we say:

(2)   For all $\alpha \in K$, E is *true in* α if and only if $\alpha \in |E|$.

Since it is traditional to say that what a sentence means or expresses is a proposition, subsets of K are henceforth called *propositions*. So each connective must correspond to an operation on propositions, that is, a map of (n-tuples of) subsets of K into subsets of K. I shall now list a number of connectives with the corresponding Equation (1) and then make some comments on that list.

(3)   Conjunction $|A \ \& \ B| = |A| \cap |B|$

(4)   Necessity $|\Box A| = \{\alpha: \beta \in |A| \text{ for each } \beta \text{ such that } \alpha R \beta\}$

(5)   Implication
       (a) material $|A \supset B| = (K - |A|) \cup |B|$
       (b) counterfactual $|A > B| = \{\alpha: f(\alpha, |A|) \in |B|\}$
       (c) strict $|A \dashv B| = \{\alpha: \{\beta: \alpha R \beta \ \& \ \beta \in |A|\} \subseteq \{\beta: \alpha R \beta \ \& \ \beta \in |B|\}\}$

The variables α and β range over the set K here. In example (3) we read '$A \ \& \ B$' as 'A and B', and the interpretation of 'and' is that the conjunction of A and B is true (in a given world in K) exactly if both A and B are true (for α is in the intersection $|A| \cap |B|$ exactly if it is in $|A|$ and in $|B|$).

The relation R which occurs in (4) and (5c) is called an *accessibility* or *alternative possibility* or *relative possibility* relation. So we can read (4) as saying that $\Box \ A$ (read as 'Necessarily, A') is true in a given world α exactly if A is true in each world β which is possible relative to α. (Slogan formulation: necessary truth is truth in all possible worlds.)[3] In (5) we have a series of varieties of implication. They are ordered: $A \supset B$ is true if $A > B$ is true if $A \dashv B$ is true (when the relation R in (5c) and the function f in (5b) are related in the usual way). So $|A \supset B|$ is the largest of the three implicative propositions expressed; it is also the least interesting.[4]

Since implication will play a role in a later section, I'll list a few properties of these three connectives. *Strict* or necessary implication was introduced by C. I. Lewis, and it is fairly easy to see that it could be defined:

(6) $|A \dashv B| = |\Box (A \supset B)|$

Counterfactual or subjunctive implication has several explications; the one given by (5b) is due to Robert Stalnaker[5]. The function f which maps pairs of worlds and propositions into worlds is called a *selection* function or sometimes a *Stalnaker* function. Intuitively, we read '$f(\alpha,|A|)$' as 'the world nearest α in which A is true'. Hence '$A > B$', which is officially read as 'If it were the case that A, it would be the case that B' (or with 'be' for 'were' and for 'would be'; this sounds less idiomatic but conveys that we are not assuming that A is definitely not the case) is given the gloss 'B is true in the world nearest the actual world in which A is true'.

There are certain laws that any implication must fulfil. To explain these I must draw a distinction between implication as an *operation* on propositions and as a *relation* among propositions. There is a map

(m:) $|A|, |B| \rightarrow |A \supset B|$

which is exactly what $\supset$ expresses: each connective corresponds in this way to an operation. So this map *is* $|\supset|$ and we can write

(7) $|A \supset B| = |\supset|(|A|, |B|)$.

And there are similar maps $|>|$ and $|\dashv|$. These are all implication operations. But in addition there is (in each interpretation, in each model structure) a *relation* which may be called implication too: the relation that if a given premise is true (in any given world α), so is a given conclusion. This is clearly the relation $|A| \subseteq |B|$. Now the basic properties that any conective $\rightarrow$ must have to be reasonably called an implication connective, are:

(8)   (a) If $|A| \subseteq |B|$ then $|A \rightarrow B| = K$
      (b) $|A| \cap |A \rightarrow B| \subseteq |B|$.

That is, if A implies B then $A \rightarrow B$ is true (in *any* world α in K, in the model structure $M = \langle K, R \rangle$ under discussion); and if A and $A \rightarrow B$ are both true (in given α) then B is true (in α).

That material implication is thus is immediate. But to ensure that strict and counterfactual implication are so, the following restrictions on R and

f are imposed:

    (9)   (a) R is reflexive
           (b) $f(\alpha, |A|) \in |A|$
           (c) If $\alpha \in |A|, f(\alpha, |A|) = \alpha$.

In (9b), there is a little technical problem: suppose A is an absurdity so that presumably $|A| = \Lambda$, the empty set. Then what is $f(\alpha, |A|)$? Stalnaker made the decision to have K include, as a technical convenience, an *impossible world* λ which belongs to all propositions, so that if A is an absurdity, $|A|$ is not Λ but $\{\lambda\}$.

It is clear from (8 a) that, for any implication connective $\rightarrow$, $A \rightarrow A$ is always true. So reflexivity is a law for implication. But there are notable examples of theses that have been accepted as laws for implication which do not follow from (8). One such example is transitivity: the thesis that $|A \rightarrow B| \cap |B \rightarrow C| \subseteq |A \rightarrow C|$. In fact, this holds for material and strict implication, but not for counterfactual implication. Many informal examples can be found to show that transitivity should not be accepted as a general law. I can assure you that if the police were to arrest me for the brutal axe-murder of a parakeet, it would be a case of false arrest. On the other hand I could hardly deny that if I *had* perpetrated this evil and the police had witnesses to that effect, they would arrest me.

3. *Transition to Quantum Logic*

When discussing physical systems of a given kind, we do not talk in terms of possible worlds, but of possible *states*. In the discussion of quantum logic, the set of states play exactly the role of the set of possible worlds in the discussion of modal logic. And so, quantum logical propositions are sets of states. (In some treatments, the propositions are identified independently, but there is then a one-to-one mapping of them into sets of states.) However, not all sets of states are equally significant: in quantum logic only certain propositions are studied (sometimes called 'empirical propositions'; often, confusingly, just 'propositions'). And a main aim of the quantum logician is to describe the mathematical structure of this set of (significant) propositions, given certain assumptions (which, in our terminology, are assumptions concerning the model structure).

The description of the set of propositions is usually quite abstract, the relation to quantum mechanics being left to the imagination of the reader.

When motivating considerations are presented, these are sometimes in terms of the actual quantum theory, and sometimes instead in terms of a general 'phenomenological' picture of measurement. The latter is preferred especially to give the axiomatic development greater generality, with a view to the possibility of alternatives to orthodox quantum theory. But how, exactly, this phenomenological account used as motivation coheres with extant quantum theory is in fact very problematic. I think it would be a mistake *either* to accept such intuitive motivation as the correct touchstone for general axiomatics or to dismiss it a *pons asinorum*.

In order to give the reader some concrete guide to the intentions underlying the more abstract development, I shall here note some elementary aspects of quantum mechanics.

We begin the representation of a quantum mechanical system with a Hilbert space. The set of states of the system is (represented by) the set of *statistical operators* on this space. What exactly these operators are we need not specify just now, but we may note that a certain subset of them is in one-to-one correspondence with the unit vectors of the Hilbert space, and these are called the *pure* states. (Because of the one-to-one correspondence in question, those vectors are also referred to as pure states, a convenient identification.) Each state is a mixture of a set of pure states, but not generally of a unique set.

Besides the states, we recognize the *observables*, which are (represented by) the *self-adjoint* operators on the Hilbert space. Among these, the *projections* on subspaces are in one-to-one correspondence with the subspaces of the Hilbert space. These projections are observables with only two possible values, 0 and 1; the values of any observable are real numbers.

States and observables are related trough the use of a scalar function, the *trace* (written 'Tr'), in the following way:

(10)   The expectation value of observable A in state W equals
        Tr (WA).

When A is a projection, the expectation value of A in state W is exactly the probability that the value found upon measurement of A in that state is 1. But this probability itself equals 1 exactly if W is a mixture of pure states corresponding to vectors in the subspace upon which A projects. Hence the sentence 'The value found upon measurement of A will certainly be 1' is true in exactly those states: the proposition it expresses is the set of those states. Here we have an example of the kind (indeed the only kind) of proposition which is considered in quantum logic.

II. DEVELOPMENT OF QUANTUM LOGIC

1. *Orthodoxy*

There is a variety of efforts to throw light on the structure of quantum theory through reference to non-classical logics. These efforts have resulted in the display of a number of divergent logical systems: examples are those of Strauss, Février, Reichenbach, Birkhoff and von Neumann, Kochen and Specker. But the work of Birkhoff and von Neumann engendered a specific line of research (continued by Mackey, Jauch, Piron, Varadarajan and others) which may presently be considered the 'mainstream'. In this line, adequate axioms yield logics isomorphic to the lattice of subspaces of a Hilbert space. In the research closest to this, the axioms accepted are weaker but not incompatible to this (examples are Gudder and Greechie) with a view to the development of theories alternative to orthodox quantum theory.

In this section I shall trace a development of quantum logic that is potentially of the orthodox variety, with special attention to the basic philosophical assumptions concerning the structure of a physical theory which underlie this.[6]

2. *General Aspects of Orthodox Quantum Logic*

With a particular kind of physical system there is associated a set of possible states (the state-space H), and a set of physical magnitudes or parameters (the set O of observables). Observables take values, and these values are real, and probabilities may be assigned to the taking of values given specific states. Thus we have for each α in H and each observable m a probability measure P_m^α on the real line.

For convenience we shall regard the pairs (m, E) with m in O and E a Borel set, as sentences. The propositions which are here our sole concern are those given by

(1) $|(m, E)| = \{\alpha : P_m^\alpha(E) = 1\}.$

We shall call the set of these propositions P. By set-theoretical inclusion, P is a partially ordered set ('poset').

It is a deliberate decision made in quantum logic to concentrate on propositions to the effect that a certain probability equals one. This procedure of ignoring intermediate probabilities seems justified only given the following assumption.

(2)   (ASSUMPTION) If $P_m^\alpha(E) = 1$ implies $P_{m'}^\alpha(E') = 1$ for all states α, then $P_m^\alpha(E) \leqslant P_{m'}^\alpha(E')$ for all states α.

That is, there is a natural ordering of propositions via set inclusion (*semantic entailment*) and also a natural ordering via the probabilities P^α, and ignoring the latter would be unjustified if it did not coincide with the former. The assumption could be doubted; however, it holds in orthodox quantum theory.

3. Proposition-Ranges of Observables

By the *proposition-range* $[m]$ of an observable m we shall mean the set $\{|(m, E)| : E \text{ is Borel set}\}$. Clearly P is the union of the proposition-ranges of the observables. Also, the proposition-ranges share a highest and lowest element:

(3) $\Lambda = |(m, \Lambda)| \subseteq p \subseteq |(m, R)| = H$

for all propositions p in P. For a little while now we shall explore the structure of a single proposition-range $[m]$. To begin we note that if $P_m^\alpha(E_1) = 1$ and $E_1 \subseteq E_2$, then $P_m^\alpha(E_2) = 1$ also. And secondly, $P_m^\alpha(E_1 \cup E_2) = P_m^\alpha(E_1) + P_m^\alpha(E_2) - P_m^\alpha(E_1 \cap E_2)$. From this it follows that $P_m^\alpha(E_1) = P_m^\alpha(E_2) = 1$ only if $P_m^\alpha(E_1 \cap E_2) = 1$ since $P_m^\alpha(E_1 \cup E_2)$ could not possibly be greater than one. To sum up:

(4) (a) If $E_1 \subseteq E_2$ then $|(m, E_1)| \subseteq |(m, E_2)|$
     (b) $|(m, E_1 \cap E_2)| = |(m, E_1)| \cap |(m, E_2)|$.

So intersection may be regarded as a kind of conjunction or meet, and since the Borel sets are a Boolean sigma-algebra, (4b) can be generalized to countable intersections.

In the similar definition of complement,

(5) $|(m, E)|' = |(m, R-E)|$

we may doubt uniqueness: if $|(m, E)| = |(m', E')|$, does it follow that $|(m, R-E)| = |(m', R-E')|$? The answer is affirmative in view of our Assumption (2). For suppose the antecedent. By (2), $P_m^\alpha(E) = P_{m'}^\alpha(E')$ for all α. Hence for all α, $P_m^\alpha(R-E) = P_{m'}^\alpha(R-E')$. Note that this uniqueness is not relative to $[m]$: if p belongs to $[m]$ and to $[m']$, so does its complement, and its complement is the same whenever defined.

For this complement we find that, for all propositions p, q

(6) (a) $p'' = p$
     (b) if $p \subseteq q$ then $q' \subseteq p'$
     (c) $p' \cap p = |(m, \Lambda)|$.

To prove (b), Assumption (2) is needed. For suppose $p \subseteq q$ and $p = |(m, E)|$, $q = |(m', E')|$. Then for all α, $P_m^\alpha(E) \leqslant P_{m'}^\alpha(E')$, hence for all α $P_{m'}^\alpha(R-E') \leqslant P_m^\alpha(R-E)$. By (2) again we conclude that $q' \subseteq p'$.

We can introduce a join by a De Morgan definition:

(7) $p \cup q = (p' \cap q')'$

and obtain the last clause needed to show that our complement is an orthocomplement:

(8) $p' \cup p = (p'' \cap p')' = |(m, \Lambda)|' = H$.

It must be noted here, however, that $p \cap q$, and hence $p \cup q$ is guaranteed to be a proposition in P only if p and q belong to the proposition-range of a single observable (for see (4b)).

The join we have defined is also the least upper bound in $[m]$ regarded as ordered by set-theoretic inclusion. Note that the following is equivalent to (7) if p, q are in $[m]$:

(9) $|(m, E)| \cup |(m, F)| = |(m, E \cup F)|$.

To show this, suppose that $p = |(m, E_1)|$ and $q = |(m, E_2)|$ have an upper bound $r = |(m, F)|$. Now if $P_m^\alpha(E_1) = 1$, then $P_m^\alpha(F) = 1$, so $P_m^\alpha(E_1 \cap F) = 1$; similarly if $P_m^\alpha(E_2) = 1$ then $P_m^\alpha(E_2 \cap F) = 1$. Hence if either $P_m^\alpha(E_1)$ or $P_m^\alpha(E_2)$ equals 1 then so does $P_m^\alpha(E_1 \cap F \cdot \cup \cdot E_2 \cap F) = P_m^\alpha(E_1 \cup E_2 \cdot \cap F)$. Clearly $s = |(m, E_1 \cup E_2 \cdot \cap F)| \subseteq r$. So it would suffice to show that $s = p \cup q$. Let us abbreviate $(E_1 \cup E_2) \cap F$ to G. Since s is an upper bound of p and q we have $p \cap s = p$ and $q \cap s = q$, so

$$p \mathbin{\underline{\cup}} q = (p \cap s) \mathbin{\underline{\cup}} (q \cap s)$$
$$= |(m, E_1 \cap G)| \mathbin{\underline{\cup}} |(m, E_2 \cap G)|$$
$$= |(m, (E_1 \cap G) \cup (E_2 \cap G)|$$
$$= |(m, (E_1 \cup E_2) \cap G)|$$
$$= |(m, G)|$$

because $G = (E_1 \cup E_2) \cap F \subseteq E_1 \cup E_2$. Thus $p \mathbin{\underline{\cup}} q = s$.

Finally distributivity is immediate from (4b) and (9). Therefore $[m]$ forms a Boolean algebra, and indeed, since (4b) and (9) can be generalized to countable meets and joins due to the sigma structure of the Borel sets, $[m]$ is a Boolean sigma algebra.

4. Relations Among Incompatible Observables

A proposition-range $[m]$ is a Boolean algebra. Classical logic is Boolean. But is the proposition-range really a classical logical structure? The answer is decidedly *no*.

The most important feature to note is that if state α is in $p \mathbin{\underline{\cup}} q$ (so that a sentence expressing $p \mathbin{\underline{\cup}} q$ is *true* in α), it does not follow that α is either in p or in q. For $P_m^\alpha(E \cup E') = 1$ does not entail that $P_m^\alpha(E)$ or $P_m^\alpha(E')$ equals 1. Most specifically, the law of excluded middle holds: $p \mathbin{\underline{\cup}} p'$ is valid, that is, true in every state. But the law of bivalence does not hold: in a given state, both p and p' may fail to be true. The classical appearance of the structure of $[m]$ is therefore mainly a matter of formal analogy.

When we look at P, the set of all propositions (of the indicated sort), the union the proposition-ranges of the observables, even that formal analogy breaks down. Let us look specifically at the intersection of two ranges, $[m, m'] = [m] \cap [m']$.

What does it mean for p to belong to $[m, m']$? It means that there must be Borel sets E, E' such that $p = |(m, E)| = |(m', E')|$. This is certainly possible; for example, $|(m, R)| = |(m', R)|$. It may not cohere with our intuitive notion of what a proposition is: intuitively we might say that 'the probability that upon measurement of m a value is found in E, equals 1' does not express the same proposition as 'the probability that upon measurement of m' a value is found in E', equals 1', even if it is a matter of empirical law (or a logical consequence of our axioms) that the one is the case when and only when the other is the case. But if there are significant differences, they are not differences that could be handled without some change in the semantic framework we are presently utilizing.

What propositions belong to $[m, m']$? We showed already that our complementation is unique regardless of the observable used to define it. Hence we have that if $p = |(m, E)| = |(m', E')|$ then $p' = |(m, R-E)| = |(m', R-E')|$.

(10) (a) If $p \in [m, m']$, so is p'
    (b) If $p, q \in [m, m']$, so is $p \cap q$
    (c) If $p, q \in [m, m,]$, $p \:\cup\: q$ is defined and is in $[m, m']$.

In other words, $[m, m']$ is a subalgebra of $[m]$ and of $[m']$, and must itself be Boolean accordingly.

Is there an observable m'' such that $[m''] = [m, m']$? Nothing we have said implies this, but in the orthodox case that is true.[7]

There is a much discussed result for the orthodox case to the effect that P cannot be mapped homomorphically into the two-element Boolean algebra $\{0, 1\}$ that is used in the semantics of classical logic.[8] That is, we cannot assign 1 (true) and 0 (false) to every proposition in P in such a way that

(a) $|(m, E)|$ has value 1 if and only if $|(m, R-E)|$ has value 0
(b) $|(m, E \cap E')|$ has value 1 if and only if $|(m, E)|$ and $|(m, E')|$ both have value 1.

for *all* observables m. Since each range $[m]$ is Boolean, *it* can be mapped into $\{0,1\}$ in this way; but there can be no single map for the union P of all ranges $[m]$. The reason lies obviously in the existence of non-trivial intersections $[m, m']$, i.e. cases in which $[m, m']$ has elements other than the absurd $|(m, \Lambda)|$ and the tautological $|(m, R)|$.

The set P is ordered by set-inclusion and orthocomplemented by the complement we defined. Even in efforts to generalize upon orthodoxy, it is furthermore asserted to be *orthomodular*. For this, a new assumption is needed. This new assumption relates distinct observables, just as does Assumption (2) and it does this in the same way: it asserts a sufficient abundance of different states to separate propositions concerned with different observables. Let us call p and q in P *orthogonal* exactly if $p \subseteq q'$.

(11)   ASSUMPTION. If X is a set of mutually orthogonal propositions in P then there is an observable m such that $X \subseteq [m]$.

Because of this assumption we can prove orthomodularity:

(12) P is an orthomodular orthocomplemented poset, i.e.
         (a) $p \,\underline{\cup}\, q$ exists if $q \subseteq p'$
         (b) if $p \subseteq q'$ and $q \,\underline{\cup}\, p = H$ then $p = q'$.

Clearly (a) follows at once from (11) and the existence of the join in each
proposition-range. For (b), suppose $p \subseteq q'$ and $q \,\underline{\cup}\, p = H$. By (11), the
former entails that for some m, E_1, and E_2, $p = |(m, E_1)|$ and $q = |(m, E_2)|$.
Now suppose α is in q'. Then $P_m^\alpha(R - E_2) = 1$, and $P_m^\alpha(E_2) = 0$. Because
$E_1 \cap E_2 \subseteq E_2$, $P_m^\alpha(E_1 \cap E_2) = 0$ also. And since $q \,\underline{\cup}\, p = H$, $P_m^\alpha(E_1 \cup E_2)$
$= 1$. So $P_m^\alpha(E_1 \cup E_2) = P_m^\alpha(E_1) + P_m^\alpha(E_2) - P_m^\alpha(E_1 \cap E_2)$, hence $P_m(E_1) =$
$= 1 - 0 + 0 = 1$. Therefore α is in p. This being general in α, $q' \subseteq p$ as well
as $p \subseteq q'$.

What we have now deduced, namely that the propositions form an
orthomodular, orthocomplemented poset, is a basic result considered
general within orthodox quantum logic. Indeed, it is considered so general
that departures from orthodoxy almost always leave it intact. In the next
subsection I shall describe how the same two basic assumptions lead
to that conclusion by a shorter route, the more usual one; the only
purpose being to show how the usual deduction coheres with what I
have just done.[9]

5. Probabilities of Propositions

States can be regarded as assigning probabilities directly to propositions,
by the definition.

(13) $\mu_\alpha |(m, E)| = P_m^\alpha(E)$.

This assumes that if $|(m, E)| = |(m', E')|$, then for each state α, $P_m^\alpha(E) = P_{m'}^\alpha$
(E'). But that is guaranteed by our Assumption (2). Let us call these
functions μ_α measures on P; they can be called probability measures but
only in virtue of their derivation from the probability measures P_m^α, not

strictly speaking. We define $\underline{M} = \{\mu_\alpha : \alpha \in H\}$. We derive:

(Q1)   If $p \neq q$ in P, then there is at least one element μ of $\underline{M}$ such that $\mu(p) \neq \mu(q)$.

For of course $p = \{\alpha : \mu_\alpha(p) = 1\}$, so indeed, if $p \neq q$, then there is a μ in m such that $1 \neq \mu(q)$ or $\mu(q) = 1 \neq \mu(p)$.

(Q2)   There is a member ϕ of P such that $\mu(\phi) = 0$ for all μ in $\underline{M}$

(Q3)   For each proposition p in P there is a proposition p' in P such that
        (a) if $p \neq \phi$ then there is a μ in $\underline{M}$ such that $\mu(p) \nleq \mu(p')$
        (b) for every μ in $\underline{M}$, $\mu(p) + \mu(p') = 1$.

Only (Q3a) needs discussion: $\phi = \Lambda$, hence if $p \neq \phi$ then for some α in H, $\mu_\alpha(p) = 1$; then $\mu_\alpha(p') = 0$, hence $\mu_\alpha(p) \nleq \mu_\alpha(p')$.

From (Q1)–(Q3) it follows that P is an orthocomplemented poset, the partial ordering being given by

(14) $p \leq q$ iff $\mu(p) \leq \mu(q)$ for all μ in $\underline{M}$

which of course also follows from (2) directly. To achieve orthomodularity we need

(Q4)   If X is a countable orthogonal subset of P (i.e. $p \leq q'$ for p, q in X), then there is a proposition $r = VX$ in P such that
$$\mu(VX) = \sum \mu(p)$$
for all μ in $\underline{M}$.

But this follows of course from our Assumption (11). It is clear that (2) and (11) are somewhat stronger than (Q1)–(Q4), but it is from the former that the latter receive their intuitive justification. On the other hand, it does not seem that (2) and (11) can yield any stronger conclusion about the structure of P as a whole than the orthomodularity and orthocomplementation implied by (Q1)–(Q4).

6. *Operations and Counterfactuals*

If the observables are those of orthodox, elementary quantum theory, P is isomorphic to the lattice of subspaces of a Hilbert space. This is indeed an orthocomplemented orthomodular poset; but it is much more.

It is a lattice ($p\cup q$, $p\cap q$ are in P for any p and q in P), it is atomic, it is semi-modular.[10]

To say that P is atomic means that if p is in P there is an atom or atomic proposition $q \subseteq p$ in P: an atom is a non-absurd proposition which is not implied by any other distinct non-absurd proposition. In the orthodox case we have the following: we call state α pure if $\mu_\alpha = c\mu_\beta + (1-c)\mu_\gamma$ for $0 < c < 1$ entails that $\alpha = \beta = \gamma$; and the *atoms* of P are exactly the sets $a_\alpha = \{\alpha\}$ for pure states α. So in the orthodox case, the atomic propositions can actually be read as 'the system is certainly in pure state α'. Or, more precisely, for each pure state α there is an observable m_α which has value 1 in state α alone, and the atomic propositions are the propositions $|(m_\alpha, \{1\})|$ for pure states α. It will be no surprise that this is not implied by our previous assumptions.

The most interesting problem however, is probably that of giving some physical content to the assertion that P is a lattice. And some exciting work has been done on this problem fairly recently, though there is far from universal agreement on its significance, simply because the main departures from orthodoxy involve the rejection of the lattice structure of P.[11] What I shall now attempt to display in the perspective of our present approach are the assumptions whereby J. C. T. Pool derives that P is a lattice.[12]

As an aid to the imagination, we have the following phenomenological account of measurement processes. In a *yes-no* measurement of proposition p, systems are separated into two beams: in the first beam we find only systems in states in which p is true, and in the other only systems in states in which p' is true. If a given system, whose initial state is α, were to leave in the first beam, the p-beam, would it also be such as to satisfy proposition q? Following Stalnaker's theory (explained in Section I) of conditionals, we construe the preceding question as one asking whether the counterfactual conditional 'Were the end-state to satisfy p, it would satisfy q' is true. Recall that in Stalnaker's account of conditionals we use a selection function $f(\alpha, p)$ which yields the end-state in question. Writing $f_p(\alpha) = f(\alpha, p)$, and accepting Stalnaker's conditions, we have

(15)   (Assumptions)

    (a) $f_p(\alpha) \in p$

    (b) if $\alpha \in p$, $f_p(\alpha) = \alpha$.

(16)   (DEFINITION)

$$p > q = \{\alpha: f_p(\alpha) \in q\},$$

where we shall also henceforth stipulate that there is a null-state λ, belonging to all propositions (the physically impossible state) as a technical convenience.

(17)   (ASSUMPTION) There is a state λ such that
    (a) $\lambda \in p$ for all $p \in P$
    (b) $f_p(\alpha) = \lambda$ if and only if $\alpha \in p'$.

It is clearly not to be taken for granted that $p > q$ belongs to P. Nor must it be thought that the function f_p could be defined in quantum mechanics, except in very special cases. For a non-atomic proposition p there are clearly distinct pure, non-null states β such that $f_p(\alpha) = \beta$ is a possibility. (This was pointed out in discussion by Professor J. Stachel.) In the general theory of conditionals there is a similar ambiguity: if p is the proposition expressed by 'Bizet and Verdi are compatriots', and α is the real world, does $f_p(\alpha)$ satisfy 'Bizet and Verdi are both French' or 'Bizet and Verdi are both Italian'?

However that may be, it will be instructive to look at the properties of these counterfactual conditionals before displaying their relations to Pool's assumptions. Writing H for the set of all states and Λ for $\{\lambda\}$, here are some immediate corollaries

(18) (a) $H = (p > p)$
    (b) $p = \{\alpha: f_p(\alpha) = \alpha\} = (H > p)$
    (c) $p' = \{\alpha: f_p(\alpha) = \lambda\}$
    (d) $f_p f_p = f_p.$

These are all immediate; the following shows how the ordering of the propositions is reflected in the selection function:

(19) $p \subseteq q$ if and only if $f_q f_p = f_p.$

Suppose first that $p \subseteq q$. Then $f_p(\alpha)$ is in q just because it is in p. Hence $f_q f_p(\alpha) = f_p(\alpha)$. On the other hand, suppose that $f_q f_p = f_p$ and let α be in p. Then $f_p(\alpha) = \alpha$. Hence $f_q f_p(\alpha) = f_p(\alpha)$ implies that $f_q(\alpha) = \alpha$. But $f_q(\alpha)$ is in q; hence α is in q. So, since α is here an arbitrarily chosen member of p, $p \subseteq q$.

By the notion of the null-space of an operator we arrive at a direct characterization of the conditionals in terms of the selection function:

(20)   (Definition) $N(f_{p_1}\cdots f_{p_n}) = \{\alpha: f_{p_1}\cdots f_{p_n}(\alpha) = \lambda\}$.

(21) $p > q = N(f_{q'} f_p)$.

For α is in $(p > q)$ exactly if $f_p(\alpha)$ is in q (which is q'') exactly if $f_{q'}(f_p(\alpha)) = \lambda$. More generally,

(22) $p_n > (p_{n-1} > \cdots (p_1 > q)\cdots) = N(f_{q'} f_{p_1} \cdots f_{p_n})$.

We may call these conditionals *right counterfactuals*.

Now in Pool's work, the first and basic assumption is the following:

(*Pool)   For all $p_1, \cdots, p_n \in P$, $N(f_{p_1}\cdots f_{p_n}) \in P$.

(For readers of Pool this may be obscured somewhat since he does not use the technical convenience of a null-state λ, but leaves $f_p(\alpha)$ undefined exactly if $\alpha \in p'$.) This means then that he assumes that all right counterfactuals formed from propositions in P are again in P. But then of course $f_{p>q}$ is also defined, namely as $f_{N(f_{q'} f_p)}$ so *all* counterfactuals are in P:

(*)   If p, q are in P, so is $p > q$

is an assumption equivalent to (*Pool).

Pool adds only one other assumption to yield the conclusion that P is a lattice, namely that if $f_{p_1}\cdots f_{p_n} = f_{q_1}\cdots f_{q_m}$ then $f_{q_m}\cdots f_{q_1} = f_{p_n}\cdots f_{p_1}$. The case for its physical motivation I shall leave for the reader to consider.

III. INTERPRETATION OF QUANTUM-LOGICAL PROPOSITIONS

1. *Problems About Mixtures*

While I shall continue to pursue a general logical approach, the problems discussed in this part are motivated mainly in terms of extant quantum theory. So shall now continue the sketch of quantum mechanics begun at the end of Section I. In a standard[13] quantum logic, the poset P is isomorphic to the lattice of subspaces of a Hilbert space H. This space has as elements the (vectors representing) *pure* states only, so that it would be wrong to identify the propositions in P with corresponding subspaces. However, if α is any state, it is (represented by) a statistical operator on H. A proposition p in P is represented by a subspace S_p of H, or equivalently by the projection E_p on S_p. And we have

$$\mu_\alpha(p) = \text{Tr}(\alpha E_p),$$

where Tr, the trace, is the function already referred to in Section I.

As my formulation suggests, I shall identify states with the statistical operators that represent them, although I eschew most other such identifications. A *pure* state α is an idempotent statistical operator ($\alpha\alpha = \alpha$), equivalently, it is the projection on a 1-dimensional subspace $[\phi]$ spanned by (unit) vector ϕ, and I shall also write E_ϕ for $E_{[\phi]}$. Any state β can be given as a weighted sum of pure states

(1) $\qquad p = \sum d_i \alpha_i, \qquad 0 < d_i \leqslant 1$
$$\sum d_i = 1$$

and then we also have, equivalently,

(2) $\qquad \mu_\beta = \sum d_i \mu_{\alpha_i}.$

This will explain the general definition of 'pure state': α is pure if $\mu_\alpha = c\mu_\beta + (1-c)\mu_\gamma$ for $0 < c < 1$ entails that $\alpha = \beta = \gamma$. A state which is not pure is called a *mixed state* or *mixture*.

We say that Equation (1) or (2) gives a decomposition of β into the set $\{\alpha_i\}$. We must carefully distinguish mixtures from superpositions: given Equation (1) we say β is a mixture of the set $\{\alpha_i\}$, and also a superposition thereof, by the definition:

(3) $\qquad \beta$ is a *superposition* of the set of states X exactly if $\mu_\beta(p) = 0$ for all propositions p such that $\mu_\alpha(p) = 0$ for all α in X.

Because of the existence of our orthocomplement the condition is clearly equivalent to:

(4) $\qquad$ if $X \subseteq q$ then $\beta \in q$, for all $q \in P$.

While all mixtures are superpositions, the converse holds only for the classical (and not the quantum-mechanical) case.

In quantum mechanics, superpositions which are not mixtures can be formed as follows: if $\phi = c\psi + d\zeta$ then E_ϕ is a superposition of E_ψ and E_ζ, while, say, $\frac{1}{2}E_\psi + \frac{3}{4}E_\zeta$ is a mixture of E_ψ and E_ζ.

There is no doubt that mixtures can be used conveniently to represent subjective ignorance. We might know that a system is in fact in state E_ψ

with probability $\frac{1}{4}$ and in state E_ζ with probability $\frac{3}{4}$. In such a case we would base our predictions about future measurement outcomes on the premise that the system is in state $\frac{1}{4}E_\psi + \frac{3}{4}E_\zeta$. However, as a general interpretation of mixed states this will not do: the assertion that a system is in state $cE_\psi + (1-c)E_\zeta$ does not imply that the system is really either in state E_ψ or in state E_ζ. The problems with that interpretation, and some variants, are discussed in the Appendix. The first problem is that it is possible to have distinct states $\alpha, \beta, \gamma, \delta, \zeta$ such that

$$\zeta = c\alpha + (1-c)\beta = d\gamma + (1-d)\delta,$$

where $0 < c, d < 1$, and $c \neq d$. Decompositions of mixed states are not unique.

The second problem, more serious than the first, concerns the states of interacting systems. If two systems X and Y have pure states α on space H_1 and β on space H_2, we assign the complex system $X + Y$ the tensor-product $\alpha \otimes \beta$ on space $H_1 \otimes H_2$. As they interact, $X + Y$ will go into some superposition of states $\{\alpha_i \otimes \beta_j\}$, say Ξ. We can now ask what state is to be assigned to components X, Y separately at that time. There can be an answer (not entirely unproblematic) to this question (see Appendix), to the effect that X is in a mixture of $\{\alpha_i\}$ and Y in a mixture of $\{\beta_j\}$. If we now assumed that being in a mixture implies *really* being in one of the component pure states, then by the practice described above we would have to say that $X + Y$ is really in one of the states $\alpha_i \otimes \beta_j$. But that contradicts the assertion that $X + Y$ is in Ξ.

What I have described is curious enough in itself, but it is also the core of the measurement problem in quantum mechanics. For the process described could have been a measurement interaction, in which X is a system being subjected to measurement, and Y an apparatus which is in state β_j exactly if it shows pointer-reading r_j, say. Surely we would like to say that, at the end, Y really does show *some* pointer reading, and hence is *really* in one of the states β_j! And many writers have taken exactly the assertion that Y is in a mixture of the set $\{\beta_j\}$ as justification for this conclusion.

My aim is to show that we can have our cake and eat it too. That is, I shall maintain the conclusion described in the preceding paragraph, while avoiding the inconsistency pointed out in the paragraph before that. To this end I have previously proposed a 'modal interpretation'.[14] I shall now re-develop this interpretation in a general quantum-logical context.

2. *Logical Aspects of Mixtures*

To begin, I shall define an alternative possibility or *access* relation, which will play an important role in the remainder of this discussion

(5)   (DEFINITION) $\alpha \, R \, \beta$ exactly if there is a real number $c \, (0 < c \leqslant 1)$, and state γ such that $\mu_\alpha = c\mu_\beta + (1-c)\mu_\gamma$.

Clearly, if α is pure, then there is access from α only to α; if α is a mixture there is access from α to all states which can appear in a decomposition of the form of Equation (2).

Recall that in modal logic, α is in $\square p$ exactly if $\{\beta : \alpha R\beta\} \subseteq p$. If we call p a *necessary proposition* exactly if $\square p = p$, then this amounts for necessary propositions to: α is in p exactly if $\{\beta : \alpha R\beta\} \subseteq p$. I shall now show that all quantum-logical propositions have this character. Writing $U_\alpha = \{\beta : \alpha R\beta\}$, that means:

(6) $\alpha \in p$ if and only if $U_\alpha \subseteq p$

To show this, we must prove that if $\mu_\alpha = c\mu_\beta + (1-c)\mu_\gamma$ for $0 < c < 1$, then $\mu_\alpha(p) = 1$ if and only if $\mu_\beta(p) = \mu_\gamma(p) = 1$. The 'if' part is immediate. For the 'only if' part, suppose that $\mu_\beta(p) \neq 1$ (or that $\mu_\gamma(p) \neq 1$, the argument is similar). Then $c\mu_\beta(p) < c$. Now $(1-c)\mu_\gamma(p) \leqslant (1-c)$; therefore $\mu_\alpha(p) = (c\mu_\beta(p) + (1-c)\mu_\gamma(p)) < (c + (1-c)) = 1$.

The set U_α has a number of important subsets, which I shall call *bases* of α.

(7)   (DEFINITION) A set X of states is a *base* of α if there are real numbers $0 < c_\beta \leqslant 1$, for $\beta \in X$ such that
        (a) $\sum\{c_\beta : \beta \in X\} = 1$
        (b) $\sum\{c_\beta \mu_\beta : \beta \in X\} = \mu_\alpha$
        (c) Every member of X is pure.

So to display a base is to give a decomposition. The numbers ('weights') c_β are definable in a special case. Call a base X of α orthogonal if any two of its elements are orthogonal states (by the definition $\mu_\beta(a_\gamma) = 0$, equivalently $a_\gamma \subseteq a'_\beta$, which is symmetric in β and γ). In such a base, $\mu_\alpha(a_\beta) = c_\beta$ of course. (Recall that $a_\alpha = \{\alpha\}$, an atomic proposition.)

Since $\sum\{c_\beta\mu_\beta: \beta\in X\}=c_\gamma\mu_\gamma+\sum\{c_\beta\mu_\beta: \beta\in X \ \& \ \beta\neq\gamma\}=c_\gamma\mu_\gamma+$
$+(1-c_\gamma)\sum\{(c_\beta/(1-c_\gamma))\mu_\beta: \beta\in X \ \& \ \beta\neq\gamma\}$, we see that a base X of α is indeed a subset of U_α. As a quick corollary we have

(8) $\alpha\in p$ if and only if $X\subseteq p$, for any base X of α.

Because we are only concerned with the standard case here, we have

(9)   (ASSUMPTION) Every state has at least one base.

In addition, standard quantum logics are complete lattices, so we have

(10)   The least proposition to which α belong is $p_\alpha=\bigcup\{a_\beta: \beta\in X\}$, where X is any base of α.

If α is pure, $p_\alpha=\{\alpha\}=a_\alpha$; but if α is not pure, p_α has many members distinct from α. For example, if $\mu_\alpha=\frac{1}{2}\mu_\beta+\frac{1}{2}\mu_\gamma$ and $\mu_\delta=\frac{1}{4}\mu_\beta+\frac{3}{4}\mu_\gamma$ then $p_\alpha=p_\delta$.

The sets U_α and p_α can be related to some extent: for any base X of α we have $X\subseteq U_\alpha\subseteq p_\alpha$. Nancy Delaney Cartwright has shown me a proof by D. J. Collup to the effect that in quantum mechanics U_α is the image space of α (i.e., $\{\alpha(\phi): \phi\in H\}$) at least if α has a finite base. It is tempting therefore to introduce the assumption that $p_\alpha=U_\alpha$ but whether this assumption could be justified (either with respect to quantum mechanics or on more general grounds) is not clear at this point. Without assuming it, we number this thesis

(11) $p_\alpha=U_\alpha$

for future reference. (See note added in proof, page 113.)

What is clear is that distinctions among mixtures with a common base cannot be made in terms of the quantum-logical propositions that are made true by them. Nor does 'the system is in mixture $\sum c_i E_{\gamma_i}$' express a quantum-logical proposition (member of P), which certainly shows significant limits on expressibility within quantum logic. As further corollary we may point out that if proposition p is represented by subspace S, then p cannot be read as 'the system is in a state represented by a vector in S' but only as 'the system is in a mixture of states represented by vectors in S'. And finally, it is essentially the pure states alone that order the quantum-logical propositions: $p\subseteq q$ if and only if all the pure states in p belong to q.

3. *The Modal Interpretation*

We have seen that, formally speaking, each quantum-logical proposition is a necessary proposition, in the sense that $p = \{\alpha: U_\alpha \subseteq p\}$, where U_α, the set of states possible relative to α, is the set of states β such that α is a mixture of β and some other state(s). This is perhaps not too surprising since p is expressed by some sentence (m, E) which reads 'Upon measurement of m, the value found is *certainly* in E'. The formal relation between necessity-statements and statements that a certain probability equals one, has often been noted before.

The question is now: are there also contingent propositions, related to quantum-logical propositions in the way that $|A|$ is related to $|\Box A|$ in modal logic generally? Certainly many writers have spoken, implicitly, in this way. Jauch draws a distinction between states and events; quantum-logical propositions describe states; what propositions describe events?

My argument that there are contingent propositions that play an important role vis à vis quantum theory is as follows. Quantum mechanics does not yield exact predictions of measurement outcomes. Let Φ_0 be a proposition giving exactly the state of system and measurement apparatus at a time t_0. From this, by Schrödinger's equation, we derive a proposition Φ_t giving exactly the state at t. Let $A(t)$ be a proposition describing the measurement outcome at t. Then, by hypothesis, Φ_t does not imply $A(t)$. Could $A(t)$ be a proposition attributing a state, or kind of state? No, for then it is, if true, either implied by Φ_t or in contradiction with Φ_t. Therefore $A(t)$ must be, unlike Φ_t, a proposition which is not a state-attribution. Hence there are two kinds of proposition.

One thing may be confusing. Say the sentence (m, E) expresses proposition Φ_t. This sentence is naturally read as relating an observable to a Borel set. The sentence expressing $A(t)$ will also relate an observable to a Borel set: it will read something like 'observable m_1 (actually) has a value in E_1.' But the similarity is deceptive. For (m, E) is true exactly if the state α of the system satisfies a certain condition, namely $P_m^\alpha(E) = 1$. The distinction can be made carefully in the reading: Let us read (m, E) always as 'the observable m has its value certainly in E' and introduce a new notation, $\langle m, E \rangle$, to be read as 'the observable m has its value actually in E'. I am suppressing any reference to measurement in the readings at this point; but I shall return to its role below.

In our stylized example we found examples of propositions whose truth-value is not determined by the state. In that case, an actual situation cannot be modelled by a state alone, since presumably what is true depends exactly on the actual situation. So I shall introduce an extra factor, a parameter λ, and say that a *model* is a couple $M = \langle \alpha, \lambda \rangle$, where α is a state and λ this other parameter, presently not further specified. (It might only be an index, a mathematical convenience, or it might have some physical significance; this question I leave open.)

And we have two sets of sentences: let L_1 be the set of sentences (m, E) and L_2 the set of sentences $\langle m, E \rangle$, where m is an observable and E a Borel set.[15] All we have to do now to complete the description of this compound language is to give truth-conditions to specify which sentences are true in which models. But we cannot do that precisely yet, because λ is not exactly specified.

However, I shall lay down four conditions

(12) (a) (m, E) is true in $\langle \alpha, \lambda \rangle$ exactly if $P_m^\alpha(E) = 1$;

      (b) $\langle m, E \rangle$ is true in $\langle \alpha, \lambda \rangle$ if (m, E) is true in $\langle \alpha, \lambda \rangle$;

      (c) for each observable m there is a non-empty Borel set E such that $\langle m, E' \rangle$ is true in $\langle \alpha, \lambda \rangle$ if and only if $E \subseteq E'$;

      (d) the sentences of L_2 true in $\langle \alpha, \lambda \rangle$ are a maximal satisfiable subset of L_2; that is, if all sentences of L_2 which are true in $\langle \alpha, \lambda \rangle$ are true in $\langle \alpha', \lambda' \rangle$, then all sentences of L_2 true in $\langle \alpha', \lambda' \rangle$ are true in $\langle \alpha, \lambda \rangle$.

Because of condition (a), the proposition *now* being expressed by (m, E) is essentially that proposition $|(m, E)|$ it expressed in the earlier parts of the paper, namely $\{\langle \alpha, \lambda \rangle : \alpha \in |(m, E)|\}$. It will do no harm therefore to keep the same notation: that is, $|(m, E)| = \{M : (m, E)$ is true in model $M\}$ and $|\langle m, E \rangle| = \{M : \langle m, E \rangle$ is true in $M\}$. Condition (b) says that certainly implies actuality; condition (c) does not contradict (b), because there is a least Borel set E such that $P_m^\alpha(E) = 1$, and is surely necessary if we are to read $\langle m, E \rangle$ as relating observables to their values at all. We have as consequences:

(13)   if $\langle m, E \rangle$ is true in M, $\langle m, R - E \rangle$ is not true in M

(14) $\langle m, E \cap E' \rangle$ is true in M if and only if $\langle m, E \rangle$ and $\langle m, E' \rangle$ are true in M.

But it is not ruled out at this point that $\langle m, E \rangle$ and $\langle m, R-E \rangle$ may neither be true, nor that $\langle m, E' \cup E \rangle$ may be true while neither $\langle m, E \rangle$ nor $\langle m, E' \rangle$ is true. For we cannot, in my opinion, either assert or deny at this point that the logical structure of L_2 is different from that of L_1.

For future convenience, I shall henceforth write $S_i(M)$ for the set of sentences of L_i that are true in M, $i = 1,2$. And I shall adopt the terminology that state α *makes* both (m, E) and $\langle m, E \rangle$ *true* if $\langle \alpha, \lambda \rangle \in |(m, E)|$ for any parameter λ. In virtue of condition (12b) this usage will be consistent. Conditions (a), (b), (d) may be summed up as

(15) (a) $\{(m, E) : P_m^\alpha(E) = 1\} = S_1(M) \subseteq \{(m, E) : \langle m, E \rangle \in S_2(M)\}$
      (b) if $S_2(M) \subseteq S_2(M')$ then $S_2(M) = S_2(M')$.

Several questions have been left open because λ was not specified. I shall now state one more such question, which it is essential to answer in the affirmative if the term 'modal interpretation' is to be justified. Is there a significant relation R^* among models such that

(16) $M \in |(m, E)|$ if and only if $\{M' : MR^*M'\} \subseteq |\langle m, E \rangle|$

so that (m, E) could be written as $\Box \langle m, E \rangle$? In the remainder of this paper I shall explore two positions concerning the contingent propositions expressed in L_2, for which these questions have definite answers.

4. *A Copenhagen Variant*

If two observables are incompatible, they cannot have simultaneous definite values. Thus phrased, the thesis sounds tautological. What else could 'incompatible' mean? But the statement that if two observables are represented by non-commuting operators, they cannot *actually* have simultaneous definite values, is definitely not tautological. That it is true nevertheless is Copenhagen doctrine. And the doctrine is supported by discussions of measurement. The statement that a given observable actually has a certain value has as typical use the description of a measurement-outcome. But then its truth requires (as necessary condition), the actuality of a certain measurement set-up. And measurement set-ups appropriate to observables represented by non-commuting operators are themselves incompatible: they cannot be simultaneously realized.

I shall certainly not attempt to evaluate here either the position or the supporting argument. What is important is that this position implies a definite answer to the questions left open in the preceding section. For it asserts that L_2 statements can be true together only if (as well as if) they can have probability one together. Precisely,

(17)   (Copenhagen I). There is a model M in which all of set X of L_2 sentences are true only if there is a state β which makes all of X true.

This means that for any model M there is a state β such that

$$S_2(M) \subseteq \{\langle m, E\rangle : P_m^\beta(E)=1\}.$$

This fulfills condition (12c) since there is a least Borel set E such that $P_m^\beta(E)=1$. But condition (12d) adds that $S_2(M)$ is a maximal satisfiable subset of L_2. This implies that β is pure; for if β is a mixture of γ and δ, then $P_m^\beta(E)=1$ implies $P_m^\gamma(E)=1$, and in the present (orthodox) context, every state is a mixture of pure states (every state has a base). To remain consistent with conditions (12a) and (12b) this state β must make true whatever statements are made true by α if $M=\langle\alpha, \lambda\rangle$; hence β is in p_α. So we have the following consequence:

(18)   For each model $M=\langle\alpha, \lambda\rangle$ there is a pure state $\beta=\alpha_\lambda$ such that $\beta \in p_\alpha$ and $S_1(M)=\{(m, E):P_m^\alpha(E)=1\}$ and $S_2(M)= = \{\langle m, E\rangle : P_m^\beta(E)=1\}$.

If Thesis (12) were assumed, the following assumption could be deduced (essentially):

(19)   (Copenhagen II). The parameters λ are the mappings of all states into pure states such that $\lambda(\alpha)=\alpha_\lambda \in U_\alpha$.

Now (18) and (19) together give an exact specification of the models with complete truth-definition for both L_1 and L_2. And if we define R^* by $\langle\alpha, \lambda\rangle R^* \langle\beta, \lambda'\rangle$ if and only if α R β and β is pure (so that $\lambda'(\beta)=\beta$) then Equation (16) holds (since U_α will contain a base X of pure states, and for each proposition q, α is in q if and only if $X \subseteq q$ if and only if $U_\alpha \subseteq q$).

Before turning to an anti-Copenhagen position, I shall explain briefly how this account needs to be modified for multibody systems. If more than one system is being described, reference to these systems must appear in the sentences. So we must write for example $\langle m, S_i, E\rangle$ rather than $\langle m, E\rangle$, to say that m has, in system S_i, a value in E. The total system

$S_1 + \cdots + S_n$ is assigned a state α and mixed states β_i can be calculated for the components S_i. We will get in a perfect analogue to (18) the consequence that for each S_i there is a pure state γ_i such that $\beta_i R \gamma_i$ and $\langle m, S_i, E \rangle$ is true exactly if $P_m^{\gamma_i}(E) = 1$. But this is not good enough. For if $\alpha = \sum a_{ij}(\phi_i \otimes \psi_j)$ and $a_{km} = 0$, then although γ_k could be ϕ_k and γ_m could be ψ_m, we must rule out that *both* of these possibilities obtain. This is a very non-classical feature of quantum mechanical systems; it concerns the kind of correlation that appears in the Einstein-Podolski-Rosen paradox. So if our account is modified for this more complex case, it is clear that an extra condition, of the kind just indicated, must be imposed on the models. The function λ must yield an assignment of pure states to all subsystems of the total system, but in such a way that these non-classical correlations are not violated.

Something like the ignorance interpretation of mixtures is going on here. If a system is in a mixture of eigenstates of m with corresponding values v_i, then, I am saying, the observable m really has one of the values v_i in the actual situation. But notice that I am *not* saying that the system is really in one of the corresponding eigenstates! In the model $M = \langle \alpha, \lambda \rangle$, we find that α has a decomposition $\alpha = \sum w_i E_{\phi_i}$ such that $\lambda(\alpha)$ is indeed one of the set $\{E_{\phi_i}\}$, say $\lambda(\alpha) = E_{\phi_k}$. But $\lambda(\alpha)$ is used only to assign truth or non-truth to L_2- statements, to contingent propositions. And the principles of quantum theory apply only to L_1-statements, to necessary propositions. For that reason the modal interpretation is not subject to the inconsistencies of the ignorance interpretation.

5. *Hidden Variables*

When I said that truth-values in an actual situation are not entirely determined by states, and introduced a further parameter λ to summarize the additional information, the reader must at once have thought of hidden variable theories. (In an earlier paper I presented the modal interpretation (essentially the Copenhagen variant of Section 4), and Dr. J. Dorling commented that I was really introducing hidden variables.) But it can also be seen that without restrictions on the parameter λ, the predictions of quantum theory would not be affected. And in the version just developed, λ does no more than implement something analogous to the ignorance interpretation, which is familiar from the Copenhagen writings as well.

Indeed, there are close relations to hidden variables, but the Copenhagen variant is not a hidden variable theory, *either* in the senses given to that term by von Neumann, Kochen and Specker, and Jauch and Piron, *or* in the sense of Bohm and Bub ('contingent parameters')[16]. Bub has proposed a general scheme for hidden variable theories.[17] It can be briefly described as follows:

> A hidden variable theory defines for each maximal Boolean sublattice B of P a set of maps $h_B: S \times X \to V_B$, where S is the set of states, X the set of hidden variables, and V_B the set of bivalent valuations on B.

By a bivalent valuation of B is meant a complete homomorphism of B into the two-element Boolean algebra $\{0, 1\}$. That is, if v is in V_B, and $p \in P \subseteq B$, then $v(p) = 1$ if and only if $v(p') = 0$, and $v(\cap P) = 1$ exactly if $v(q) = 1$ for each member q of P. Or, more perspicuously, v is a classical truth-table assignment to B. The significance of this is that each proposition range of an observable is a Boolean sublattice of P, as we saw in Section II; the converse holds also in the standard case; and a maximal Boolean sublattice is the proposition range of a maximal observable; each observable can be defined as a function of a maximal observable.

Now in the variant of the modal interpretation that we presented in Section 4, it is not true that bivalent valuations are constructed for each maximal Boolean sublattice of P. Indeed, let α be pure; then in any model $\langle \alpha, \lambda \rangle$ we must have $\lambda(\alpha) = \alpha$. And the valuation $v_\alpha(p) = 1$ exactly if $\alpha \in p$, $v_\alpha(p) = 0$ if $\alpha \in p'$, is the only valuation available in this case. It will be a bivalent valuation on *some* maximal Boolean sublattice, but all the others remain undiscussed.

But clearly all this is because we did not take advantage of the enormous latitude the introduction of L_2-statements gives us. Could we not jettison the Copenhagen assumption, extend v_α in the obvious way to

$$v_\alpha(\langle m, E \rangle) = 1 \text{ if } \alpha \in |(m, E)|$$
$$v_\alpha(\langle m, E \rangle) = 0 \text{ if } \alpha \in |(m, R - E)|.$$

and *then extend* it to a complete truth-table assignment to all of L_2?

Before we try to answer this question, we need to see exactly why Bub specified that a hidden variable theory gives separate valuations for each maximal Boolean sublattice, and not to P as a whole. The reason is the

Kochen and Specker proof concerning limitations on hidden variable theories.[18]

The results are as follows: there is no map h of P into $\{0, 1\}$ such that if p and q belong to the proposition-range of the same observable then

$$h(p \cap q) = h(p) \cdot h(q)$$
$$h(p \cup q) = h(p) + h(q) - h(p \cap q)$$
$$h(I) = 1,$$

where I is the maximal element, i.e. $|(m, R)|$. As a corollary, there can, in the standard case, be no map f of the observables into the real numbers such that

$$f(mA + nB) = mf(A) + nf(B)$$
$$f(AB) = f(A) \cdot f(B)$$
$$f(I) = 1$$
$$f(A) \text{ is in the spectrum of } A,$$

where I is now the observable which has value 1 in every state and the spectrum of A is the set of values v such that $P_A^\alpha(\{v\}) = 1$ for some state α. The reason for this corollary is not hard to see: if each observable m could be assigned a specific value $v = f(m)$, then $|(m, E)|$ could be called true exactly if $v \in E$ and false otherwise, thus yielding the kind of homomorphism of P into $\{0, 1\}$ which was stated to be impossible.

And why exactly is such a homomorphism impossible? Each proposition-range can be mapped homomorphically into $\{0, 1\}$ because it is a Boolean algebra. But the union of such homomorphisms will not be a homomorphism: it will not be even a function, since the proposition-ranges overlap, and hence some propositions may (indeed, in this case, must) get more than one value. And so, in Bub's scheme, propositions do get more than one value, but relative to different hidden parameters. And these different parameters correspond, in a very Copenhagen way, to different, incompatible, measurement set-ups.

But look for a moment at the set of maps $\{h_B(\alpha, \lambda) : B \text{ a maximal}$ Boolean sublattice of $P\}$. The information given by this set of maps can be summed up by a single map: $h^*(B, \alpha, \lambda) = h_B(\alpha, \lambda)$ for all B. Suppose we give to each L_2 sentence $s = \langle m, E \rangle$ a value (in a purely indexical sense) which is a triple $\langle B_s, \alpha, \lambda \rangle$ and say that in model $\langle \alpha, \lambda \rangle$ the truth-value of $\langle m, E \rangle$ is $h^*(B_s, \alpha, \lambda)$. Then we have a truth-table assignment to all of L_2. In the next and final section I shall explain the uses of this trivial bit of conjury.

6. *An Anti-Copenhagen Variant*

It is quite possible for the quantum-logical sentences 'The value of m is certainly in E' and 'The value of m_1 is certainly in E_1' to be true in exactly the same states, although $m \neq m_1$ and $E \neq E_1$. This is at the root of the Kochen and Specker proof. And as long as we regard propositions as members of P, we must then say that these so different looking sentences express the same proposition. In a different, more ordinary sense they do not, *just because* they are about different observables. And certainly, considering such sentences as 'the value of m is actually in E', we can see that $|(m, E)| = |(m_1, E_1)|$ does not need to imply that $|\langle m, E \rangle| = = |\langle m_1, E_1 \rangle|$ – although it does in the Copenhagen variant of Section 4.

Various writers, including Popper and later Margenau and Park, have argued that each observable may be assigned a definite simultaneous value. It is clear that to do so, they must say that quantum-states do not incorporate all possible information, and observables cannot be identified with the operators that represent them. But then the immediate question is: if not by the representing operators, how are observables to be identified?

I shall answer this question as follows: maximal observables can be identified with the operators that represent them.[19] All other observables are functions of maximal observables,[20] and are identified jointly by a maximal observable and a function. This is different from the usual procedure. If $P_m^\alpha = P_{m'}^\alpha$, for all α, then m and m' are represented by the same operator. It is in addition usually assumed (e.g. explicitly by Kochen and Specker, *op. cit.*, p. 63), that in that case $m = m'$. But we accept this only for the case in which m and m' are maximal observables. Now the following two facts will help us here:

(i)   for each observable m there is a maximal observable m^* and Borel function f such that
$$P_m^\alpha(E) = P_{m^*}^\alpha(f^{-1}(E))$$
for all states α and Borel subsets E of the real line, so that
$$|(m, E)| = |(m^*, f^{-1}(E))|$$
for all Borel sets E;

(ii)   the mapping $m \to [m]$ restricted to maximal observables m is
            a one-to-one mapping onto the maximal Boolean sublattices
            of P.

With reference to (i) we may note that, if m^* is represented in quantum
mechanics by the operator A, m is represented by $f(A)$. And now we
can give the new postulate:

(20)   (anti-Copenhagen I) For each observable m there is a maximal
            observable m^* and Borel function f_m such that
            (a) for all states α and Borel sets E, $P_m^\alpha(E) = P_{m^*}^\alpha(f^{-1}(E))$,
            (b) for observables m_1 and m_2, $m_1 = m_2$ if and only if $m_2^* = m_1^*$
            and $f_{m_1} = f_{m_2}$.

If β is any state, there is a least Borel set $E(\beta, m)$ such that β makes
$(m, E(\beta, m))$ true. This set $E(\beta, m)$ is always non-empty, and, in the
standard case, part of the spectrum of m. The actual values of observables should always fall inside the sets $E(\beta, m)$ for some given state (so
that nothing that has zero probability everywhere can be true). To stay
as close to the usual gloss on mixtures as possible, we must choose this
state β for model $\langle \alpha, \lambda \rangle$ in such a way that $\alpha \, R \, \beta$. So we add as postulate

(21)   (anti-Copenhagen II). For each model $\langle \alpha, \lambda \rangle$, there is a pure
            state β such that $\alpha \, R \, \beta$ and λ maps each maximal observable
            m into a member $\lambda(m)$ of $E(\beta, m)$.

Now we can give a complete truth-definition for these anti-Copenhagen
models.

(22) (a) (m, E) is true in $\langle \alpha, \lambda \rangle$ exactly if $\alpha \in |(m, E)|$
     (b) If m is a maximal observable, then $\langle m, E \rangle$ is true in $\langle \alpha, \lambda \rangle$
          exactly if $\lambda(m)$ belongs to E
     (c) If m is not a maximal observable, then $\langle m, E \rangle$ is true in
          $\langle \alpha, \lambda \rangle$ exactly if $\langle m^*, f_m^{-1}(E) \rangle$ is true in $\langle \alpha, \lambda \rangle$.

It is clear that each observable m receives a definite value, namely
$f_m(\lambda(m^*))$. The only danger of inconsistency one might suspect lies in
our requirement that if (m, E) is true so is $\langle m, E \rangle$. But if (m, E) is made
true by α then it is made true by β if $\alpha \, R \, \beta$. So the values $\lambda(m)$ are chosen
from sets $E(\beta, m) \subseteq E(\alpha, m)$. and therefore conditions (12a–b) are not

contradicted. This is obvious in the case in which m is maximal, and follows for the other cases because of fact (i) above.

It is very important to see that in this anti-Copenhagen variant each observable receives a definite value, and values of functions of observables are indeed functions of the values of those observables, without contravening the Kochen and Specker result.

Finally we must show that Equation (16) of Section 3 is satisfied. We define R^* as at the end of Section 4. Then the equation is satisfied if, when v is in E_α^m, the set of values specified above Equation (21), there is some function λ of the kind allowed by (21) such that observable m is assigned value v in $\langle \alpha, \lambda \rangle$ in the manner described. And this seems trivially true.[21]

APPENDIX. ENSEMBLE-MODELS OF MIXTURES[22]

For each statistical operator W there is an orthonormal base $\{\phi_i\}$ and weights w_i such that $W = \sum w_i E_{\phi_i}$. (Recall that E_ϕ is the projection on subspace $[\phi]$, a pure state which is equally represented by the vector ϕ itself.) The weights are non-negative and sum to 1; we say that W is *diagonal* in the base $\{\phi_i\}$. If all the weights are distinct, there is only one such base, but if not (degeneracy), that base is not unique.

If p is a proposition then $\mu_W(p)$ is the weighted sum $\sum w_i \mu_i(p)$, where μ_i is the measure determined by pure state E_{ϕ_i}. So the weights look like classical probabilities, and this gives rise to the ignorance interpretation: a system is in state W if and only if it is really in state ϕ_i, with probability w_i. But classical probabilities can be represented by measures on an ensemble X of the form

(i) $\mu(X) = 1, \ \mu(\Lambda) = 0$

(ii) $\mu(X_1 \cup X_2) = \mu(X_1) + \mu(X_2) - \mu(X_1 \cap X_2)$.

So if the ignorance interpretation is correct there is, for a system in state W, in principle an ensemble X of systems and measure μ on X as above such that

(a)   each member of X is in some state ϕ_i

(b) $\mu(\{x \in X : x \text{ is in state } \phi_i\}) = w_i$.

We would say that the ensemble X represents a system in state W. Unfortunately, because of the equal weights problem noted above, the representing ensemble cannot be unique. For if it were, some of its members would have to have two states at once which is impossible.

This is not just a technical problem. If the ignorance interpretation is to be consistent it must assert *not* that to be in a state $\sum c_i E_{\phi_i}$ is to be really in some state ϕ_i, but that a system is in a state W if and only if the system is really in some state ϕ (with probability w) such that $W = = w E_\phi + w' U$, where U is some other state. But in this amended form it is consistent.

There is a different interpretation, the *statistical interpretation*, according to which a system in a given mixed state can be represented by a unique ensemble. The principle of this interpretation (maintained in various forms by Margenau and his students and recently by Ballantine) is that a state cannot be assigned to an individual system, but only to a (sub)-ensemble. So to say that a system is in state W can make sense at most as shorthand for the assertion that it is a member of an ensemble X for which there is a measure μ as described above, and assignment of states to subensembles of X such that

(a′)   there is a complete partition of X into subensembles $\{X_i\}$ such that X_i has the state ϕ_i

(b′) $\mu(X_i) = w_i$.

Of course, this holds in general: (a′) and (b′) give the meaning of the assertion that X has state W. For the same ensemble X however there might be another partition $\{X_i'\}$ such that X_i' has state ψ_i and $\mu(X_i') = p_i$, in which case W must also be equal to $\sum p_i E_{\psi_i}$. This is perfectly consistent *provided* not every subensemble of X need be assigned a state. For if $X_i \cap X_j'$ were assigned a state, and also $X_i - X_j'$ and $X_j' - X_i$, we would have to say that $X_i \cap X_j'$ is assigned both ϕ_i and ψ_j, which is absurd unless $\phi_i = \psi_j$. So such adages, which come naturally tripping to the tongue, as that subensembles of pure ensembles are pure, are false on the statistical interpretation.

A very different problem is raised by many-body systems. Let system $S_3 = S_1 + S_2$ be in pure state $\phi = \sum a_{ij}(\phi_i \otimes \psi_j)$ then we can find mixed states W_1 and W_2 to characterize S_1 and S_2. This is discussed by Jauch and is rejected by Hooker; I shall here accept its correctness. Now let $W_1 = \sum c_i E_{\phi_i}$ and $W_2 = \sum d_j E_{\psi_j}$; the bases in question can be chosen so that this holds. Then the ignorance interpretation would say that S_1 is really in some state ϕ_k and S_2 really in some state ψ_m. However, by the usual principle for combining pure states of components of complex systems, we would have that $S_1 + S_2$ is in state $\phi_k \otimes \psi_m$ – and this contradicts that $S_1 + S_2$ is in state Φ (in general). I must note that this reasoning is based on two principles concerning states of complex systems and their components, which could be disputed.

I do not see how the statistical interpretation can fare better here. This interpretation must specify exactly how the ensemble representing $S_1 + S_2$ is related to the ensembles representing S_1 and S_2. If this is done by a one-one pairing, and all the assumptions made so far carry over, the same problem recurs. (For if $S_1 + S_2$ is represented by a one-one map $X_1 + X_2$ of X_1 onto X_2, then a subensemble Y_1 of X_1 in state ϕ_k plus a subensemble Y_2 of X_2 in state ψ_m will give a subensemble $Y_1 + Y_2$, in state $\psi_k \otimes \psi_m$, of $X_1 + X_2$, and so $X_1 + X_2$ will be assigned a mixture of states $\phi_i \otimes \psi_j$ which contradicts its being in state Φ.)

I suppose an adherent of the statistical interpretation will dispute my assumptions about how the ensemble representing $S_3 = S_1 + S_2$ must be related to the ensembles representing S_1 and S_2. But then he must answer, with equal precision, how they are to be related; and until this is done, the dispute is idle.

In a previous paper, I presented a modal interpretation by means of an ensemble model also. To see how this is done, it must be noted that in the example above in which W_1 is diagonal in the base $\{\phi_i\}$, we can find factors n_i and unit vectors ζ_i such that $\Phi = \sum a_{ij}(\phi_i \otimes \psi_j) = \sum n_i(\phi_i \otimes \zeta_i)$ This was shown by Everett. And then we have in fact (see the Appendix to that previous paper) that $W_1 = \sum n_i^* n_i E_{\phi_i}$ and $W_2 = \sum n_i^* n_i E_{\zeta_i}$. I then implemented what I have here called a Copenhagen variant of the modal interpretation.[23] The system $S_1 + S_2$ is represented by the set of possible worlds $Q = \{\alpha_i = \langle \alpha_i^1, \alpha_i^2, \alpha_i^3 \rangle = \langle \phi_i, \zeta_i, \Phi \rangle\}$ and statement $\langle m, S_i, E \rangle$ is true in α_j exactly if for $\beta = \alpha_j^i$, $P_\beta^m(E) = 1$, and (m, S_i, E) is true in α_j exactly if $\langle m, S_i, E \rangle$ is true in all elements of Q. (I wrote $\square(m, E)$ and $O(m, E)$ rather than (m, E) and $\langle m, E \rangle$ at that time.) The principles relating states of complex systems and their components are to be applied only to quantum-logical sentences expressing necessary propositions, and not those expressing contingent propositions.

The ensemble models so constructed do not fall prey to the problem that apparently invalidates the ignorance and statistical interpretations applied to many-body systems. But they have the usual drawback of non-uniqueness: in many cases, the orthonormal base in which W_1 is diagonal is not unique. Besides, why is W_1 treated in this preferred way? We could get equally good models if we began by taking a base in which W_2 is diagonal. And why are orthogonal decompositions given a preferred status? Because of these questions, of which discussions with Nancy Delaney Cartwright made me very conscious, I chose in the present paper to develop the modal interpretation in a very general manner, eschewing the over-specificity that tends to come with model building.

University of Toronto

NOTES

* The research for this paper was supported by the John Simon Guggenheim Memorial Foundation and Canada Council grant S71-0546. While I have many debts, some of which are indicated at various points below, my debts to Nancy Delaney Cartwright in Part III and in the Appendix are so pervasive that they cannot be adequately chronicled. I hereby acknowledge them gratefully.
1 *List of special symbols.*
$|E|$ Section I.1.
$\&$, $\Box$, $\supset$, $\dashv$, (Section I.2
Tr Section I.3
H, (m, E), P_m^α, P Section II.2
$[m]$,' $\cup$ section II. 3.
$[m, m']$ Section II.4
μ_α Section II.5
a_α, f_p, λ, N Section II.6
$[\phi]$, E_p, E_ϕ, $\otimes$, Section II.1
R, U_α, p_α Section III.2
$\langle m, E \rangle$, λ, L_1, L_2, $S_1(M)$, $S_2(M)$, R^* Section III.3
$\langle m, S_i, E \rangle$, Section III.4
Note that the symbol λ plays different roles in Parts II and III.
2 For a general treatment see Chapter 2 of my *Formal Semantics and Logic* New York, Macmillan, 1971.
3 Cf. *Formal Semantics and Logic* (Note 3), Chapter 5, Section 2a.
4 This may be of some relevance to Professor Greechie's remarks on implication in quantum logics in this volume.
5 See R. Stalnaker and R. H. Thomason, 'A Semantic Analysis of Conditional Logic' *Theoria* 36 (1970), 23–42; for alternatives see D. Lewis, 'Completeness and Decidability of Three Logics of Counterfactual Conditionals', *Theoria* 37 (1971), 74–85.
6 For a general exposition of the actual developments in quantum logic, see the paper by Professors Gudder and Greechie in this volume. More general discussions of quantum logics of various stripe are also found in my 'The Labyrinth of Quantum Logics' and 'The Formal Representation of Physical Quantities' in *Boston Studies in Philosophy of Science*, D. Reidel Publ. Co., Dordrecht, Holland, forthcoming.

[7] Cf. J. M. Jauch, *Foundations of Quantum Mechanics* Addison-Wesley, Reading, Massachusetts, 1968, p. 100.

[8] S. Kochen and E. P. Specker, 'The Problem of Hidden Variables in Quantum Mechanics' *Journal of Mathematics and Mechanics* 17 (1968), 59–87. See also N. Zierler and M. Schlesinger, 'Boolean Embeddings of Orthomodular Sets and Quantum Logic' *Duke Mathematical Journal* 32 (1965), 251–162.

[9] The deduction I have in mind is that of G. W. Mackey, *Mathematical Foundations of Quantum Mechanics*, W. A. Benjamin, New York, (1963); see also M. J. Maczynski, 'A Remark on Mackey's Axiom System for Quantum Mechanics' *Bulletin de l'Académie Polonaise des Sciences (Série des sciences mathematique astronomique et physique)* 15 (1967), 583–7.

[10] Specifically, see the definition of 'projective logic' and Theorem 7.44 in Chapter VII of V. S. Varadarajan, *Geometry of Quantum Theory*, Vol. I, Van Nostrand, Princeton, 1968.

[11] Cf. S. Gudder, 'Hidden Variables in Quantum Mechanics Reconsidered', *Review of Modern Physics* 40 (1968), 229–31.

[12] J. C. T. Pool, 'Baer*-Semigroups and the Logic of Quantum Mechanics', *Communications of Mathematical Physics* 9 (1968), 118–41, and 'Semi-Modularity and the Logic of Quantum Mechanics' *ibid.*, 212–28.

[13] This use of 'standard' is from Varadarajan, *op. cit.* For a more detailed motivation for the problems discussed in this part see the papers by C. Hooker and myself in the volume referred to in Note 14.

[14] In Part II of my 'A Formal Approach to the Philosophy of Science' (henceforth FAPS), pp. 303–366 in *Paradigms and Paradoxes* (ed. by R. Colodny), University of Pittsburgh Press, Pittsburgh, 1972.

[15] In FAPS, these were called □-sentences and O-sentences, and written ' □ (m, E)' and 'O(m, E)'.

[16] J. Bub, 'What is a Hidden Variable Theory of Quantum Phenomena?', *International Journal of Theoretical Physics* 2 (1969), 101–23.

[17] J. Bub, 'Hidden Variables and Quantum Logic', *Boston Studies in the Philosophy of Science*, D. Reidel Publ. Co., Dordrecht, Holland, forthcoming.

[18] *Op. cit.* (Note 8).

[19] In different contexts, different Hilbert spaces are used to represent the states of the same system. The present assertion of identity can be similarly understood as having the tacit rider 'relative to all purposes in a given context.' It can be taken as a metaphysical assertion only if it is presupposed that for each system there is a state-space which is peculiarly appropriate to it.

[20] See the definition in Kochen and Specker, *op. cit.*, p. 63; note that they follow this with the assumption that if $P_m^\alpha = P_{m'}^\alpha$ for all states α then $m = m'$, which we here deny.

[21] After the completion of this paper I found ideas similar to those of what I call the modal interpretation in a paper by A. Fine presented at the *Fourth International Congress of Logic, Methodology and Philosophy of Science*, Bucharest, August 1971.

[22] Except for the remarks on the statistical interpretation this appendix is a summary of some parts of FAPS; mathematical details may be found in the appendix of FAPS.

[23] With the qualification that in FAPS a single (orthogonal) decomposition of a mixture β was always specified, with the remainder of U_β ignored.

Note added in proof. The question discussed in Section III-2 concerning Equation (11) is settled by a theorem in E. Schrödinger, 'Probability Relations Between Separated Systems', *Proc. Cambridge Philosophical Society* 32 (1936), 446–52.

E. GERJUOY*

IS THE PRINCIPLE OF SUPERPOSITION
REALLY NECESSARY?

I. INTRODUCTION

My main intention today is to examine a few well-known and a few not-so-well-known experiments, primarily with regard to their implications for the validity of the principle of superposition. In order to set my remarks in proper context, however, especially in a conference of this sort (on the Foundations of Quantum Mechanics), I think it appropriate to state first my views on the general subject of the wave function's interpretation.

Since I am a working or lumpen-proletariat physicist, it should occasion no surprise that I have no fundamental disagreements with the well-known orthodox viewpoints on the physical interpretation of the wave function; these are the viewpoints stated in most standard contemporary texts [1], usually no better expressed today than they were by competent textbook authors thirty years ago (e.g., Kramers [2]). The best succinct exposition of this orthodox view that I have encountered has been given by Heisenberg, in his 1955 article [3], 'The Development of the Interpretation of the Quantum Theory'. I quote:

An important step forward was made by the work of Born... the wave in configuration space was interpreted as a probability wave, in order to explain collision processes on Schrodinger's theory. This hypothesis contained two important new features... The first of these was the assertion that, in considering 'probability waves', we are concerned with processes not in ordinary three-dimensional space, but in an abstract configuration space (a fact which is, unfortunately, sometimes overlooked even today); the second was the recognition that the probability wave is related to an individual process. The probability wave describes the behaviour, not of a large number of electrons, but only of one system of particles whose number is finite and is given by the number of dimensions in the configuration space; the wave can be conceived as representing a statistical assembly only in so far as the experiment concerned can be repeated as often as we please.

These assertions of Heisenberg's are not universally agreed to, as you all know, but they are agreed to by me. It *is* a wonderful mystery that the wave function propagating in configuration space should yield the probabilities of events in physical space, but one can learn to live with mere mystery, which is not a disaster for theory like logical inconsistency or discrepancy between prediction and observation. Moreover, the postulate of a probability wave function evolving in configuration space cannot be controverted by observations on any individual system in physical space, unless what is observed is sufficiently improbable to be incredible. I add that some of the criticisms of these orthodox views on the interpretation of the wave function seem to be based on almost semantic confusion. For example, Heisenberg's reference to an 'individual' process seems to have been the basis for Ballantine's espousal (in a very recent article [4]) of what he terms a 'Statistical Interpretation' of quantum mechanics, in opposition to the 'Copenhagen Interpretation'. To quote Ballantine almost verbatim, "a state vector characterizes an ensemble of similarly prepared systems, but need not provide a complete description of an individual system." However, Ballantine's actual procedures for using the wave function are quite orthodox. Actually, Heisenberg's assertion 'the probability wave is related to an individual process' mainly is designed to stress that quantum mechanics applies to every individual electron being described by a wave function, not merely to the ensemble of electrons. For no single electron can one measure position accurately without being uncertain about momentum; for each electron in a double slit experiment the interference of probability amplitudes from the individual slits is significant. But each of these double slit electrons blackens but a single spot on a photographic film; neither Heisenberg nor I claim that an individual electron going through a double slit apparatus would produce an interference pattern, no matter how perfectly the pattern was enhanced to make it visible.

On the other hand, my views do not include Heisenberg's further belief (also very well expressed in this same 1955 article) that there are no difficulties with the orthodox interpretation of measurement, which difficulties (it is only fair to state) were an added inspiration for Ballantine's article. Making quite reasonable assumptions about the nature of the measuring process, Wigner, Komar [5], and more recently Fine [6], have argued convincingly that use of the conventional Schrödinger equation – to follow the evolution of the wave function throughout its interaction with the measuring apparatus – does not yield states or mixtures consistent with the requirements for a good measurement. But because the characteristic features of macroscopic apparatus still are obscure, I am not sure such arguments conclusively demonstrate the need for modifications of orthodox quantum mechanics (along lines Wigner [7], for instance, has suggested). However, granting the measuring apparatus is able to perform a good measurement, by some means we do not yet understand, there is no reason to dispute use of the conventional pre-apparatus-interaction wave function to predict post-apparatus-interaction observations.

On the basis of considerations such as the above, I am unable to take seriously those conceptual difficulties and seeming paradoxes of quantum mechanics which – from where I sit – appear to arise from failure to distinguish between the actual particles and the wave function describing them.

For example, I cannot see the problem in Schrödinger's cat paradox [8], where (to again quote Ballantine [4]):

A cat is placed in a chamber together with a bottle of cyanide, a radioactive atom, and a device which will break the bottle when the atom decays. One half-life later the state vector of the system will be a superposition containing equal parts of the living and dead cat, but any time we look into the chamber we will see either a live cat or a dead one.

To me, it is absurd to pretend that a macroscopic object like a cat can be described by a wave function, implying the possibility of interference between coherent live-cat and dead-cat amplitudes when the cat-in-the-box wave function is squared. I am not even sure the cat is describable by a mixture, i.e., by an incoherent superposition of live-cat and dead-cat states. I make these assertions about the cat even though I firmly believe that in principle quantum mechanics is perfectly valid in the domain of macroscopic bodies. After all, the cat is composed of a myriad of atoms whose total number and total energy constantly fluctuate under the influence of an external essentially random environment; in this ensemble of cat-environment systems, which quantum states retain the essence of 'cat', not to mention 'live-cat' rather than 'dead-cat', is impossible to determine. However, I can't see the paradox even if the cat in-the-box system is describable by a wave function because, as I said earlier, what is evolving continuously – and in a radioactive half-life becomes an equal-amplitude superposition of live-cat and dead-cat states – is the wave function, not the cat; the cat is entirely alive until it dies, and then it stays entirely dead. Nevertheless, in all candor I must report that deWitt [9], who is a perfectly respectable physicist, last year cited the cat paradox as evidence for a multi-valued parallel universes theory of nature advanced by Everett [10] and Wheeler [11] in 1957, of a type which will be familiar to science fiction addicts.

II. ANALYSES OF EXPERIMENTS

Having made clear (I trust) my views on how the wave function is to be interpreted, let me turn to my main subject. Conventional quantum theory is substantiated by a vast body of experiments, and I assuredly am not questioning its validity. Rather fewer experiments can be termed 'fundamental', however, where by a fundamental experiment I mean one which by and of itself seems to demand acceptance of a basic tenet of orthodox quantum mechanics, e.g., the interpretation of the wave function I favor. As used here, 'demand' is a loaded word, of course. Since one can have many alternatives [12] to conventional quantum mechanics, agreeing with conventional theory in the ordinary domains of observation, it is probable that existing experiments strictly demand neither conventional quantum theory as a whole, nor any of its basic principles. Granting the general validity of conventional quantum theory, however, it is reasonable to seek fundamental experiments, and to examine what basic tenets they 'demand', a term I will make more precise in a moment. To keep the discussion from being too diffuse, let's concentrate on the principle of superposition, which assuredly is as basic a quantum mechanical tenet as one can name. So I will pose the question: Granting the general validity of quantum mechanical ideas (e.g., wave-particle duality, the interpretation of the wave function I've described, etc.), but omitting the principle of superposition, what experiments fundamentally demand the principle of superposition, in the sense that they manifest effects immediately understandable with the principle of superposition, and hard to understand without it? As a start, I propose to examine – from the standpoint of this question – three fundamental experiments which are among those most commonly cited in the literature on the philosophic foundations of quantum mechanics. These three justly-famous experiments are the double slit experiment, the Stern-Gerlach experiment, and the Wu-Shaknov experiment.

The implications of the double slit experiment are fairly apparent.

Granting the general validity of wave mechanics, the double slit experiment shows that the probability of reaching the observing screen is the square of a sum of coherent amplitudes; or, to put it a little differently, the double slit experiment certainly does demand the principle of superposition. So far, so good, and that's that for the double slit experiment.

2.1. *Double Stern-Gerlach Experiment*

The Stern-Gerlach experiment is not quite as transparent. Of course, it immediately demonstrates the need for quantizing angular momentum. However, neither the usual Stern-Gerlach experiment nor the typical double Stern-Gerlach experiment really require the principle of superposition. Figure 1 shows a beam of hydrogen atoms from a furnace going

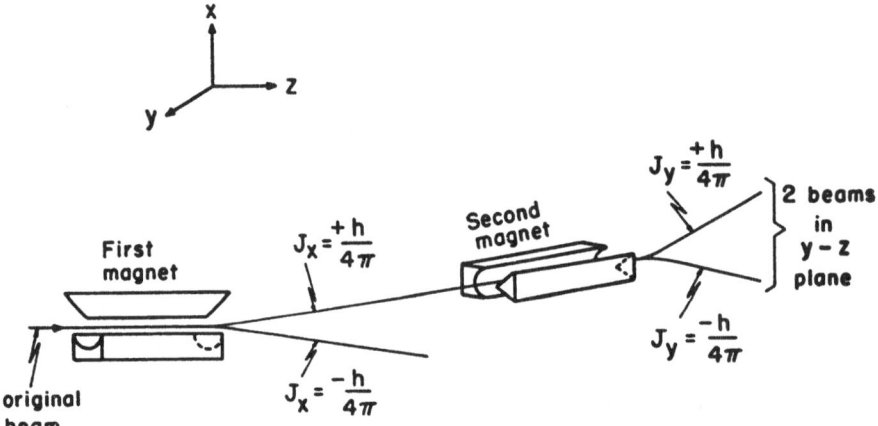

Fig. 1. Double Stern-Gerlach experiment. Atoms of total angular momentum $J = h/4\pi$ are incident on the first magnet, whose inhomogeneous field is along x. The inhomogeneous field of the second magnet is along y.

through a double Stern-Gerlach apparatus. The beam of spin $\frac{1}{2}$ atoms from the furnace is a mixture, but the atoms deflected up after passing through the first magnet (whose inhomogeneous field is along x) presumably all have their spins along x, i.e., are in the pure state $J_x = h/4\pi$. When this beam passes through the second magnet, whose inhomogeneous field is along y, two beams of equal intensity, corresponding to $J_y = \pm h/4\pi$, are produced. It is possible to infer this result on the basis that the pure state $J_x = h/4\pi$ is a coherent superposition of the pure states $J_y = \pm h/4\pi$, with equal amplitudes. But the actual observation at the exit end of the second magnet is no different than if the beam entering the second magnet were a mixture of $J_y = \pm h/4\pi$ states, with equal intensities. One could orient the second magnet at an arbitrary angle θ to the x axis, in which event the superposition principle predicts the two beams exiting from the second magnet would have intensities proportional to $\sin^2 \theta/2$ and $\cos^2 \theta/2$, whereas a completely incoherent mixture entering the second magnet still would predict equal intensities in the two beams. I presume this measurement of the beam intensities as a function of θ could be done, and would verify the superposition intensity prediction. Nevertheless, it is clear that the Stern-Gerlach experiment has much less immediate implications for the principle of superposition than does the double slit experiment.

Wigner [7] has pointed out that in the double Stern-Gerlach experiment the need for the principle of superposition would be demonstrated by bringing together the two $J_y = \pm h/4\pi$ beams exiting from the second magnet, and then observing that the combined beam corresponded once more to a pure $J_x = h/4\pi$ state. If the two $J_y = \pm h/4\pi$ beams really are coherent, and the experiment actually could be performed, it would be a marvelous demonstration of interference, consistent with the principle of superposition. However, I am not so sure the two separated $J_y = \pm h/4\pi$ beams can be considered coherent, even before detectors have begun counting the atoms in them. The question of whether or not the $J_y = \pm h/4\pi$ beams are automatically incoherent involves the measurement process, which I earlier admitted I didn't fully understand. If the whole passage of the original $J_x = h/4\pi$ beam through the second magnet is governed by the Schrödinger equation for a hydrogen atom in a fixed inhomogeneous magnetic field along y, there is no doubt that the two $J_y = \pm h/4\pi$ beams will be coherent. But the apparatus is a macroscopic system; the magnet responds to thermal and voltage fluctuations; the vacuum system has pressure fluctuations, etc. Thus once the $J_y = \pm h/4\pi$ beams begin to separate in space, maintenance of coherence between them may not be possible. I confess that if this automatic incoherence between the two $J_y = \pm h/4\pi$ beams could be demonstrated theoretically, via some very basic arguments which would hold for perfect magnets in perfect vacua, it would please me, because it would remove the need for seeking a mechanism for incoherence in the final process of actually observing the atoms. Unfortunately, I don't really believe such arguments will be found.

2.2. *Wu-Shaknov Experiment*

Analysis of the Wu-Shaknov experiment [13] also reveals some features of interest. Let me take a moment to remind you of this experiment, which is not as quite well known as the double slit or the Stern-Gerlach. The experimental configuration is illustrated in Figure 2. A positron-electron pair annihilates at the point O, and produces a pair of photons which (by energy-momentum conservation) have to leave O in opposite directions with equal energy. Theory [14] indicates that the planes of polarization of these outgoing photons are correlated, in fact are perpendicular. In other words, if the photon going to the right is polarized in the plane of the paper, the photon going to the left must be polarized in a direction perpendicular to the paper. This theoretical prediction is tested [13] by observing the rates with which these photons are scattered in dense matter, e.g., iron. A photon scattering defines a scattering plane, the plane containing the directions of propagation before and after the scattering event. Again theory [14] indicates that the probability of scattering through a given scattering angle θ (defined relative to the initial direction of propagation) depends on the angle between the plane of polarization and the scattering plane. What is observed in the experiment, therefore, is the rate at which a pair of simultaneously emitted photons are simultaneously scattered though the same scattering angle θ, as a function of the angle ϕ between the pair of scattering planes thus defined. In particular, one mea-

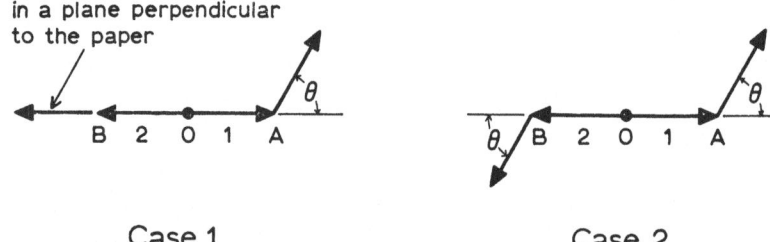

Fig. 2. Schematic representation of Wu-Shaknov experimental arrangement. Photons 1 and 2 leave the annihilation point O in opposite directions, and are scattered through the same angle θ at points A and B respectively. In case 2, the scattering plane of each photon lies in the plane of the paper; in case 1, the scattering plane of photon 2 is perpendicular to the plane of the paper, whereas photon 1 scatters as in case 2.

sures the ratio of these rates for the two cases $\phi=0$ and $\phi=\pi/2$, and finds that this observed ratio is completely consistent with the usual quantum theoretic predictions. Moreover, the whole detection apparatus can be rotated about any axis through the annihilation source O without in any way affecting the aforementioned ratios. Consequently, (after comparing the observed ratio with the results of calculations performed by Bohm and Aharonov [15]), one can infer that the Wu-Shaknov experiment is consistent with the conventional formalism describing the outgoing photon pair by a properly rotation-invariant wave function, but is inconsistent with the supposition that the system is described by an incoherent mixture of perpendicularly polarized outgoing photon pairs, uniformly distributed over all possible emission directions from O.

In effect, therefore, the Wu-Shaknov experiment – like the Stern-Gerlach experiment – indirectly verifies the principle of superposition, via quantitative observations which inferentially rule out the possibility that the outgoing pair of photons could be described by a mixture instead of by a coherently evolving wave function. However, having gone into this much detail about the Wu-Shaknov experiment, I should like to take the opportunity to examine some of its other important implications. For an individual photon, the direction of polarization can't be inferred from the single observation that the photon has been scattered at some chosen scattering angle. In fact, the meaning of the term 'direction of polarization' has to be very carefully prescribed for individual photons, as has been discussed in a second very admirable paper by Bohm and Aharonov [16] on the Wu-Shaknov experiment. Let us suppose, for the sake of argument, that there exists the analogue of double refraction for 500 keV positron annihilation photons. When a macroscopic beam of plane polarized light enters a properly cut doubly refracting crystal, like calcite, the beam breaks up into two distinct beams, the ordinary and extraordinary rays, which are plane polarized along perpendicular directions with intensities proportional to $\cos^2 \theta$ and $\sin^2 \theta$ respectively, where θ is the angle between the original plane of polarization and the plane of polarization in the ordinary ray. However, any individual photon in the original beam will be counted only in the ordinary ray or in the extraordinary ray, not in both. The probability that an individual photon in the original beam will be found in the ordinary ray is $\cos^2 \theta$. Evidently, using a doubly refracting crystal to separate plane polarized light into two beams is very analogous to the previously discussed double Stern-Gerlach experiment; in fact, as explained earlier, to have the same $\cos^2\theta$ and $\sin^2\theta$ relative intensities in the pure spin-polarized atomic beams exiting from the second magnet, the second magnet merely need be turned through an angle 2θ relative to the first.

Now, let's pretend two such 500 keV photon polarization analysers are available, and let's grant that the Wu-Shaknov experiment demonstrates the conventional theory of positron annihilation to be correct. Then when the photon emitted to the right produces a count in the ordinary ray channel of the analyser on the right, the photon emitted to the left will produce a count *only* in the extraordinary ray channel of the analyser on the left, assuming the analysers are similarly oriented. This is the precise meaning of the assertion that the outgoing pair of positron annihilation photons are polarized along perpendicular directions; of course, the polarization measurements here being discussed are supposedly performed before the photons have been scattered by matter, as they are scattered in the Wu-Shaknov experiment. Thus we may additionally conclude that it has become possible to infer the polarization of a photon emitted to the left from a polarization measurement on the by now very distant photon emitted to the right, without letting the leftward photon scatter, or in any other way disturbing it. Furthermore, the pair of perpendicularly oriented analysers could have been simultaneously rotated through an arbitrary angle (30° say), which – now speaking very loosely – means we could have inferred the polarization amplitudes of the leftward photon along directions 30° apart, without in any way disturbing it. For photons, polarization amplitudes along directions differing by any angle other than 90° are non-commuting observables. [16] Therefore, photon pair production by positron annihilation often is said to provide an illustration in nature of the Einstein-Podolsky-Rosen [17] claim that it is possible to arbitrarily accurately measure the values of non-commuting observables in a system (here the leftward-propagating photon) without disturbing the system.

On the other hand, this last conclusion stemmed from loose language, as I tried to stress. These is no conflict with the uncertainty principle. The uncertainty principle merely forbids arbitrarily accurately measuring the simultaneous values of non-commuting observables in a single system for the purposes of future prediction. There are no restrictions for two different systems; there is no objection to measurements which enable us to infer values of non-commuting observables that a system must have had sometime in the past. To illustrate, when an electron goes through a narrow slit at some carefully measured instant, and is observed on a small distant screen at a carefully measured later instant, it is no conflict with the uncertainty principle to assert – after the fact – that the electron left the slit with an essentially exactly known velocity, just right to bring it to the observing screen at the measured time of arrival.

Getting back to the photons under present consideration, if the photon traveling to the right is observed in a vertical polarization channel, this very observation alters the probability of observing it in a channel oriented 30° to the vertical. Consequently, if the same rightward-traveling photon is successively observed in vertical and in 30°-from-vertical polarization channels, no information contradicting the uncertainty principle can be learned concerning its leftward-traveling twin's polarization amplitudes along horizontal and 30°-from-horizontal directions. Accurate measurements of vertical polarization for different rightward-propagating photons do not yield horizontal and off-horizontal polarizations for the same leftward-traveling photon. In other words, these Einstein-Podolsky-Rosen type polarization measurements provide no contradiction with the uncertainty principle, as I asserted.

Nor is there any conceptual paradox. There is no implication that observations on the same leftward-traveling photon using a fixed polarization analyser can be modified in any way by changing the orientation of the polarization analyser being used to observe the rightward-moving photon. Suppose, for some given pair of photons, simultaneous counts are observed with the right analyser vertical, and with the left analyser horizontal. These particular photons cannot be regained for the purpose of repeating the measurement, but there is every reason to expect that for this pair the left analyser would have registered a count even if the right analyser had been rotated through 30°. Of course, there might not be a coincidence, because there is no guarantee that the right analyser will register a count in its new orientation, but this is completely understandable. We also know that when the right analyser registers a count in its orientation 30° from the vertical, there will be zero probability of the left analyser registering a simultaneous count when it is parallel to the right analyser (i.e., when it is 60° from the horizontal). On the other hand, when the right analyser was in its original vertical orientation, there was a finite probability of registering a coincidence between the 60°-from-horizontal left analyser and the vertical right analyser. But there still is no paradox. When the analyser on the left is rotated through 60°, we are doing a new experiment on the left. Irrespective of what is happening on the right, a leftward moving photon registering a count in a horizontal analyser had (before doing so) probability $\cos^2 60°$ for being counted, and probability $\sin^2 60°$ for not being counted – in an analyser oriented 60° from the horizontal. In no case is it possible to conclude that the same leftward photon was caused to behave differently merely by reorienting the polarization analyser on the right.

2.3. *Electron Capture Experiment*

Returning to my main theme, the foregoing considerations have led me to wonder how many other experiments 'demand' – in the sense previously explained – the validity of the principle of superposition. In practice, the principle of superposition will be demanded in this sense only when the experiment manifests obvious interference effects. By 'obvious' interference effects I mean here a series of regular oscillations as a function of some controllable experimental parameter, e.g., the frequency (as in the double slit experiment); otherwise the need for the principle of superposition, though real, is not likely to be immediately apparent – as we've seen in the Stern-Gerlach and Wu-Shaknov experiments. For instance, to give a very simple illustration of this last assertion, whenever an experimental collision cross section is azimuthally symmetric about the incident z-direction, but is asymmetric under reflection in the x–y 90°-scattering plane, coherent interference between the even parity and odd parity components of the wave function must be occurring. If the even parity and odd parity components of the wave function were incoherent, the experimental cross section would involve only the squares of these components, not their cross terms, and (with this azimuthal symmetry) would be invariant under the reflection cited. Thus the mere presence of the very common forward peak in observed cross sections (which are almost invariably azimuthally symmetric) is evidence for the principle of superposition. On the other hand, although the immediately preceding assertion is correct, I can get forward peaks in cross sections computed via classical mechanics (e.g., the Coulomb scattering cross section) wherein the principle of superposition is never mentioned.

It would be ridiculous for me to say that I have examined any significant fraction of the experimental literature from the above standpoint. However, I have given some thought to various types of commonly performed experiments, and it is surprising how rarely interference oscillations are unmistakably seen, excluding the domain of quantum optics, where coherent interference usually is the whole experimental point. However, quantum optics phenomena – unless no more than a glorified version of the double slit experiment – tend to be difficult to understand and to explain simply; indeed the interpretations of many such experiments still are in dispute. Consequently, it seems worthwhile to describe here some recent non-quantum-optics experiments – one reported just within the past year – which manifest interference fluctuations, and which are simply and incontestably interpretable in terms of the principle of superposition. More specifically, these experiments indicate that the superposed components of the wave function evolve in time independently but coherently (precisely as the Schrödinger equation predicts) until the moment before an observation is undertaken, at which time squaring the wave function produces interfering cross terms between the independently evolved superposed components. It is emphasized that – despite the aforementioned relative rarity of experiments manifesting interference oscillations – the experiments I am about to discuss are not the only ones that could have been selected, even excluding the domain of quantum optics. My particular selections have been influenced by a desire to avoid experiments whose interpretations could be criticized on the grounds (surely valid in high energy particle physics, and perhaps even in low energy nuclear physics) that they involved domains where quantum theoretic predictions still are quantitatively unsatisfactory.

The first of these relatively recent experiments I shall discuss involves the cross section for resonant electron capture. The simplest reaction of this kind is electron capture by protons incident on atomic hydrogen in its ground 1s state.

$$H^+ + H(1s) \rightarrow H(1s) + H^+ \tag{1}$$
$$\text{fast B} \quad \text{slow A} \quad \text{fast B} \quad \text{slow A}$$

In the laboratory the hydrogen atoms are essentially at rest initially, compared to the velocities of the incident protons. When electron capture occurs a fast ion is converted to a slow ion, as Equation (1) indicates; alternatively, one sees a fast H(1s) atom where originally there were only slow H(1s) atoms. In this way, the experimenter is able to distinguish electron capture from elastic scattering of protons by atomic hydrogen.

$$H^+ + H(1s) \rightarrow H^+ + H(1s)$$

fast B   slow A   fast B   slow A
$$\tag{2a}$$

Of course, there are some knock-on elastic collisions between the incident protons (which I label B) and the atomic protons (labeled A); these knock-on collisions, symbolized by the equation

$$H^+ + H(1s) \rightarrow H^+ + H(1s)$$

fast B   slow A   slow B   fast A
$$\tag{2b}$$

also produce fast H(1s) atoms. Your attention is directed to the differences between (2b) and (2a) and between (2b) and (1). The essential feature of (1) is that the electron originally attached to proton A ends up attached to proton B. It is stressed that for our present purposes and for the purposes of the actual observations at slow velocities – the fact that protons A and B actually are indistinguishable particles is quite irrelevant; each of the reactions (1), (2a) and (2b) would continue to take place even if particle B were a deuteron rather than a proton. With incident deuterons the knock-on reactions (2b) become physically distinguishable from electron capture; with incident protons, the knock-on reactions (2b) are physically indistinguishable from electron capture. However, the knock-on reactions (2b) are relatively very improbable at the low incident velocities of the measurements I shall report. Consequently, the protons B and A can be – and will be – treated as distinguishable particles in the following discussion. For similar reasons, the proton spins can be ignored, i.e., the protons will be treated as if spinless.

The protons, we said, are going to be moving at comparatively low velocities. Then it is reasonable to examine first the case of protons at rest. The solutions to the Schrödinger equation for an electron in the field of two stationary protons are well-known. In particular, at any finite interproton distance ρ there are precisely two eigenfunctions which at infinite ρ correspond to an isolated proton and a hydrogen atom in its ground 1s state. Because the Hamiltonian is symmetric in the proton coordinates, one of these eigenfunctions, termed $\omega_+(\rho)$, is symmetric under interchange of the protons, while the other eigenfunction $\omega_-(\rho)$ is anti-symmetric under proton interchange; I repeat, this has nothing to do with particle indistinguishability. The eigenenergies $\omega(\rho)$ of these hydrogen molecular ion eigenfunctions are sketched in Figure 3; the symmetric $\omega_+(\rho)$ lies lower, and is responsible for the molecular ion H_2^+ being stable; the antisymmetric $\omega_-(\rho)$ is a so-called repulsive branch, in which the protons are not kept together by the presence of the electron.

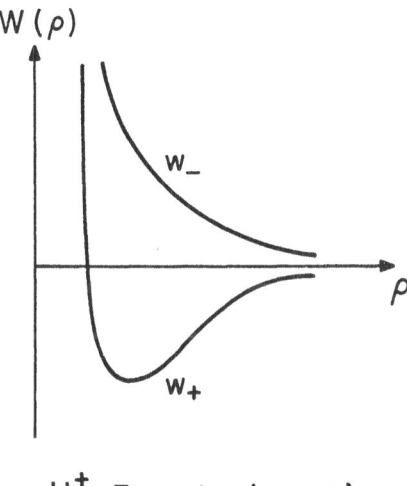

H_2^+ Energies (sketch)

Fig. 3.   Crude sketch of the interaction energy $W(\rho)$ vs. internuclear spacing ρ, for the lowest symmetric (W_+) and antisymmetric (W_-) states of H_2^+.

Now consider the actual collision (1), wherein the protons are moving slowly relative to each other. Here 'slowly' means proton velocities much smaller than typical electron velocities in these ω_+ and ω_- states, which electron velocities are of the order of 10^8 cm s^{-1}. Then it is a very good approximation to suppose that at each instant the electron wave function is a superposition of the eigenfunctions it would have if the protons were held fixed at their instantaneous separation ρ. Thus, to very good approximation, we can write the time-dependent electron wave function as a superposition

$$\psi(t) = c_+(t)\,\omega_+ + c_-(t)\,\omega_-, \tag{3}$$

where $\omega_+(\rho)$, $\omega_-(\rho)$ depend on time t through the dependence of ρ on t, and where $c_+(t)$, $c_-(t)$ are numerical coefficients, which change with time as the protons come together and separate. The wave function $\psi(t)$ satisfies the usual time-dependent Schrödinger equation

$$ i\hbar \frac{\partial\psi}{\partial t} = \mathscr{H}(\rho)\,\psi, \qquad (4) $$

where the Hamiltonian $\mathscr{H}(\rho)$ depends of course on ρ, and therefore on t for specified $\rho(t)$. For what follows, the particular dependence of ρ on t is almost irrelevant, but usually one assumes the incident proton moves with constant velocity υ, which is a quite good approximation even at very low incident energies. The geometry under this assumption is sketched in Figure 4; we see

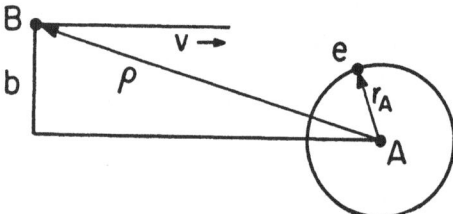

Fig. 4. Sketch of geometry for collisions of incident protons B with hydrogen atoms (nucleus labeled A). The electron, initially bound to proton A, is labeled e. The vector ρ is drawn from A to B; Proton B is moving along positive z with speed υ; b is the impact parameter; r_A is the position of the electron relative to A.

$$ \rho(t) = \sqrt{b^2 + \upsilon^2 t^2}, \qquad (5) $$

where b is the impact parameter, and the origin of time is adjusted so that at $t=0$ the incident proton – which moves in the positive z direction – lies in the plane $z=0$.

With the foregoing very valid simplifications, the theory of this electron capture reaction is almost trivial. Substitute Equation (3) into the Schrödinger Equation (4). We find

$$ i\hbar \left\{ \frac{\partial c_+}{\partial t}\,\omega_+ + c_+\,\frac{\partial\omega_+}{\partial t} + \frac{\partial c_-}{\partial t}\,\omega_- + c_-\,\frac{\partial\omega_-}{\partial t} \right\} = $$
$$ = c_+\mathscr{H}(\rho)\,\omega_+ + c_-\mathscr{H}(\rho)\,\omega_-. \qquad (6) $$

But, by definition of ω_+, ω_-

$$\mathcal{H}(\rho)\,\omega_+ = W_+(\rho)\,\omega_+$$
$$\mathcal{H}(\rho)\,\omega_- = W_-(\rho)\,\omega_-, \tag{7}$$

where W_+, W_- are the energies plotted in Figure 3. Moreover, using (5)

$$\frac{\partial\omega_+}{\partial t} = \frac{\partial\omega_+}{\partial\rho}\frac{\partial\rho}{\partial t} = \frac{v^2 t}{\sqrt{b^2 + v^2 t^2}}\frac{\partial\omega_+}{\partial\rho} < v\,\frac{\partial\omega_+}{\partial\rho}. \tag{8}$$

The right side of (8) is proportional to the incident proton velocity, and thus approaches 0 as $v \to 0$; in fact, it can be seen that the ratio of $\hbar(\partial\omega_+/\partial t)$ to $W_+\omega_+$ in (6) is proportional to v/v_e, where v_e is the electron velocity, a result obvious on dimensional grounds. At the incident energies of present interest, therefore, $\partial\omega_+/\partial t$ can be neglected in (6), and similarly for $\partial\omega_-/\partial t$. Consequently, since ω_+, ω_- are orthogonal at every ρ, Equation (6) reduces to

$$i\hbar\,\frac{\partial c_+}{\partial t} = W_+ c_+$$
$$i\hbar\,\frac{\partial c_-}{\partial t} = W_- c_- \tag{9}$$

whose solutions are

$$c_+(t) = c_+(-\infty)\exp\left\{(-i/\hbar)\int_{-\infty}^{t} dt\,W_+(t)\right\}$$

$$c_-(t) = c_-(-\infty)\exp\left\{(-i/\hbar)\int_{-\infty}^{t} dt\,W_-(t)\right\} \tag{10a}$$

implying

$$c_+(\infty) = c_+(-\infty)\exp\left\{(-i/\hbar)\int_{-\infty}^{\infty} dt\,W_+(t)\right\}$$

$$c_-(\infty) = c_-(-\infty)\exp\left\{(-i/\hbar)\int_{-\infty}^{\infty} dt\,W_-(t)\right\} \tag{10b}$$

In Equation (10), it is understood that $W_+(t)$, $W_-(t)$ are functions of time through $\rho = \rho(t)$.

When the protons are infinitely separated, it is obvious that (except for inconsequential phase factors)

$$\omega_+ = \frac{1}{\sqrt{2}}\left[u_{1s}(\mathbf{r}_A) + u_{1s}(\mathbf{r}_B)\right] \equiv \frac{1}{\sqrt{2}}\left[u_A + u_B\right]$$

$$\omega_- = \frac{1}{\sqrt{2}}\left[u_{1s}(\mathbf{r}_A) - u_{1s}(\mathbf{r}_B)\right] \equiv \frac{1}{\sqrt{2}}\left[u_A - u_B\right], \tag{11}$$

where $\mathbf{r}_A$, $\mathbf{r}_B$ are the positions of the electron relative to protons A, B respectively, and where $u_{1s}(\mathbf{r})$ is the electronic wave function of ground state atomic hydrogen. Naturally, at finite ρ the eigenfunctions ω_+, ω_- are much more complicated functions than (11), but the whole point of this treatment of the problem is that there is no need to know ω_+, ω_- at finite ρ. Moreover, we postulated that at $t = -\infty$, when the collision began, the electron was attached to proton A, i.e., we know

$$\psi(-\infty) = u_{1s}(\mathbf{r}_A) \equiv u_A = \tfrac{1}{2}(u_A + u_B) + \tfrac{1}{2}(u_A - u_B)$$

$$= \frac{1}{\sqrt{2}}\omega_+ + \frac{1}{\sqrt{2}}\omega_-. \tag{12}$$

Comparing with Equation (3), we see that the initial conditions determine

$$c_+(-\infty) = c_-(-\infty) = \frac{1}{\sqrt{2}}.$$

Thus, using (10b) in (3)

$$\psi(+\infty) = \frac{1}{\sqrt{2}}\omega_+ \exp\left\{(-i/\hbar)\int_{-\infty}^{\infty} dt W_+(t)\right\}$$

$$+ \frac{1}{\sqrt{2}}\omega_- \exp\left\{(-1/\hbar)\int_{-\infty}^{\infty} dt W_-(t)\right\} \tag{13a}$$

which, using (11) in (13a), becomes

$$\psi(+\infty) = \tfrac{1}{2}u_A \left[\exp\left\{ (-i/\hbar) \int_{-\infty}^{\infty} dt\, W_+(t) \right\} \right.$$

$$+ \exp\left\{ (-i/\hbar) \int_{-\infty}^{\infty} dt\, W_-(t) \right\} \right] + \tfrac{1}{2}u_B \left[\exp\left\{ (-i/\hbar) \int_{-\infty}^{\infty} dt\, W_+(t) \right\} \right.$$

$$\left. - \exp\left\{ (-i/\hbar) \int_{-\infty}^{\infty} dt\, W_-(t) \right\} \right] \tag{13b}$$

The coefficient of u_B in (13b) is $c_B(\infty)$, the amplitude for finding the electron attached to proton B at $t = \infty$ after starting with $c_A(-\infty) = 1$, $c_B(-\infty) = 0$, representing an electron initially attached to proton A. If $W_+(t)$ were equal to $W_-(t)$ at all t, i.e., if the energies of the symmetric and antisymmetric H_2^+ eigenfunctions were identical at all internuclear spacings, $c_B(\infty)$ would vanish and there would be no electron capture; capture occurs only because the superposed components ω_+ and ω_- evolve at somewhat different rates. The specific value of $c_B(\infty)$ readily can be computed from the accurate energies sketched in (3), for any given impact parameter b and speed v in (5). For our present purposes, however, there is no need to make this computation. Quite generally, the probability of electron capture at given b and v is, from (13b),

$$|c_B(\infty)|^2 = \tfrac{1}{4}\left\{ 2 - 2\cos\left(\frac{1}{\hbar} \int_{-\infty}^{\infty} dt\, [W_+(t) - W_-(t)] \right) \right\}$$

$$= \sin^2\left(\frac{1}{2\hbar} \int_{-\infty}^{\infty} dt\, [W_+(t) - W_-(t)] \right). \tag{14}$$

The expected behavior of (14) as a function of impact parameter for fixed velocity is sketched in Figure 5. At very large impact parameters, ρ is always very large, and $W_+(t)$ very nearly equals $W_-(t)$ at all t; this explains the rapid decrease of $|c_B(\infty)|^2$ at very large b. For small b, the capture probability oscillates between zero and one as b increases, reflecting the variation with b of the phase difference between the coherently evolved components ω_+, ω_- in Equation (13a) for $\psi(\infty)$.

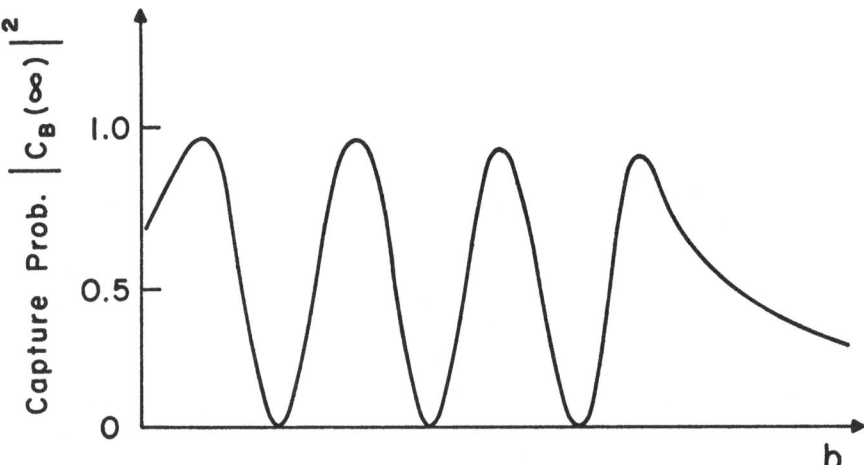

Fig. 5. Sketch of resonant electron capture probability as a function of impact parameter b, for a not too large incident velocity.

The oscillations in capture probability predicted by (14) and sketched in Figure 5 have been observed, though not quite as functions of impact parameter. It is easier experimentally to vary the velocity. Using (5), Equation (14) can be rewritten as

$$|c_B(\infty)|^2 = \sin^2 \left(\frac{1}{2\hbar v} \int\limits_{\rho(t=-\infty)}^{\rho(t=+\infty)} d\rho \, \frac{\rho}{\sqrt{\rho^2 - b^2}} [W_+(\rho) - W_-(\rho)] \right). \quad (15a)$$

If b is small enough, the integral (15a) can be approximated by

$$|c_B(\infty)|^2 \cong \sin^2 \left(\frac{1}{2\hbar v} \int\limits_{\rho(t=-\infty)}^{\rho(t=+\infty)} d\rho \, [W_+(\rho) - W_-(\rho)] \right) \quad (15b)$$

which depends only on the incident v. Now the outgoing electron-capturing protons B are observed at all angles, the angle depending on the impact parameter and incident v. However, electron capture observation at wide angles is only possible if the incident impact parameter was small, so that the internuclear forces were large. Estimates indicate that at inci-

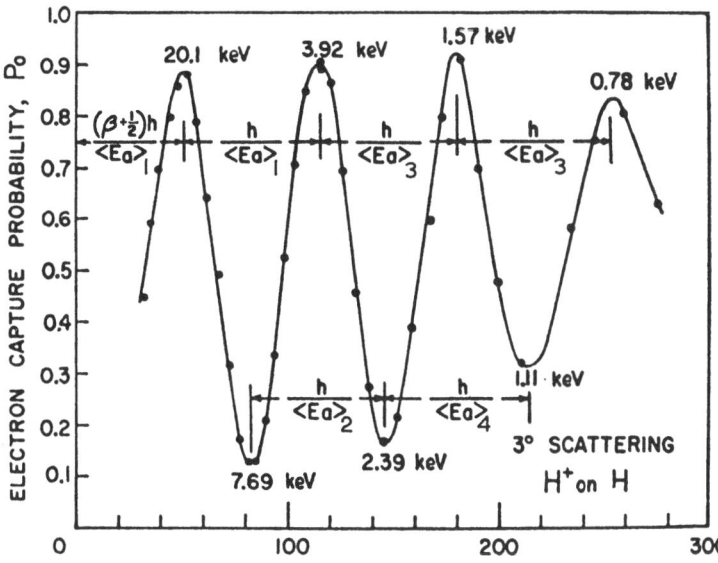

Fig. 6.   Observed electron capture probability P_0 in $H^+-H(1s)$ collisions plotted vs. the reciprocal incident velocity, at a fixed scattering angle of 3°, from Reference [18].

dent energies of about a keV (where the proton velocities still are small compared to the electron velocity), deflection angles as small as a few degrees still are large enough for (15b) to be a reasonable approximation to (15a). In other words, our theory predicts that at observation angles greater than a few degrees the electron capture probabilities will depend only on incident velocity, will oscillate between 0 and 1, and will be equally spaced when plotted against reciprocal velocity.

Figure 6 shows a plot of the electron capture probability $P_0 = |c_B(\infty)|^2$ for the reaction (1), plotted against reciprocal velocity, as measured by Lockwood and Everhart [18] at a fixed 3° deflection angle. The agreement with the predictions of (15b) is remarkable. The energies of maximum and minimum capture probability are shown on the Figure; the other symbols on the Figure ($h/\langle E_a \rangle$, etc.) may be ignored for our present purposes. Figure 7, from Helbig and Everhart [19], shows a plot of electron capture probability vs energy at various deflection angles θ from $\theta = 1.2°$ to $\theta = 6°$. Although not so indicated on the figure, the observations at $\theta = 4°$, 5° and 6° are reduced to equivalent hydrogenic energies from experiments with

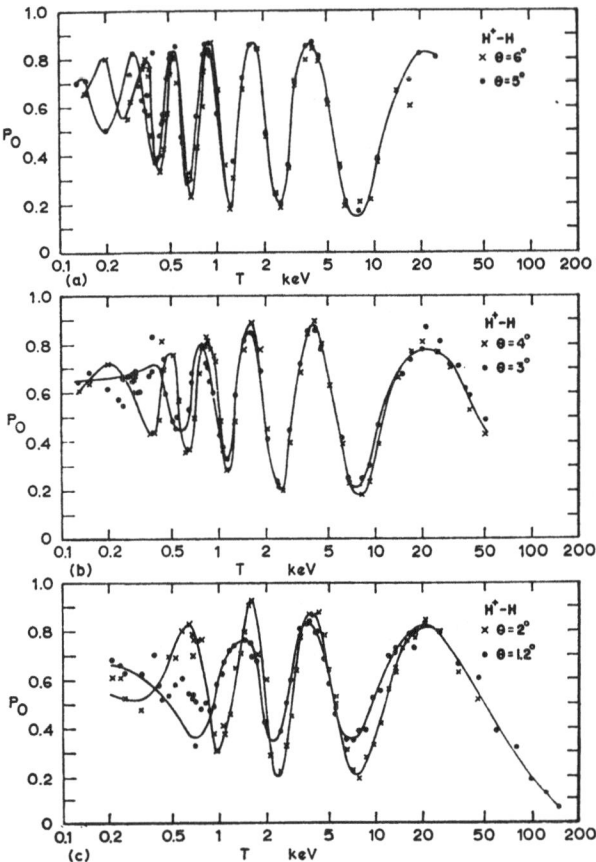

Fig. 7. Observed probabilities P_0 for electron capture in $H^+ - H(1s)$ collisions, plotted vs incident energy at several different angles of scattering, from Reference [19].

incident deuterons on ground state deuterium atoms. The capture probability oscillations are seen to be quite independent of deflection angle, again in remarkable agreement with (15b). Figure 8, from Ziemba and Everhart [20] is a plot of resonant electron capture in atomic helium

$$He^+(1s) + He(1s^2) \rightarrow He(1s^2) + He^+(1s) \qquad (16)$$
$$\text{fast} \qquad \text{slow} \qquad \text{fast} \qquad \text{slow}$$

The observed capture probabilities P_0 are plotted against reciprocal velocity at a fixed scattering angle of 5°. Again one sees regularly spaced inter-

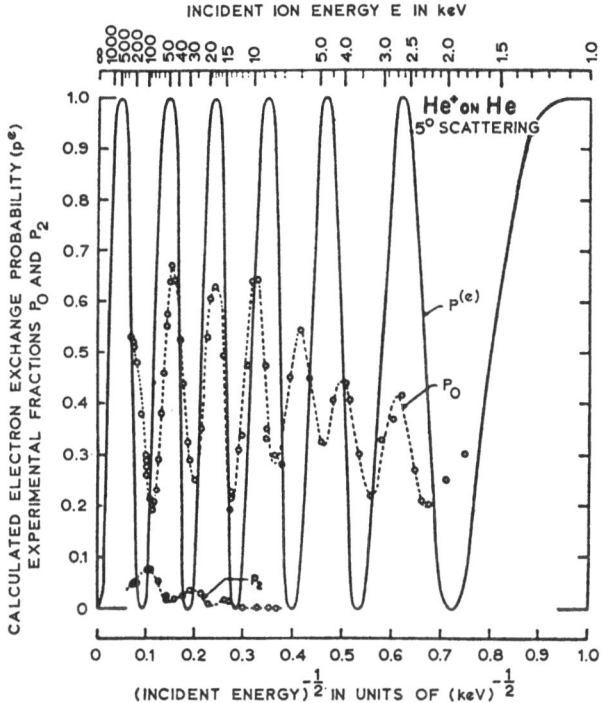

Fig. 8.   Observed probabilities P_0 for electron capture in $He^+ - He(1s^2)$ collisions, plotted vs. reciprocal incident velocity, at a scattering angle of 5°, from Reference [20]. The solid line shows the estimated theoretical probability $P^{(e)}$; also shown are the observed probabilities P_2 for ionization of the incident He^+ ion.

ference oscillations of P_0, although – presumably because this He^+ reaction is much more complicated than is capture in hydrogen – the observed probabilities are not quite as close to the full 0 to 1 oscillations predicted by (15b) as were Figures 6–7. The solid line in Figure 8 shows the theoretical predictions, based on best estimates of He_2^+ energy curves (at that time, 1959). One possible reason for disagreement with theory in Figure 8 is the detectable probability P_2 of ionizing the incident ion, a process which was ignored in the theory of (15b), and which of course cannot occur in electron capture by protons.

It seems fair to conclude that these observations by Everhart and co-workers (Figures 6–8) demand the interpretation that the wave function in resonant electron capture is a linear superposition of appropriate molecular ion states, independently but coherently evolving as the collision progresses, and eventually interfering coherently to produce the actual capture probability. As a matter of fact, the resonant capture process is quite analogous to the exchange of excitation between two classical oscillators which are slowly being coupled and decoupled as a function of time. The oscillator winding up with the major share of the energy will vary as the rate of decoupling is varied. On this picture, the states u_A and u_B correspond to the uncoupled oscillators; the coupling is provided by the interaction between a proton and atomic hydrogen. For strong interactions, however, as in the present electron capture collisions when the two nuclei are close together, the model of two independently-existing oscillators breaks down. But for slow changes in coupling, one can introduce the instantaneously exact normal modes; these are the solutions ω_+ and ω_-.

2.4. *Quantum Beats Experiments*

Figure 9 shows remarkable data reported very recently by Andra [21]. Plotted in Figure 9 are the observed intensities of radiation emitted as a function of time in a fast beam that has emerged from a foil. To make plain what is going on, consider the top three curves in Figure 9, which correspond to an allowed transition in normal helium, between the 3^3P and 2^3S levels. The excited He (3^3P) atoms were produced by the so-called beam-foil technique [22]. He^+ ions were accelerated to 475 keV, and sent through a very thin carbon foil. Some of the ions capture electrons in the foil, and emerge as neutral He(3^3P) atoms. These are the atoms which are being observed. What actually is measured is the intensity of radiation from a very thin section of the beam, as a function of distance along the beam downstream from the foil. These intensities can be replotted as a function of time via the known velocity of the ions, which are barely slowed down in passing through the foil. All observations were made perpendicular to the beam axis using photon counters. Other He lines were filtered out, but the filters were not sharp enough to distinguish between the various possible fine structure transitions. In curve A, Andra measured the radiated intensity polarized in the plane parallel to the beam axis, as indicated on the figure. In curve B the same transition was observed with detectors measuring the polarized component perpendicular to the beam. Curve C shows observed intensities without polarization analysers. All three curves A, B, C show regular oscillations having the same

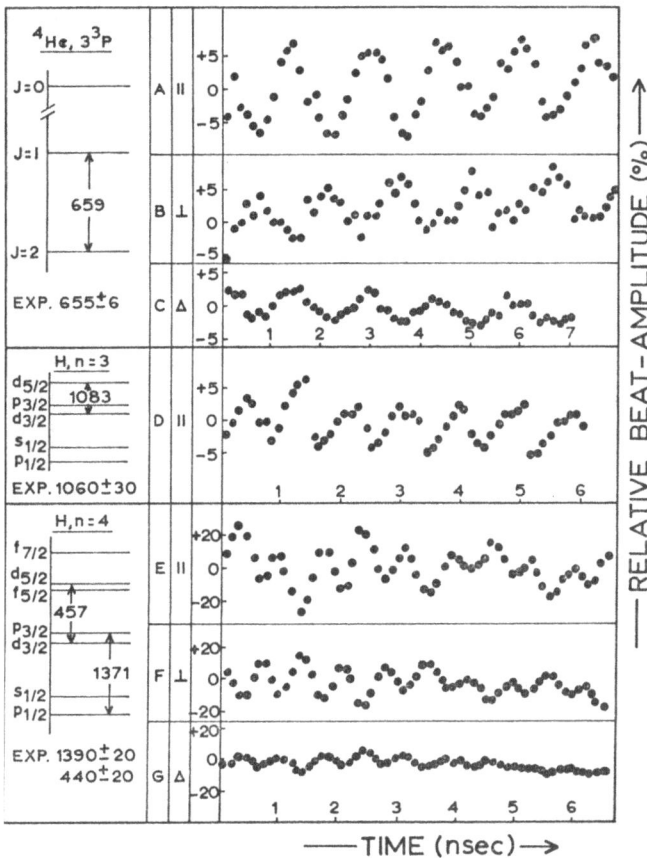

Fig. 9. Observed beats in radiation emitted from electron-capturing beams, from Reference [21]. Curves A, B, C, radiation from He 3^3P levels; curve D, $H\alpha$ radiation; curves E, F, G, $H\beta$ radiation. The designations $\parallel$, $\perp$, $\triangle$ refer respectively to measurements of intensities plane polarized parallel to the beam, perpendicular to the beam, and unpolarized. All observations were made at 90° to the beam axis. All frequencies are given in MHz.

frequency, but differing in phase and amplitude. The observed frequency of oscillation was $655 \pm 6 \times 10^6$ Hz, which can be considered identical with 658.55×10^6 Hz, the known fine-structure separation of the He 3^3P_1 and the 3^3P_2 levels.

The other curves in Figure 9 were obtained with atomic hydrogen using an incident proton beam of 133 keV, but are otherwise basically similar to the He curves. Curve D shows observations (parallel polarization only) on the Balmer line Hα; curves E, F, G are for the Balmer line Hβ. The oscillations in Hα and Hβ look a little more complicated than single frequency, but because of experimental difficulties it is not possible to make a complete resolution of the frequency components in these lines. However, it was determined that Hα had a component about equal to the $n = 3$ $d_{5/2}-d_{3/2}$ separation, while in Hβ two components, corresponding respectively to the $n = 4$ $d_{5/2}-d_{3/2}$ and $p_{3/2}-p_{1/2}$ separations, could be identified.

What is one to make of these oscillations in Figure 9? If the atoms were moving in external electric or magnetic fields, the oscillations could be understood via arguments resembling those discussed in the electron capture case. For weak external fields, the situation generally prevailing in practice, the zero-field stationary states legitimately can be regarded as independently-existing oscillators, being coupled by the external field. The coupled oscillators would be the nearly degenerate levels in the upper radiating state (assuming for the sake of simplicity that the lower final state is not split in the external field). As the excitation oscillates back and forth between the upper radiating levels, the total radiative transition probability generally also oscillates, with an oscillation or beat frequency close to the separation of the zero-field nearly degenerate levels. Thus the predicted radiation in the circumstances that there are external fields would resemble the observations in Figure 9. Furthermore, such external-field-induced beats are observable in actual practice, as Figure 10 shows. This figure shows observed beats in Balmer Hδ produced in a beam-foil experiment like Andra's of Figure 9. In Figure 10, however, the measurements (by Bickel [23]) are in a region where the beam is traversing a field of 40 volts/cm; the observed beat frequency was 1.14×10^9 Hz, which was in the range of theoretical expectation taking into account Stark shifts of the levels, and recognizing that the vast multiplicity of sublevels in the upper $n = 6$ hydrogenic state makes detailed comparison between theory and experiment very difficult. Therefore, in Figure 10 we again have an experiment which is reasonably well understood, and which provides strong evidence for the validity of the principle of superposition and subsequent coherent interference.

On the other hand, Andra – realizing this possibility of external-field-induced beats – took pains to eliminate such fields in the observations of Figure 9. Assuming his precautions were well taken, Figure 9 represents

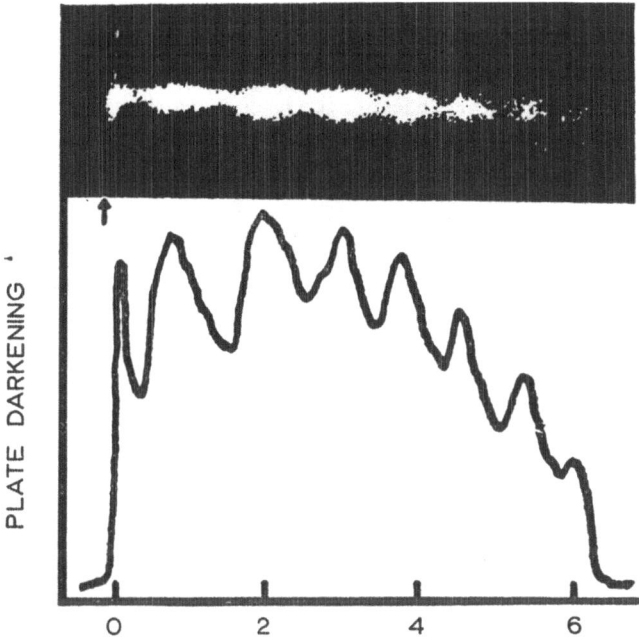

PLATE DARKENING

0   2   4   6

Fig. 10.   Oscillations in Hδ radiated in a beam-foil experiment, with the beam in a 40 V cm⁻¹ electric field, from Reference [23]. The upper portion of the figure is the original densitometer tracing of the beam emissions.

oscillations in zero external fields; or, if you will, oscillations without external coupling. This is an entirely new notion – not at all analogous to the electron capture case we have discussed – but its possibility was predicted recently by Macek [24]. According to Macek (if I may be allowed to put words in his mouth), electron capture into more than one state can occur in the foil; in particular, one can have capture into several fine structure levels. Thus the wave function of the emerging neutral beam can be a superposition of fine structure levels. Normally we would regard the foil as a macroscopic apparatus, I guess, and would suppose that multiple scatterings, stray field fluctuations, etc., inside the foil would cause the emergent atoms to be described by an incoherent mixture of fine structure levels. But in the present beam-foil experiment the beam traverses the very thin foil so rapidly (in the order of 10^{-13} s, in fact) that fine structure levels don't have time to get out of phase before the atoms emerge from the foil. Hence the radiating excited atom beam observed by Andra is described by a coherent superposition of fine structure levels. Each of these levels is individually a perfectly normal stationary state, but the superposition is not stationary, because the levels have slightly different energies. To calculate the probability of radiation, Macek proposes to compute simply the square of the usual dipole matrix element between the final state and this superposed initial state. In this way one gets beat frequencies equal to fine structure spacings in the higher radiating state.

There are several remarks to be made concerning this explanation of Figure 9 which – if it stands up – will be a very profound confirmation of quantum mechanics in general, and of the principle of superposition in particular. In the first place, as Macek points out, the external-field-interference beats – when observed with and without polarization effects – should behave qualitatively sufficiently differently from the zero-field beats to be distinguishable thereform, although this may be a difficult test to carry out in practice. Andra made some tests of this kind, but in view of the fundamental importance of his experiment, I don't think his tests should be considered conclusive as yet. Secondly, I don't think there is extant any really satisfactory derivation of Macek's hypothesis that the transition probability is to be computed from what may be termed the instantaneous interfering dipole strength. (Professor E. P. Wigner, and (independently) Professor Philip Stehle, have suggested that the derivation could be carried through by generalizing the Wigner-Weisskopf theory of radiation from a single level[2] (cf. W. Heitler [14] *ibid.*, p. 182.)). I have seen semi-classical (by which I mean non-quantum-electrodynamic) derivations of the probability for photoelectric emission [25] that confirm this plausible hypothesis, and I made the same hypothesis myself many years ago when I estimated the intensity of beats in the photoelectric current from two simultaneously incident completely incoherent sources [26]. Yes, Virginia, these incoherent sources will produce photoelectric beats, at a frequency precisely equal to the difference between the two independent incident frequencies, as Forrester *et al.* [27] first demonstrated in 1955.

However, the foregoing considerations really are not germane to the present problem; the interaction of an atom with a macroscopic anharmonic field does not immediately give us information about the spontaneous emission from an atom in a superposition of non-degenerate states.

But I will claim – as I also would claim in the case of the aforementioned photoelectric beats – that Macek's predicted effect is only possible because the measurement is unable to distinguish between the frequencies of the various photons that can be emitted; you will remember that Andra's filter didn't distinguish between his various permissible fine structure lines. I make this claim without having proved it, because it is consistent with my feelings as to what complementarity is all about. If we know which one of two slits a photon has passed through, we wipe out the interference between different possible paths; if we know what frequency a particular photon count corresponds to, we wipe out the interference between different possible frequencies. These intuitions of mine are reflected in the difficulty one has in deriving Macek's hypothesized spontaneous emission probability by conventional quantum electrodynamic techniques. The trouble is that the whole usual formalism is designed to determine the transition probability between states of well-defined energy. But as soon as we know the energies of the initial and final states, we know the frequency of the emitted photon and therefore – on my view – automatically lose the ability to discover interference between photons of different frequency.

Department of Physics,
University of Pittsburgh

NOTES

* Supported by the National Aeronautics and Space Administration under Contract No. NGL-39-011-035.
[2] Note added in proof. I am indebted to Professor Macek for pointing out that indeed he has carried through such a derivation. J. Macek, *Phys. Rev.* A1 (1970) 618.

BIBLIOGRAPHY

[1] Cf., e.g., Schiff, L. I., *Quantum Mechanics*, McGraw Hill, New York, 1955; Bohm, D., *Quantum Theory*, Prentice Hall, New York, 1951; Messiah, A., *Quantum Mechanics*, Wiley, New York, 1962.
[2] Kramers, H. A., *Quantum Mechanics*, North-Holland Publ. Co., 1957, transl. by D. ter Haar of the original 1937 edition.
[3] Heisenberg, W., in *Niels Bohr and the Development of Physics* (ed. by W. Pauli), Pergamon, New York, 1955, p. 12.
[4] Ballantine, L. E., *Reviews of Modern Physics* 42 (1970), 358.
[5] Komar, A., *Physical Review* 126 (1962), 365.

[6] Fine, A., *Physical Review* D2 (1970), 2738.
[7] Wigner, E. P., *Symmetries and Reflections*, Indiana University Press, Bloomington, Indiana, 1967, Part IV, *Epistemology and Quantum Mechanics*.
[8] Schrödinger, E., *Naturwissenschaften* 23 (1935), 807.
[9] deWitt, B. S., *Physics Today* 23 (1970), p. 30.
[10] Everett III, H., *Reviews of Modern Physics* 29 (1957), 454.
[11] Wheeler, J. A., *Reviews of Modern Physics* 29 (1957), 463.
[12] Cf. e.g., Bohm, D., *Causality and Chance in Modern Physics*, Van Nostrand, New York, 1957; Bohm, D., *Physical Review* 85 (1952), 166, 180; Bohm, D., *Physical Review* 89 (1953), 459; Bohm, D. and Vigier, J. P., *Physical Review* 96 (1954), 208; de Broglie, L., *The Current Interpretation of Wave Mechanics*, Elsevier, Paris, 1964. For a proposed experiment to distinguish between conventional and hidden variable quantum theories, see Clauser, J. F., Horne, M. A., Shimony, A., and Holt, R. A., *Physical Review* 23 (1969), 880.
[13] Wu, C. A. and Shaknov, I., *Physical Review* 77 (1950), 136.
[14] Heitler, W., *Quantum Theory of Radiation*, Oxford, 1954, pp. 268–72.
[15] Bohm, D. and Aharanov, Y., *Physical Review* 108 (1957), 1070.
[16] Bohm, D. and Aharanov, Y., *Il Nuovo Cimento* 17 (1960), 964.
[17] Einstein, A., Podolsky, B. and Rosen, N., *Physical Review* 47 (1935), 777.
[18] Lockwood, G. J. and Everhart, E., *Physical Review* 125 (1962), 567.
[19] Helbig, H. F. and Everhart, E., *Physical Review* 140 (1965), A715.
[20] Ziemba, F. P. and Everhart, E., *Physical Review Letters* 2 (1959), 299.
[21] Andra, H. J., *Physical Review Letters* 25 (1970), 325.
[22] Bashkin, S., *Nuclear Instrumentation and Methods* 28 (1964), 88.
[23] Bickel, W. S., *Journal of the Optical Society of America* 58 (1968), 213.
[24] Macek, J., *Physical Review Letters* 23 (1969), 1.
[25] Mandel, L., Sudarshan, E. C. G. and Wolf, E., *Proceedings of the Physical Society (London)* 84 (1964), 435.
[26] Forrester, A. T., Parkins, W. E., and Gerjuoy, E., *Physical Review* 72 (1947), 728.
[27] Forrester, A. T., Gudmundsen, R. A., and Johnson, P. O., *Physical Review* 99 (1955), 1691.

R. J. GREECHIE AND STANLEY P. GUDDER

QUANTUM LOGICS

PREFACE

Although the authors have great interest in physics and philosophy we are not 'experts' in these fields; moreover our primary interest (and training) is in mathematics. For these reasons our presentation is essentially mathematical in nature. We realize that, when one presents an approach which purports to deal with physical situations, physical justifications should be given for one's assumptions. If further this approach has philosophical ramifications then one ought to discuss these ramifications. In this paper we shall attempt to motivate the assumptions made; however, we shall minimize discussion of the philosophical import of these assumptions.

Our main aim is to present some mathematical tools necessary for the study of axiomatic quantum mechanics, in particular the quantum logic approach. We feel that the mathematical tools, techniques and theorems must be first understood before physical and philosophical discussions about the subject can take place. It is our opinion that the quantum logic approach can give one, if nothing else, deeper insight into the understanding of quantum phenomena.

I. INTRODUCTION

There are many advantages of axiomatic formulations for physical theories. First of all, by stating one's axioms carefully one is fully aware of the assumptions made in the theory and it is clear what hypotheses must be physically justified. These hypotheses may then be tested in the laboratory as a check of the theory. Of course, a physical theory can never be proved to be correct; the conclusions from the theory can only be compared with experimental results to test whether it is an approximate description of some small isolated portion of 'reality'. If the theory fails to compare favorably with the experimental results it must be abandoned as a theoretical description for that portion of reality. Otherwise it may be retained

Hooker (ed.), Contemporary Research in the Foundations and Philosophy of Quantum Theory, 143–173. *All rights reserved*
Copyright © 1973 by D. Reidel Publishing Company, Dordrecht-Holland

until a better theory is discovered. Secondly, an axiomatic approach gives a common 'universe of discourse' in which ideas may be discussed and conjectures formulated. Many of the great controversies in physics seem to result from difficulties in semantics. For example, it is our feeling that the controversy over hidden variables in quantum mechanics is caused to a certain extent by a failure in laying out the 'ground rules' for the game. Many arguments seem to result from the fact that the debators have different underlying physical formulations in mind and it is never clearly stated exactly what assumptions one is making. If at the beginning a common universe of discourse in terms of an axiomatic formulation were established these types of problems might vanish. Thirdly, if one operates under a consistent axiomatic model for a physical theory one is assured that no mathematical contradictions will be encountered. For example, the difficulties in quantum electrodynamics stemming from the occurrence of divergences and infinities might be avoided if a consistent axiomatic model were constructed. Indeed this is one of the reasons for the introduction of axiomatic quantum field theory by Wightman and his co-workers.

There are several axiomatic approaches to the foundations of quantum mechanics available in the physical and mathematical literature. One of these is the quantum logic approach initiated by Birkhoff and von Neumann in 1936 [4]. This study has been continued by Mackey [54, 55], and Varadarajan [80], and has been refined and altered by Jauch, Piron [38, 39, 40, 45, 46, 64] and others [1, 5, 14, 18, 28, 30, 31, 32, 42, 56, 58, 62, 65, 67, 68, 69, 76, 78, 79, 82, 84]. Another is the algebraic approach first conceived by Jordan *et al.* [47] in 1934, developed further by Segal [71, 72] and others [50, 73, 74] culminating in the elegant theories of Haag [37], Wightman [77], and their co-workers. Another formulation has been proposed by Ludwig and his collaborators [51, 52] and has been recently refined by Mielnik [59, 60], Davies and Lewis [11, 12] and others [15, 16, 36]. These models present different approaches to what appears to be essentially the same underlying theory. In fact there have been studies made comparing these different approaches [27, 35, 63]. Now it may seem, at first sight, to be wasteful and redundant to proliferate the literature with different approaches to the same subject. However, it has turned out that each approach adds new insights and different viewpoints which have led to fruitful results and contributed to a deeper understanding of quantum theory.

Of course, most working physicists do not use and are probably even unaware of the above formulations. The majority of physicists rely upon the von Neumann [81] and/or Dirac [13] formulations of quantum mechanics and in so doing have achieved many extraordinary successes. We have no quarrel with these researchers. We only contend that a knowledge of some of the basic mathematical tools involved in a consistent axiomatic model grounded upon physically justified assumptions may prove both useful and rewarding.

One of the aims of the present paper is to present one of these models, namely the quantum logic approach, in some detail. This approach has evolved in several slightly different directions. One of these directions we attribute to Mackey and another to Jauch-Piron. We single out these researches only for expediency and, although they have had profound influence on the subject, it must be realized that many others have made equally important contributions.

Having treated these two major approaches to quantum logic in some detail we discuss several attempts to define a conditional, $a \supset b$, in quantum logic. The impact of our discussion is that there has been no successful attempt to define a conditional, $a \supset b$, which is an element of the logic and behaves in a fashion similar to that of classical logic.

We conclude with a discussion of combinatorial quantum logic. Orthogonality spaces are introduced in order to present a model which distinguishes between the two possible orderings induced by states. Other applications of this promising approach are given.

II. OBSERVABLES, STATES AND QUESTIONS [55]

Although, as we have stated in the introduction, there are several axiomatic formulations for the foundations of quantum mechanics, they all seem to involve in some degree the three basic, primitive notions of observable, state, and question. In the laboratory, the experimental physicist makes different measurements. He measures physical observables which have traditional names such as energy, momentum, position, spin, charge, magnetic moment, etc. He makes these measurements by subjecting a physical system to a measuring apparatus, and he is concerned with the outcomes of these interactions when the system is in some specified state or condition.

Let us suppose that we have some fixed physical system which may exist in any one of a collection of states $S = \{s, s_1, s_2, \ldots\}$ and that in this system one may measure the observables $\mathcal{O} = \{x, y, z, \ldots\}$. Now the result of measuring an observable x can usually be formulated as a number; e.g., the spin of the particle was $+2$, the energy of the electron was 3 erg. Of course, in practice, one repeats the experiment many times (keeping the state as fixed as possible) and obtains only a statistical distribution for the values of x. Thus, given an observable x and a state s, the experimentalist obtains a probability distribution $p(x, s)(\cdot)$. By this we mean that given any set E of real numbers (mathematicians usually only consider Borel sets but we will not concern ourselves with such technicalities now) $p(x, s)(E)$ is a number between 0 and 1 representing the probability that the observable x has a value in the set E when the system is in the state s. If we denote the set of all probability distributions by M we may now formulate our first axiom.

AXIOM 1. There is a map $p: \mathcal{O} \times S \to M$ denoted by $p(x, s)(\cdot)$. (Of course $\mathcal{O} \times S$ denotes the set of all ordered pairs (x, s) where $x \in \mathcal{O}$ and $s \in S$.)

If two observables have the same probability distribution in every state then there is no experimental way to distinguish them so they must be the same observable. Similarly if two states give the same probability distributions for all observables they must be equal. We are thus led to our next axiom.

AXIOM 2. If $p(x, s)(E) = p(y, s)(E)$ for all $s \in S$ and all sets E of real numbers then $x = y$. If $p(x, s_1)(E) = p(x, s_2)(E)$ for all $x \in \mathcal{O}$ and all sets E of real numbers then $s_1 = s_2$.

If we can measure an observable x then it is just as easy to measure the observable x^2; simply take the measured values of x and square them. Now the probability that x^2 has a value λ is the same as the probability that x has the values $\pm\sqrt{\lambda}$, and more generally the probability that x^2 has a value in a set E of real numbers is the probability that x has a value in the set $\pm\sqrt{E} = \{\lambda : \lambda^2 \in E\}$. In the same way, if f is a real valued function and x is an observable then $f(x)$ is an observable and the probability that $f(x)$ has a value in the set E is the probability that x has a value in the set $f^{-1}(E) = \{\lambda : f(\lambda) \in E\}$. Usually mathematicians consider only Borel functions but again we omit the technicality. We are now ready for our next axiom.

AXIOM 3. If $x \in \mathcal{O}$ and f is a real valued function then there is a $y \in \mathcal{O}$ such that $p(y, s)(E) = p(x, s)(f^{-1}(E))$ for every $s \in S$ and every set of real numbers E.

It follows from Axiom 2 that the observable y in Axiom 3 is unique. We denote this observable by $y = f(x)$.

Now there is a particular type of observable which is extremely simple. These are the observables with only two possible values, say 0 and 1. We call such observables 'questions'. For example, a counter is a question since it gives a measurement with only two possible outcomes: unactivated (or 0) and activated (or 1). We thus define a *question* to be any observable x that satisfies $p(x, s)(\{0, 1\}) = 1$ for all $s \in S$; that is, x has the value 0 or 1 with certainty in every state. It is easy to show that $x \in \mathcal{O}$ is a question if and only if $x^2 = x$. There is another convenient way to describe questions. If $E \subseteq R$ (R denotes the real line) then the *characteristic function*
$$\chi_E(\lambda) = \begin{cases} 1 \text{ if } \lambda \in E. \\ 0 \text{ if } \lambda \notin E. \end{cases}$$
Now it is easy to show that $x \in \mathcal{O}$ is a question if and only if $x = \chi_E(y)$ for some $E \subseteq R$, $y \in \mathcal{O}$. In particular if $x \in \mathcal{O}$ then we can associate with each $E \subseteq R$ a question $\chi_E(x)$. This question has the value 1 if x has a value in E and the value 0 if x has a value not in E. Notice if $\chi_E(x) = \chi_E(y)$ for all $E \subseteq R$ then

$$p(x, s)(E) = p(x, s)(\chi_E^{-1}\{1\}) = p(\chi_E(x), s)(\{1\})$$
$$= p(\chi_E(y), s)(\{1\}) = p(y, s)(E)$$

for all $s \in S$, $E \subseteq R$ and hence $x = y$. We thus see that not only can we associate with any observable x a collection of questions $\{\chi_E(x) : E \subseteq R\}$ but that this associated collection of questions determines x.

Denote the set of questions by Q. We have seen that the system (Q, S) contains all the information given in $(\mathcal{O}, S)$. Since questions are far simpler than general observables it appears that we can make a more fundamental study by considering (Q, S) instead of $(\mathcal{O}, S)$. Let us now try to discover some of the mathematical properties of the system (Q, S). If $\alpha \in Q$ and $s \in S$ we define $s(\alpha) = p(\alpha, s)(\{1\})$. Now $s(\alpha)$ may be interpreted as the probability that α has the value 1 (or α has the answer 'yes') in the state s. Notice if $\alpha, \beta \in Q$ and $s(\alpha) = s(\beta)$ for all $s \in S$ then by Axiom 2, $\alpha = \beta$. It also follows from Axiom 2 that if $s_1(\alpha) = s_2(\alpha)$ for all $\alpha \in Q$ then $s_1 = s_2$ so there are sufficiently many questions to determine the state. If $\alpha_1, \alpha_2 \in Q$ we define $\alpha_1 \leqslant \alpha_2$ if $s(\alpha_1) \leqslant s(\alpha_2)$ for all $s \in S$. Thus $\alpha_1 \leqslant \alpha_2$ if α_1 has a smaller probability of having an answer 'yes' than α_2 in every state. It is easy to check that $\leqslant$ is a partial order relation on Q; that is $\alpha \leqslant \alpha$ for all $\alpha \in Q$, $\alpha \leqslant \beta$ and $\beta \leqslant \gamma$ implies $\alpha \leqslant \gamma$, $\alpha \leqslant \beta$ and $\beta \leqslant \alpha$ implies $\alpha = \beta$. Thus $(Q, \leqslant)$ is a *partially ordered set* or *poset*. Let f be the function $f(\lambda) = 1 - \lambda$. If $\alpha \in Q$ we define the observable α' by $\alpha' = f(\alpha)$. Notice

$$p(\alpha', s)(\{0, 1\}) = p(\alpha, s)(f^{-1}\{0, 1\}) = p(\alpha, s)(\{0, 1\}) = 1$$

for every $s \in S$ so $\alpha' \in Q$. Also

$$s(\alpha') = p(\alpha', s)(\{1\}) = p(\alpha, s)(f^{-1}\{1\}) = p(\alpha, s)(\{0\})$$
$$= 1 - p(\alpha, s)(\{1\}) = 1 - s(\alpha).$$

Thus α' corresponds to the negation of the question α. If f_0 and f_1 are the functions that are identically zero and one respectively and $x \in \mathcal{O}$ we define the observables 0 and 1 by $0 = f_0(x)$ and $1 = f_1(x)$ respectively. Notice

$$p(0, s)(\{0, 1\}) = p(x, s)(f_0^{-1}\{0, 1\}) = p(x, s)(R) = 1.$$

Hence $0 \in Q$ and also

$$s(0) = p(0, s)(\{1\}) = p(x, s)(f_0^{-1}\{1\}) = p(x, s)(\varphi) = 0$$

for all $s \in S$. Similarly $1 \in Q$ and $s(1) = 1$ for all $s \in S$. Hence $0 \leqslant \alpha \leqslant 1$ for all $\alpha \in Q$. We may interpret 0 and 1 as the questions whose answers are always 'no' and 'yes' respectively. In the poset $(Q, \leqslant)$ we say that γ is the *least upper bound* of α, $\beta \in Q$ if α, $\beta \leqslant \gamma$ and whenever α, $\beta \leqslant \delta$ we have $\gamma \leqslant \delta$. The least upper bound need not exist, but when it does it is unique. We denote the least upper bound (or sup) of α and β by $\alpha \vee \beta$ when it exists. We define the greatest lower bound (or inf) of α and β dually and denote it by $\alpha \wedge \beta$ when it exists. The following lemma is easily proved.

LEMMA 2.1. The operation $\alpha \to \alpha'$ is an orthocomplementation on Q. That is, $\alpha'' = \alpha$ for all $\alpha \in Q$, if $\alpha \leqslant \beta$ then $\beta' \leqslant \alpha'$, and $\alpha \vee \alpha'$ always exists and equals 1.

Thus $(\alpha, \leqslant, ')$ is an *orthocomplemented poset*. This section has served only as an introduction to the physical and mathematical notions involved in an axiomatic approach to quantum mechanics. The theory has not been carried out far enough to give a mathematical model for a physical system. In the next section we will start anew on a slightly different tack and present a more detailed and complete model. The framework that we have developed in the present section can be extended further [55] to obtain a model equivalent to that of the next section. However it is more common and possibly more instructive to begin with the questions as the axiomation elements as is done in the succeeding section.

III. THE BIRKHOFF-VON NEUMANN-MACKEY APPROACH [80]

In the last section we formulated an axiomatic theory based on the observables and states of a physical system. We then derived the notion of questions which turned out to be more elementary than the observables. For this reason we now formulate an axiomatic model in which the questions are the sole primitive axiomatic elements. We then derive the concepts of states and observables in terms of these primitive elements.

Let $Q = \{\alpha, \beta, \gamma, ...\}$ be the set of questions for a quantum system. A question may be interpreted as corresponding to a measurement or experiment leading to two alternatives which we call 'yes' and 'no'. The above measurement or experiment consists of a procedure to be carried out with the physical system under consideration, and a rule for interpreting the possible results in terms of 'yes' and 'no'. It is well-known that there are measurements in quantum mechanics that interfere with each other such as position and momentum measurements. Suppose α and β are noninterfering questions (i.e. performing the experiment α does not change the answers of β and vice-versa). If whenever the answer to α is 'yes' it follows that the answer to β is 'yes' we write $\alpha \leqslant \beta$. Notice the relation $\leqslant$ has no apparent connection with the order relation in Section II. We cannot use that order relation here since our only axiomatic elements are questions and we want to derive the states from these. The relation $\leqslant$ should satisfy:

(Q1) $\alpha \leqslant \alpha$, for all $\alpha \in Q$;
(Q2)   if $\alpha \leqslant \beta, \beta \leqslant \alpha$ then $\alpha = \beta$;
(Q3)   if $\alpha \leqslant \beta, \beta \leqslant \gamma$ then $\alpha \leqslant \gamma$;
(Q4)   there are questions 0,1 such that $0 \leqslant \alpha \leqslant 1$ for all $\alpha \in Q$.

Thus Q is a poset with universal bounds 0,1. A fundamental problem is whether Q is a *lattice*, i.e. does $\alpha \vee \beta$ and $\alpha \wedge \beta$ exist for all $\alpha, \beta \in Q$? For example if $\alpha \wedge \beta$ exists it would (in this setting) be interpreted as the question whose answer is 'yes' if and only if α and β both have the answer 'yes'. If α and β are interfering questions it is, to some researchers, doubtful that an experimental apparatus can be constructed corresponding to such a question except under special conditions. For this reason we do not assume Q is a lattice. In Figure 1 we diagram an example of a poset which is not a lattice. In reading such diagrams, a rising line from α to β means $\alpha \leqslant \beta$ and there is no $\delta \neq \alpha, \beta$ such that $\alpha \leqslant \delta \leqslant \beta$.

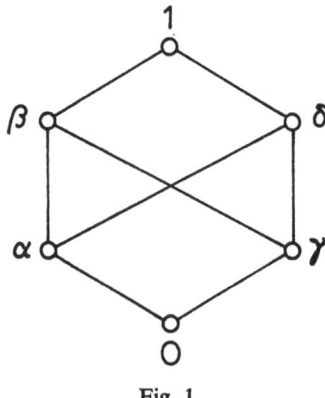

Fig. 1

Given $\alpha \in Q$ we define α' as the question whose alternatives are reversed; that is, we consider α' to have a 'yes' answer if and only if the result of α is 'no'. It is clear that the following conditions are satisfied:

(Q5) $\alpha'' = \alpha$ for all $\alpha \in Q$;

(Q6)   if $\alpha \leqslant \beta$ then $\beta' \leqslant \alpha'$;

(Q7) $\alpha \vee \alpha'$ exists and equals 1 for all $\alpha \in Q$.

Thus $(Q, \leqslant, ')$ forms an orthocomplemented poset. If $\alpha \leqslant \beta'$ we say α and β are *orthogonal* are write $\alpha \perp \beta$. Notice $\alpha \perp \beta$ if and only if $\beta \perp \alpha$. We may interpret $\alpha \perp \beta$ to mean that α and β are non-interfering and that β has a 'no' result if α has a 'yes' result. In this case it *is regarded as* physically resonable to assume that $\alpha \vee \beta$ exists. We extend this conclusion, mainly for mathematical convenience, to a countable number of questions.

(Q8)   If α_i is a sequence of mutually orthogonal questions then $\bigvee \alpha_i$ exists, i.e. Q is σ-orthocomplete.

We now make this into a statistical theory by introducing states. A state should completely describe the system as far as is physically possible. Experimentally the most we can determine about the questions of a physical system is the probabilities of getting 'yes' (and hence also 'no') results.

Thus given a question α, the state should determine the probability that α has a 'yes' result. We therefore define a *state* s as a probability measure on Q; that is, s is a map from Q to the unit interval $[0,1] \subset R$ such that (S1) $s(1) = 1$; (S2) $s(\vee \alpha_i) = \sum s(\alpha_i)$ if $\alpha_i \perp \alpha_j$, $i \neq j = 1, 2, \dots$.

A set of states S on Q is full if $s(\alpha) \leqslant s(\beta)$ for all $s \in S$ implies $\alpha \leqslant \beta$. One can give examples of systems satisfying (Q1)–(Q7) which have no states at all [22]. Physically, one would expect to have enough states to capture the ordering on the questions. We therefore postulate:

(Q9)   There is a full set of states S on Q.

If (Q, S) satisfies (Q1)–(Q9) we call it a *quantum logic*. An orthocomplemented poset is called an *orthomodular poset* if $\alpha \leqslant \beta$ implies $\alpha \vee (\beta \wedge \alpha')$ exists and equals β. The next lemma has a straightforward proof.

LEMMA 3.1. *If* (Q, S) *is a quantum logic then* Q *is an orthomodular poset.*

Now there are quantum logics which are not lattices. Let $\Omega = \{1, 2, 3, 4, 5, 6\}$ and let Q be the collection of subsets of Ω with an even number of elements. Order Q by inclusion and let $'$ be the usual set complementation. For $A \in Q$, $i = 1, \dots, 6$ define $s_i(A) = \begin{cases} 1 \text{ if } i \in A \\ 0 \text{ if } i \notin A \end{cases}$. Then $S = \{s_i : i = 1, \dots, 6\}$ is a full set of states and (Q, S) is a quantum logic. However Q is not a lattice since, for example, $\{1, 2, 3, 4\} \wedge \{2, 3, 4, 5\}$ does not exist.

We say that two questions α, β are *compatible* (written $\alpha C \beta$) if there are mutually orthogonal questions $\alpha_1, \beta_1, \gamma$ such that $\alpha = \alpha_1 \vee \gamma$, $\beta = \beta_1 \vee \gamma$. We shall see that compatible questions are ones that can be answered simultaneously; that is, questions which do not interfere. In fact, the words compatible, simultaneously answerable, and non-interfering are frequently used synonymously. Notice if $\alpha \perp \beta$ then $\alpha C \beta$ and $0 C \alpha$, $1 C \alpha$ for all $\alpha \in Q$. Physically, our interpretation of $\alpha \leqslant \beta$ demands that $\alpha C \beta$ if $\alpha \leqslant \beta$. This need not happen if there is no full set of states which is another reason we insist upon (Q9).

LEMMA 3.2. *If* (Q, S) *is a quantum logic then* $\alpha C \beta$ *whenever* $\alpha \leqslant \beta$.

Dynamical variables are very important in classical physics and they are equally important in quantum mechanics. In classical mechanics dynamical variables are defined as functions on phase space. Since we have no phase space in our present setting, we must define them differently. To distinguish these objects from their classical counterpart (our definition will actually be a generalization of dynamical variables) we shall call them 'observables'. An observable should be an object associated with our physical system which can be measured. That is, it determines a set of real numbers, the values of the observable. On the other hand, given any set of real numbers E an observable x gives us the question: 'Does x have a value in E?' Now it would be cumbersome mathematically to consider all subsets of R. For this reason mathematicians usually consider a class of subsets of R which is at the same time large enough to contain all the physically important subsets of R and yet small enough to make the theory manageable. Now open intervals are certainly important subsets of R (these intervals correspond to inexact results of measurements such as 'the result is between 2.03 and 2.05 degrees centigrade') and we would surely like to be able to take set complements and countable unions of sets and still remain within our class of subsets. The class of *Borel subsets* $B(R)$ of R is defined to be the smallest collection of subsets of R that contains the open intervals and that is closed under set complementation and countable unions [38, 48]. It is easy to show that open sets and closed sets are Borel sets and that $B(R)$ is closed under countable intersections. Now if x is an observable and $E \in B(R)$ we have the corresponding question $x(E)$: 'Does x have a value in the set E?'. We thus define an observable x as a map from $B(R)$ to Q that satisfies the following physically plausible conditions:

($\mathcal{O}$1) $x(R) = 1$;

($\mathcal{O}$2)   if $E \cap F = \emptyset$, then $x(E) \perp x(F)$;

($\mathcal{O}$3) $x(\bigcup E_i) = \bigvee x(E_i)$ if $E_i \cap E_j = \emptyset$, $i \neq j = 1, 2, \ldots$.

It follows that $x(\emptyset) = 0$ and denoting the complement of $E \in B(R)$ by E' we have $x(E') = x(E)'$. To give an example of an observable, let α_i be a sequence of mutually disjoint questions such that $\bigvee \alpha_i = 1$ and let λ_i be a sequence of distinct real numbers. Defining the map x by $x(E) = \bigvee \{\alpha_j : \lambda_j \in E\}$, $E \in B(R)$, it is easily checked that x is an observable.

Two observables x, y are *compatible* (written $x C y$) if $x(E) C y(F)$ for every $E, F \in B(R)$. We shall show later that observables which are compatible may be thought of physically as being observables which are simultaneously measurable. It can also be shown that a collection of compatible observables may be identified with a collection of dynamical variables.

The reader should notice that we have constructed a generalized probability theory. Instead of being a Boolean σ-algebra of subsets of a set, our events (questions) which are more general, form a logic with less structure than a Boolean σ-algebra. The probability measures are replaced by states and the random variables by observables. Notice if x is an observable and s a state then the probability that x has a value in $E \in B(R)$ is $s[x(E)]$. Thus $s[x(E)]$ corresponds to $p(x, s)$ (E) in Section II. Before proceeding further, let us consider two examples of quantum logics.

Example 1. Let Ω be a phase space and let $B(\Omega)$ be the Borel subsets of Ω (defined in a similar way as $B(R)$ above). $B(\Omega)$ may be thought of as the set of mechanical events. Now $B(\Omega)$ satisfies (Q1)–(Q8). A state is now a probability measure on $B(\Omega)$ and from the existence of measures concentrated at points we see that $B(\Omega)$ has a full set of states and is thus a quantum logic. If x is an observable it follows from a theorem of Sikorski-Varadarajan [80] that there exists a (measurable) function $f: \Omega \to R$ such that $x(E) = f^{-1}(E)$ for every $E \in B(R)$. Thus observables are just inverses of dynamical variables. We thus see that the quantum logic generalizes classical mechanics and also the conventional Kolmogorov formulation of probability theory [48]. It is easily checked that all events (questions) and observables are compatible in this example.

Example 2. Let H be a separable complex Hilbert space and let P be the collection of all closed subspaces of H. Ordering P by inclusion and defining the complement of a subspace as its orthocomplement it is easily seen that P satisfies (Q1)–(Q8). If $\alpha \in P$ we denote the unique orthogonal projection on α by P_α. Now if $\varphi \in H$ and $\|\varphi\| = 1$ then the map $\alpha \to \langle \varphi, P_\alpha \varphi \rangle$ is a state. If $\alpha \not\leq \beta$ choose a unit vector φ_0 in α which is not in β. Then $\langle \varphi_0, P_\alpha \varphi_0 \rangle = 1$ and $\langle \varphi_0, P_\beta \varphi_0 \rangle \neq 1$ so P has a full set of states and is thus a quantum logic. It is an interesting and important fact that every state is a convex combination of states of the above form. Indeed, Gleason [21] has shown that any state s on P has the form $s(\alpha) = \sum_1^\infty \lambda_i \langle \varphi_i, P_\alpha \varphi_i \rangle$, $\lambda_i \geq 0$, $\sum \lambda_i = 1$, φ_i is an ortho-normal set of unit vectors. Identifying closed subspaces with their orthogonal projections, an observable may be thought of as a projection-valued measure. Since, using the spectral theorem [38], there is a one-one correspondence between projection-valued measures and self-adjoint operators, we may identify observables with self-adjoint operators. It is straightforward to show that $\alpha, \beta \in P$ are compatible if and only if P_α and P_β commute. It follows that two observables are compatible if and only if they commute. Of course, the present example gives the usual framework of conventional quantum mechanics. We thus see that the quantum logic is a generalization of conventional quantum mechanics.

Let us now return to general quantum logics. If x is an observable we call $\{x(E): E \in B(R)\}$ the *range* of x.

LEMMA 3.3 [80]. Two questions α, β are compatible if and only if they are in the range of a single observable.

This last lemma justifies the fact that compatible questions are physically non-interfering questions since to measure two compatible questions we need measure only a single observable.

A function $f: R \to R$ is said to be a *Borel function* if $f^{-1}(E) \in B(R)$ for every $E \in B(R)$. Again we consider Borel functions instead of arbitrary functions for mathematical manageability. It can be shown that the Borel functions form the smallest class of functions which contains the continuous functions and which is closed under pointwise convergence. Now if x is an observable and u a Borel function on R then there is an operational significance for $u(x)$. That is, if x has the value $\lambda \in R$ then $u(x)$ has the value $u(\lambda)$. This is equivalent to saying that the question 'Does $u(x)$ have a value in $E \in B(R)$?' is the same as the question 'Does x have a value in $u^{-1}(E)$?'. Motivated by this we define $u(x)$ as $u(x)(E) = x(u^{-1}(E))$ for all $E \in B(R)$. It is easily checked that $u(x)$ is an observable and that $u(x) C x$.

THEOREM 3.4 [80]. Two observables x, y are compatible if and only if there is an observable z and Borel functions u, v such that $x = u(z)$ and $y = v(z)$.

This last theorem shows that, physically, compatible observables are measurements that can be performed simultaneously (i.e., non-interfering) since to measure compatible observables one need only measure a single observable.

One can continue this approach to a considerable extent and introduce such notions as time evolution, spectral theory, symmetry, superposition principle, superselection rules, scattering theory, and many others [18, 27, 30, 31, 32, 40, 41, 45]. However we refer the reader, interested in further study, to the literature and hope that we have conveyed some of the flavor of this subject.

IV. THE JAUCH-PIRON APPROACH [42, 65]

In the last section we gave an approach to quantum logics and in this section we present a slightly different approach. Each approach has its advantages and disadvantages. One of the advantages of the present approach is that it gives a much richer (and hence more specific) mathematical structure than that of Section III. For example one is able to derive the existence of sups and infs in a reasonable physical manner and thus show that in this case Q is a lattice. A disadvantage of this approach is that the probabilistic interpretation seems to disappear (although it can be partly recovered later under certain conditions); this is at the same time an advantage since states can be defined without recourse to probabilistic statements and therefore no difficulties arise attributing a state to an individual system.

In this approach the questions $Q = \{\alpha, \beta, \gamma, \ldots\}$ are again taken as the primitive axiomatic elements where the questions are interpreted exactly the same as in Section III. If $\alpha \in Q$ the question $\alpha^\sim$ is the question obtained by interchanging the alternatives of α. We have changed notation because we want to use ' for something else later. If $\{\alpha_i\}$ is a collection of questions (not necessarily countable) we denoted by $\pi\alpha_i$ ($\alpha \cdot \beta$ if there are two) the question defined in the following manner: Measure an arbitrary one of the α_i's and attribute to $\pi \, a_i$ the answer thus obtained. Clearly $(\pi\alpha_i)^\sim = \pi\alpha_i{}^\sim$. There exists a trivial question 1 which consists in doing anything (or nothing!) and stating that the answer is 'yes'. Let $0 = 1^\sim$.

DEFINITIONS. When the physical system is prepared in such a way that the result of a measurement of α is certain to be 'yes' then α is *true*. If whenever the physical system is prepared so that α is true we have β true also then we write $\alpha \leqslant \beta$.

This last relation expresses a physical law. This order relation is weaker than the order given in Section III and in fact this is the essential difference between the two formulations. We shall see that this difference enables one to develop a much richer structure. Clearly $\alpha \leqslant \alpha$ for all $\alpha \in Q$ and $\alpha \leqslant \beta$, $\beta \leqslant \gamma$ implies $\alpha \leqslant \gamma$. We define an equivalence relation $\alpha \sim \beta$ if $\alpha \leqslant \beta$ and $\beta \leqslant \alpha$. A *proposition* is an equivalence class of questions. We denote the equivalence class containing α by $a = \{\beta \in Q : \beta \sim \alpha\}$. We say a is *true* if any (and hence all) questions in a are true.

At this point we see that the probabilistic interpretation of propositions is lost. For example let α and β be questions. Define the question γ in the following way: Flip a fair coin; if the coin comes up heads measure α, if the coin comes up tails measure β. Define the question δ as follows: Flip a weighted coin in which the probability of heads is $\frac{2}{3}$ and the probability of tails is $\frac{1}{3}$; if the coin comes up heads measure α, if the coin comes up tails measure β. Now the two questions γ and δ are equivalent since the only way γ or δ can be true is if α and β are both true. Suppose we had a notion of the probability λ that α has a 'yes' answer and the probability μ that β has a 'yes' answer. Then the probability that γ has a 'yes' answer would be $\frac{1}{2}\lambda + \frac{1}{2}\mu$ while the probability δ has a 'yes' answer would be $\frac{2}{3}\lambda + \frac{1}{3}\mu$ which is, in general, different from $\frac{1}{2}\lambda + \frac{1}{2}\mu$. Thus there would be no unique way to associate a probability to the proposition containing γ and δ.

Let L be the set of propositions defined for a given physical system. If $a, b \in L$ define $a \leqslant b$ if $\alpha \leqslant \beta$ for all $\alpha \epsilon a$, $\beta \epsilon b$. It is easy to see that if a_i is a collection of propositions then $\bigwedge a_i = \{\beta : \beta \sim \pi\alpha_i, \alpha_i \in a_i\}$ and that $\bigvee a_i = \bigwedge \{x \in L : x \geqslant a_i$ for all $i\}$. We thus see that L is a *complete lattice*. It is clear that '$a \wedge b$ true'$\Leftrightarrow$'a true' and 'b true' so $\wedge$ plays the same role as 'and' in ordinary logic. This formulation overcomes the difficulties one has in the formulation of Section III where one gives the interpretation to $a \wedge b$ as the question which has answer 'yes' when a and b have a 'yes' answer in which case a and b must be measured simultaneously. In this formulation a and b are not measured simultaneously but one at a time.

For the sup in this formulation, however, we only have 'a true' or 'b true' $\Rightarrow$ '$a \vee b$ true'. In fact we have the following lemma.

LEMMA 4.1 [63]. If '$a \vee b$ true'$\Leftrightarrow$('a true' or 'b true') for every $a, b \in L$ then L is distributive; i.e., $a \wedge (b \vee c) = (a \wedge b) \vee (a \wedge c)$ for all $a, b, c \in L$.

The implication '$a \vee b$ true' $\Rightarrow$ 'a true' or 'b true' is an essential distinction between classical and quantum theory. This implication holds in classical theory but in general not in quantum mechanics. Thus in classical theory L is distributive and it can then be shown [3] that L is isomorphic to the set of subsets of some phase space.

Example. The linear polarization of photons. The experiment consists of placing a polarizer in a beam of linearly polarized photons. By dispatching photons one by one, this experiment leads to an alternative: Either the photon passes through or is absorbed. We define the question α_φ by specifying the orientation of the polarizer (the angle φ) and interpreting the passage of a photon as a 'yes'. By experiment one can show that in order to obtain a photon prepared so that α_φ is 'true' it is sufficient to consider photons which have passed a first polarizer oriented at this angle φ. Also experiments show that it is impossible to prepare photons capable of traversing with complete certainty a polarizer oriented at angle φ as well as another oriented at angle $\varphi' \neq \varphi \pmod{\pi}$ i.e. $\alpha_\varphi \cdot \alpha_{\varphi'} \sim 0$. The corresponding lattice of propositions is given in Figure 2.

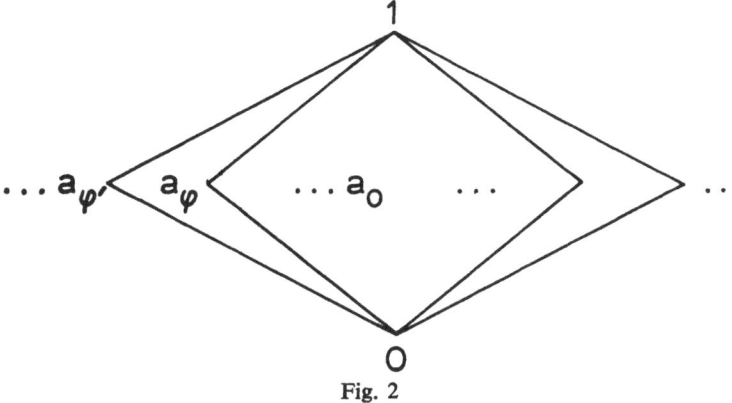

Fig. 2

One cannot define a question which is 'true' if and only if 'a_φ is true or $a_{\varphi'}$ is true' since a photon which is absorbed does not get through the polarizer so a second experiment cannot be performed. Notice in this case '$a_\varphi \vee a_{\varphi'}$ is true' $\not\Rightarrow$ 'a_φ is true or $a_{\varphi'}$ is true'. The lattice is not distributive and therefore corresponds to a quantum system.

We say that b is a *compatible complement* of a if it is a complement of a (i.e., $a \wedge b = 0$, $a \vee b = 1$) and if there is a question $\alpha \in a$ such that $\alpha^\sim \in b$. For example in the polarization experiment $a_{\varphi + \pi/2}$ and a_φ are compatible complements but no other $a_{\varphi'}$ is a compatible complement for a_φ.

Note. Every proposition has at least one compatible complement. This can be seen as follows. If $a \in L$, let $\alpha \in a$ and let b be the equivalence class containing $\alpha^\sim$. Then b is a compatible complement of a.

AXIOM P. If $a \leqslant b$ and a', b' are compatible complements of a, b respectively, then the sub-lattice generated by $\{a, b, a', b'\}$ is distributive.

We can justify Axiom P as follows. If $a \leqslant b$ then a and b are non-interfering so they may be considered part of a classical subsystem and we have seen that classical systems are distributive.

Suppose a_1, a_2 are both compatible complements of a. Then since $a \leqslant a$ by Axiom P, a, a_1, a_2 are in a distributive sub-lattice. But in a distributive lattice complements are unique so $a_1 = a_2$. Hence compatible complements are unique. We denote the compatible complement of a by a'. It follows from Axiom P that $a'' = a$, $a \leqslant b$ implies $b' \leqslant a'$ and $a \leqslant b$ implies $b = a \vee (b \wedge a')$. Thus L is a complete orthomodular lattice or CROC. We say that two propositions a, b are *compatible* if the sub-lattice generated by $\{a, b, a', b'\}$ is distributive. This definition is equivalent to the definition of compatibility given in Section III.

A *JP-state* of L is a subset $S \subset L$ satisfying:

(S_1) $0 \notin S$;

(S_2) if $a \in S$ and $a \leqslant x$ then $x \in S$;

(S_3) if $a_i \in S$ then $\wedge a_i \in S$;

(S_4) S is maximal, i.e., if a subset of L satisfies (S_1), (S_2), (S_3) and contains S, it equals S.

A JP-state corresponds physically to the set of propositions that are true for some preparation of the system. Since a preparation determines and is determined by the set of propositions that are true for that preparation, a JP-state may be thought of as a preparation of the system.

Let S be a JP-state and let $p = \wedge \{a : a \in S\}$. Then by (S_3), $p \in S$. By (S_2), $S = \{a \in L : p \leqslant a\}$. Now p is an atom (i.e. $0 \leqslant p_1 \leqslant p$ implies $p_1 = p$ or 0). Indeed, if there exists $0 \leqslant p_1 \leqslant p$ and $p_1 \neq 0$, p then $\{a \in L : p_1 \leqslant a\}$ satisfies (S_1), (S_2), (S_3) and properly contains S which contradicts (S_4). Thus every JP-state defines an atom and every atom defines a JP-state.

By definition if a proposition is different from 0, then it must be true for some preparation. We have therefore justified the next axiom.

AXIOM A_1. L is atomic (i.e. if $a \neq 0$ there is an atom $p \leqslant a$).

Axiom A_1 is equivalent to the axiom: Every proposition is contained in a JP-state.

Our last axiom is the following:

AXIOM A_2. If $a \in L$ and p is an atom of L such that $p \nleqslant a'$ then $(p \vee a') \wedge a$ is an atom.

Roughly speaking, the justification for A_2 is that if p is the JP-state of the system and if an 'ideal measurement of the first kind' of a is made [42, 65] with the resulting answer 'yes' then the smallest proposition that is 'true' after the measurement is $(p \vee a') \wedge a$. Hence the resulting JP-state should correspond to $(p \vee a') \wedge a$ so this proposition must be an atom. A similar interpretation for this axiom may be found in [67, 68].

Axiom A_2 is called the *semimodular* or *covering law*. A complete lattice satisfying Axioms P, A_1, and A_2, i.e., a complete, atomic, semimodular, orthomodular lattice is called a *propositional system*. A propositional system L is *irreducible* if the only elements of L compatible with all other elements of L are 0 and 1.

THEOREM 4.2. Any propositional system is isomorphic to a direct product of irreducible propositional systems.

One of the great achievements of this theory is the following.

THEOREM 4.3 (Piron) [64]. Every irreducible propositional system of dimension ≥ 4 is isomorphic to the lattice of all closed subspaces of a Hilbert space over a division ring with involution.

We thus see that a propositional system gives a structure which is very close to the conventional quantum mechanical formalism in terms of a complex Hilbert space. This theory can then be considered as a 'derivation' of the Hilbert space that mysteriously occupies such an important place in conventional quantum mechanics. Of course an important problem is to give more physical information so that the division ring is determined (hopefully to be the complex numbers). One result along these lines has been obtained by Gudder and Piron [34]. It states that if L admits an observable that is maximal in a certain sense then the division ring must contain the reals as a subfield. Examples of maximal observables are the position, momentum, and magnetic moment observables in conventional quantum mechanics. If further the division ring is a finite extension of R it follows from a theorem of Frobenius [66] that it must be the reals, complexes, or quaternions.

It follows from Theorem 4.3 that a propositional system L with dim $L \geq 4$ has a full set of states $\mathscr{S}$, although, as we have pointed out earlier, these states cannot be defined in terms of questions. In this way $(L, \mathscr{S})$ becomes a quantum logic which enjoys a still richer structure.

In summary, we define an *abstract quantum logic* $(\mathscr{L}, \mathscr{S})$ to be a σ-orthocomplete orthomodular poset $\mathscr{L}$ together with a full set of states $\mathscr{S}$. This structure seems to be the underlying framework for (almost) all quantum logical studies. In particular the models presented in Sections III and IV contain abstract quantum logics in which the axiomatic elements are given specific physical interpretations.

V. HIDDEN VARIABLE THEORIES

As was mentioned in the introduction, one of the advantages gained by an axiomatic formulation is that it presents a common universe of discourse in which to study deep quantum mechanical concepts. We also stated in Section III that many of the important concepts of quantum mechanics can be formulated within the quantum logic framework. Since we do not have the space to consider all these concepts, we shall attempt to illustrate the utility of the quantum logic approach by considering one concept which is not only important to this conference but has played a recurrent role throughout the history of the development of quantum theory. This is the concept of hidden variables.

One of the problems in hidden variable discussions is that they have occurred in different frameworks simultaneously and therefore investigators in many cases have been writing (and talking) about different subjects but have called them the same thing. Also it is very common that what one researcher calls hidden variables is entirely different from what another calls them so it is not surprising that some investigators are able to prove they do not exist while others prove they do. We would like to consider two hidden variable proofs in the quantum logic framework, one a proof that they do not exist and one a proof that they do.

Jauch and Piron's interpretation of hidden variables is that if hidden variables exist then there would be states for which every proposition is either true or false; that is, any question would have the answer 'yes' with certainty or the answer 'no' with certainty. They call such JP-states *dispersion-free*. Thus, in their approach to quantum logics, if an atom p corresponds to a dispersion-free state we must have $p \leqslant a$ or $p \leqslant a'$ for every proposition $a \in L$. It follows that p is compatible with every proposition. Jauch-Piron go even further than this. Precisely, they say that a propositional system L *admits hidden variables* if every JP-state is dispersion-free [44]. Thus if L admits hidden variables then each atom is compatible with every atom. It follows that each proposition is compatible with every proposition (in particular L is distributive) and hence there are no interfering experiments and L is a classical propositional system. This is a contradiction since there are noncompatible propositions for quantum mechanical systems. These concepts can also be phrased in the BVM (Birkhoff-von Neumann-Mackey) framework for quantum logic in which case we say a state is *dispersion-free* if its values are just 0 and 1, and a similar proof goes through.

We now consider another interpretation of hidden variables phrased in the BVM quantum logic context. We feel that the main difficulties in hidden variables discussions is that the investigators giving impossibility proofs [29, 44, 46, 49, 61, 81, 84] are not referring to the same thing as the hidden variable proponents. They are proving something is impossible but these things are not what hidden variable researchers such as Einstein [17], Bohm and Bub [6, 7, 8, 9], and others [2] are referring to when they speak of hidden variables. We will give a general definition in the BVM quantum logic context of what we feel an HV (hidden variables) theory is as described by HV proponents and prove that such a theory is *always* possible and is, in fact, unique in a certain sense.

We first give an English-language version of what we feel the HV researchers mean by an HV theory.

The state *s* of a quantum mechanical system is not complete in the sense that another variable ω can be adjoined to *s* so that the pair (s, ω) completely determines the system. That is, a knowledge of (s, ω) enables one to predict precisely the outcome of any single measurement. Furthermore, an average of (s, ω) over the values of ω gives the usual quantum state *s*.

We now attempt to translate the above version of an HV theory into a mathematical-language version on a quantum logic (Q, S). First a single measurement corresponds to a Boolean sub σ-algebra of Q. This is because in a single measurement there is no possibility of interference so the measurement corresponds to a distributive subsystem. To say that the results of a measurement (corresponding to a Boolean sub σ-algebra $B \subset Q$) are completely determined means that one has a dispersion-free state s_0 defined on B (*not* on Q). We denote the set of dispersion-free states on B by S_B. Recall that in probability theory a *probability space* is a triple (Ω, F, μ) where Ω is the set of elementary outcomes, F is the Boolean σ-algebra of events, and μ is a probability measure on F.

DEFINITIONS. A quantum logic (Q, S) *admits an* HV *theory* if there is a probability space (Ω, F, μ) with the following property: For any maximal Boolean sub σ-algebra $B \subset Q$ there is a map H_B from $S \times \Omega$ onto S_B such that (i) $\omega \rightarrow H_B(s, \omega)(\alpha)$ is measurable for every $s \in S$, $\alpha \in B$; (ii) $\int_\Omega H_B (s, \omega)(\alpha) d\mu(\omega) = s(\omega)$ for every $s \in S$, $\alpha \in B$. Denote the set of maximal Boolean sub σ-algebras of Q by $\mathscr{B}$. We call $((\Omega, F, \mu), \{H_B : B \in \mathscr{B}\})$ an HV *theory for* (Q, S). An HV theory $((\Omega, F, \mu), \{H_B : B \in \mathscr{B}\})$ is *minimal* if $H_B(s, \omega_1) = H_B(s, \omega_2)$ for every $s \in S$, $B \in \mathscr{B}$ implies $\omega_1 = \omega_2$.

The definition merely says that for each 'completed' state (s, ω) there is a dispersion-free state H_B (s, ω) for any single measurement B and (i), (ii) say that the average of these dispersion-free states over ω give back the quantum state s. We consider only maximal Boolean sub σ-algebras so that the theory does not become too cumbersome. This is really only a technicality since any Boolean sub σ-algebra is contained in a maximal one. The probability space (Ω, F, μ) may be thought of as the space of hidden variables. If an HV theory is minimal, there exists a minimal number of hidden variables – just enough to give all the dispersion-free states. We now state our main theorem [26].

THEOREM 5.1. Any quantum logic (Q, S) admits a minimal HV theory $((\Omega, F, \mu), \{H_B : B \in \mathscr{B}\})$. Furthermore, $((\Omega, F, \mu), \{H_B : B \in \mathscr{B}\})$ is the unique minimal HV theory in the sense that if $((\Omega', F', \mu'), \{H_B' : B \in \mathscr{B}\})$ is another HV theory, there exists a measurable map τ from Ω' into Ω such that $H_B(s, \tau\omega')(\alpha) = H_B'$ $(s, \omega')(\alpha)$ for all $B \in \mathscr{B}$, $\omega' \in \Omega'$, $s \in S$, $\alpha \in B$, $\mu'(\tau^{-1}(\Lambda)) = \mu(\lambda)$ for all $\lambda \in F$ and if $((\Omega', F', \mu'), \{H_B' : B \in \mathscr{B}\})$ is minimal then τ is one-one.

VI. THE CONDITIONAL

We begin this section with a brief sketch of the treatment of the conditional in classical logic. Let B be a Boolean algebra and $\mathfrak{F}$ a (lattice) filter (i.e., (I) $a, b \in \mathfrak{F}$ implies $a \wedge b \in \mathfrak{F}$; (II) $a \in \mathfrak{F}$, $x \in B$ with $x \geqslant a$ implies $x \in \mathfrak{F}$) in B. We regard B as a logic and $\mathfrak{F}$ as the set of true propositions of B. For any two propositions a, b in B the conditional $a \supset b$ (read 'a implies b') is the proposition $a' \cup b$. Thus $a \supset b$ is true in case $a' \cup b \in \mathfrak{F}$. By 'dividing out' the filter $\mathfrak{F}$ (or the dual filter of all false propositions) we obtain another Boolean algebra called the reduced algebra, $B_1 = B/\mathfrak{F}$. Let the elements of B_1 be denoted by $\bar{a}$ where $a \in B$. Formally $\bar{a} = \{b \in B \mid b \vee c = a \vee c$ for some $c \in B$ with $c' \in \mathfrak{F}\}$. It turns out that the following statements are equivalent: (1) $a \supset b \in \mathfrak{F}$, (2) $\overline{a' \vee b} = \bar{1}$, (3) $\bar{a} \leqslant \bar{b}$. Thus in a Boolean algebra the statement '$a \supset b$ is true' may be translated into the statement '$\bar{a} \leqslant \bar{b}$'.

More than anything else the fact which allows us to regard the Boolean algebra B as a logic is the existence of a valid inference scheme (modus ponens): If $a \in \mathfrak{F}$ and $a \supset b \in \mathfrak{F}$ then $b \in \mathfrak{F}$. The argument runs as follows: $a \in \mathfrak{F}$, $a' \cup b \in \mathfrak{F}$ so $a \wedge (a' \cup b) \in \mathfrak{F}$ by (I); but since B is a Boolean algebra

(*) $a \wedge (a' \cup b) = a \wedge b \in \mathfrak{F}$;

moreover $a \wedge b \leqslant b$ so that $b \in \mathfrak{F}$ by (II).

(*) is the key to the above computations. The equation given in (*) is equivalent to saying that aCb.

The question of whether there exists in quantum logic a conditional which is a logical proposition allowing a valid inference scheme has received some attention. So far the results are one-sided: If, for every $a, b \in \mathscr{L}$ there is a proposition $a \supset b$ behaving enough like the conditional of classical logic then $\mathscr{L}$ is classical (a Boolean algebra). Here are four examples:

(1) Fay [19] has proved the following: If a relation R is defined on an orthomodular lattice L by $(a, b) \in R$ if and only if $a' \vee b = 1$, then R is transitive if and only if L is a Boolean algebra.

(2) Skolem [75] defines an *implicative lattice* to be a lattice L with the property that for every a, b in L there exists an element $a \supset b$ in L such that

(i) $a \wedge (a \supset b) \leqslant b$

and

(ii)   if $a \wedge c \leqslant b$ then $c \leqslant (a \supset b)$.

He then proves that any implicative lattice is distributive.

(3) Working in an orthomodular lattice L, Catlin [10] defines a relation I on L by $(a, b) \in I$ if there exists an element $a \supset b$ satisfying (i) and (ii) above. (If such an element exists it is unique; in fact, $a \supset b = a' \vee b$.) Catlin proves (for example) that $(a, b) \in I$ and $(b', a') \in I$ both hold if and only if aCb and $(a' \vee b)Cx$ for all $x \in L$. Thus (again) if $a' \vee b$ behaves as the conditional for every a, b in L, then L is a Boolean algebra.

(4) Łukasiewicz [53] has developed a (classical) system in which a proposition P is assigned a truth value $[P] \in \mathbb{R}$ satisfying $0 \leqslant [P] \leqslant 1$. He defines the truth values of the conditional $P \to Q$ and negation $\bar{P}$ as follows:

$$[P \to Q] = \begin{cases} 1 & \text{if } [P] \leqslant [Q] \\ 1 - [P] + [Q] & \text{if } [Q] < [P] \end{cases}$$

and

$$[\bar{P}] = 1 - [P].$$

$[P] = 1$ is interpreted as 'P is true'. Intermediate values of $[P]$ stand for various degrees of certainty. Note that

(i) $[P] = 1$ and $[P \to Q] = 1$ implies $[Q] = 1$

and

(ii) $[P \to Q] = 1$ and $[Q \to R] = 1$ implies $[P \to R] = 1$.

Jauch and Piron [43] considered an adaptation of the infinite-valued logic as developed by Łukasiewicz [53]. They argued as follows. Reichenbach [70] has proposed that the

elementary propositions about quantum mechanical systems should admit three truth values: True, false, and undetermined. In view of the fact, however, that the state of a system attributes to each yes-no experiment a probability function $p(a)$ with $0 \leqslant p(a) \leqslant 1$, it seems more natural, once one has passed beyond the ordinary double-valued logic, to consider 'quantum-logic' as an infinite-valued logic.

The various degrees of certainty should depend on the state of the physical system. Thus they sought the existence in a quantum logic $(L, \mathscr{S})$ of a conditional $p \to q$ for each pair p, q in L. It must satisfy, *for each* $m \in \mathscr{S}$,

(*) $m(p \to q) = \begin{cases} 1 & \text{if } m(p) \leqslant m(q) \\ 1 - m(p) + m(q) & \text{if } m(q) < m(p) \end{cases}$.

They showed that in a standard quantum logic there exist p and q that admit no conditional.

Greechie and Gudder [24] generalized this result to an arbitrary quantum logic. In fact, they did more. An outline of the technical results is given below.

Let $(L, \mathscr{S})$ be an abstract quantum logic and assume that $\mathscr{S}$ is *closed under the formation of midpoints*, i.e.

$$m_1, m_2 \in \mathscr{S} \text{ implies } \tfrac{1}{2}m_1 + \tfrac{1}{2}m_2 \in \mathscr{S}.$$

Call a pair $(a, b) \in L \times L$ conditional if there exists $c \in L$ such that for all $m \in \mathscr{S}$

(**) $m(c) = \min \{1, m(a') + m(b)\}$.

If c exists it is unique, write $c = a \to b$. (Note that (*) and (**) are, in fact, the same condition.) Call $(L, \mathscr{S})$ *conditional* if $a \to b$ exists for all pairs (a, b). Then $(L, \mathscr{S})$ is conditional if and only if $L = \{0,1\}$. Moreover, if

$\mathscr{S}$ is strongly order determining, then $a \to b$ exists if and only if $a \leqslant b$ or $b \leqslant a$.

From a different point of view, Piron has argued that a conditional $a \supset b$ does not exist because 'no experimental arrangement is possible which measures the proposition $a \supset b$'.

Thus the indications are that the conditional in quantum logic must be treated as a relation, perhaps nothing more than the relation $\leqslant$. In this case the required valid inference scheme would be: a is true and $a \leqslant b$, therefore b is true. Here 'a is true' is apparently interpreted as $a \in \mathfrak{F}$ where $\mathfrak{F}$ is the filter of all propositions true in some (fixed) state.

E. L. Marsden has made an interesting observation (oral communication) on implication in a quantum logic. His approach has the advantage (or disadvantage?) of ignoring the states, i.e. of working completely within the orthomodular poset.

Let P be an orthomodular poset, S a subset of P with $1 \in S$. Modifying a notion of classical logic (cf. A. Church, *Mathematical Logic*, Princeton University Press, 1956) Marsden defines an element $d \in P$ to be *a theorem based on S*, written $S \vdash d$, in case there exist $a_1, \ldots, a_n \in P$ such that $a_n = d$ and, for all i, either $a_i \in S$ or there exist j, $k < i$ with $\{a_i, a_j, a_k\}$ a commuting set and $a_j = a'_k \vee a_i$. He also defines a *C-filter* in P to be a non-empty subset $\mathfrak{F}$ of P such that (1) if $x \in \mathfrak{F}$ and $x \leqslant y$ then $y \in \mathfrak{F}$, and (2) if $x, y \in \mathfrak{F}$ and $x C y$ then $x \wedge y \in \mathfrak{F}$.

Let $\bar{S}$ denote the C-filter generated by S. Theorem: $d \in \bar{S}$ if and only if $S \vdash d$. The difficulty with this is that $\bar{S}$ is not associated with any congruence relation on P. Thus $\bar{S}$ cannot be used to form a reduced logic.

VII. COMBINATORIAL QUANTUM LOGIC

In this section we discuss some aspects of the combinatorial approach to quantum logic.

We restrict our considerations to finite structures. Let $(P, \leqslant, ')$ be a finite orthomodular poset and A the set of atoms in P. Make A into a graph by defining, for a, $b \in A$, $a \perp b$ to mean that $a \leqslant b'$ in P. For $M \subseteq A$, let $M^\perp = \{x \in A \mid x \perp m$ for all $m \in M\}$ and $M^{\perp\perp} = (M^\perp)^\perp$. Let $\mathscr{L} = \{D^{\perp\perp} \mid D$ is an orthogonal subset of $A\}$. Then $(\mathscr{L}, \subseteq, {}^\perp)$ is isomorphic to $(P, \leqslant, ')$. Thus we may recapture $(P, \leqslant, ')$ from the *orthogonality graph* $(A, \perp)$. We now pass to the *orthogonality space* $(A, \mathscr{E})$ by defining $\mathscr{E}$ to be the set of all maximal orthogonal sets in $(A, \perp)$, that is, the set of maximal complete subgraphs (or cliques) of the graph $(A, \perp)$. It is easy to see that $(A, \mathscr{E})$ determines the graph $(A, \perp)$; thus we can recapture $(P, \leqslant, ')$ from $(A, \mathscr{E})$. The main reason for passing from the poset to the associated orthogonality space is psychological: The diagramatic representation for $(A, \mathscr{E})$ is most perspicuous.

We illustrate this process for the well-known orthomodular lattice D_{16}, the Hasse diagram of which is given in Figure 3.

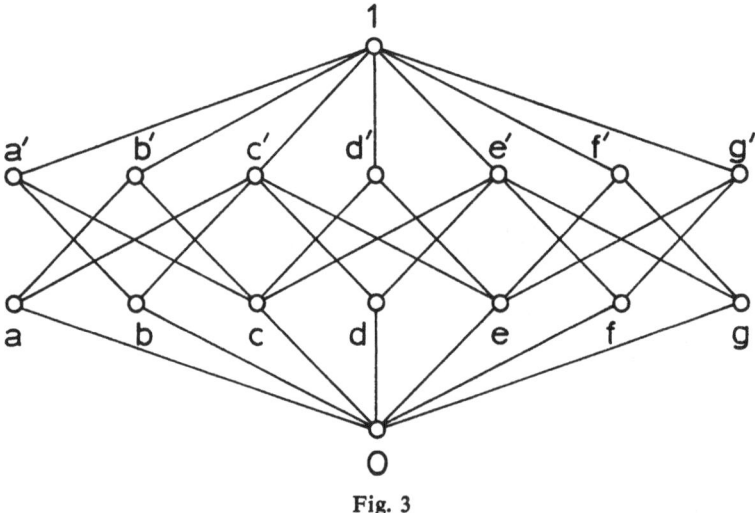

Fig. 3

Here $A = \{a, b, c, ..., g\}$, the orthogonality graph $(A, \perp)$ is given in Figure 4, and the orthogonality space $(A, \mathscr{E})$ is given in Figure 5 where each line represents a clique $E \in \mathscr{E}$.

Note that the cliques of the orthogonality space correspond to the maximal Boolean subalgebras of the poset, for example $\{c, d, e\}$ of Figure 5 corresponds to $\{0, c, d, e, c', d', e', 1\}$ of Figure 3.

By focussing on the intertwining of the maximal subalgebras of an orthomodular poset and then translating into the orthogonality space we are able, in certain instances, to create structures tailored to predescribed criteria. We illustrate this with an example of a quantum logic $(\mathscr{L}, \mathscr{S})$ in which $\mathscr{L}$ is an orthomodular poset and $\mathscr{S}$ is a full but not strongly order determining set of states; moreover $\mathscr{S}$ is sufficient. First, recall the

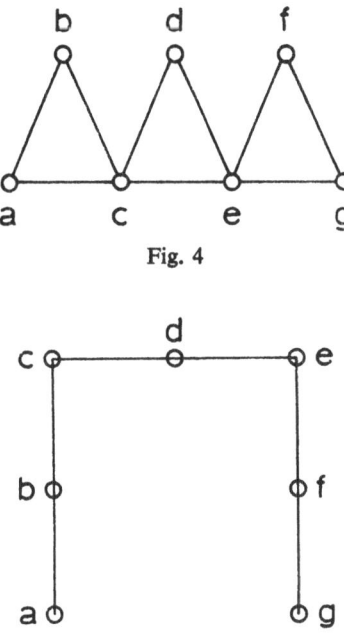

Fig. 4

Fig. 5

meaning of the terms full, strongly order determining, and sufficient. A set of states $\mathscr{S}$ on an orthomodular poset $\mathscr{L}$ is *full* (or *order determining*) in case $m(x) \leqslant m(y)$ for all $m \in \mathscr{S}$ implies $x \leqslant y$; $\mathscr{S}$ is *strongly order determining* in case $\{m \in \mathscr{S} | m(x) = 1\} \subseteq \{m \in \mathscr{S} | m(y) = 1\}$ implies $x \leqslant y$; $\mathscr{S}$ is *sufficient* (or satisfies the *projection postulate*) in case for all non-zero $x \in \mathscr{L}$ there exists a (not necessarily unique) state $m \in \mathscr{S}$ with $m(x) = 1$. (Note that in Section II we defined $\leqslant$ in such a way that $\mathscr{S}$ was full. Also, in Section III, we eventually assume $(Q9)$ that the set of states on Q is full. Because of Theorem 4.3 there is, in fact, a strongly order determining set of states on the resulting structure of Section IV.)

The example is given in Figure 6. This is a diagram of an orthogonality space $(X, \mathscr{E})$. The corresponding orthomodular poset may be obtained by the construction outlined above, by passing through the associated orthogonality graph $(X, \perp)$ and on to the poset $\mathscr{L} = \mathscr{L}(X, \perp)$. The details of the argument that the structure has the required property appear in [23].

It is easy to see that an orthomodular poset with a strongly order deter-

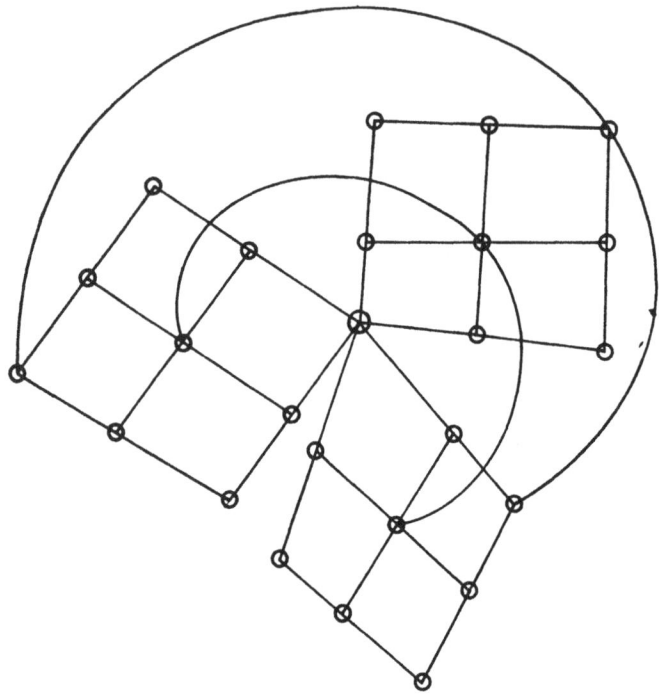

Fig. 6

mining set of states $\mathscr{S}$ is in fact a quantum logic, i.e. $\mathscr{S}$ is full. We have just proven that the reverse implication fails: There exists a quantum logic with a set of states which is (sufficient but) not strongly order determining. B. Collings has given an example of an orthomodular lattice with a state space having these same properties. The example, however, is too complicated to be given here.

There exists [25] an orthomodular poset $P_{j,k}$ with a distinguished element $x \in P_{j,k}$ such that $m(x) < j/k (0 < j < k)$ for every state m on $P_{j,k}$; moreover $P_{j,k}$ admits a full set of states if and only if $\frac{1}{2} < j/k$. Thus we see in a rather dramatic fashion that the axiom of sufficiency does not follow from the other axioms of a quantum logic.

Combinatorial quantum logic is useful for reasons other than axiomatics. For example, one can prove the following theorem using the combinatorial methods alluded to above and expounded in [22, 25].

THEOREM (Schrag). For any finite group G there exists a quantum logic $(\mathcal{L}, \mathcal{S})$ such that the full automorphism group of $\mathcal{L}$ is G.

The combinatorial approach has also yielded results of a negative character. For example, consider Finkelstein's [20] very readable presentation of what he calls a model quantum logic. Beginning with a finite set of entities he constructs a structure which is mathematically equivalent to a finite orthocomplemented projective geometry large enough to contain at least one plane, which, in turn, must be orthocomplemented. But there exists a combinatorial argument, due to R. Baer, that no such plane exists and therefore no such model quantum logic exists. This comment is not intended to be a criticism of Finkelstein's carefully presented argument but serves only to point out, once again, that combinatorics can play an important role in quantum logic.

VIII. CONCLUSION

In conclusion we wish to emphasize that quantum logic is not a closed subject. Although penetrating investigations have been made there exists no universally accepted theory. The basic framework has not been established to the satisfaction of even a majority of researchers. We feel that Jauch and Piron's axiom A_2 has not been sufficiently justified; the problem appears intricately interwoven with the philosophical problem of what is meant by 'truth with certainty' in an empirical setting, with axiom (S_3) in the definition of state, and with the 'ideal measurements of the first kind'. While in the Mackey-von Neumann presentation the axioms presented herein may be more easily accepted by some, in order to complete this development it appears necessary to make the *ad hoc* axiom that the logic, in fact, is the usual Hilbert space structure.

There are numerous lines of investigation which may provide insight into the sought-for underlying theory. We take this opportunity to list four directions which we consider of prime importance:

(1) Provide a physically meaningful interpretation for existing infima in a quantum logic. This, of course, is one of Birkhoff and von Neumann's original suggestions.

(2) Provide a connection between the structure of the state space and that of the underlying logic.

(3) Develop a general theory of group representations on an abstract quantum logic.

(4) Explain the meaning of the word 'logic' in the title of this paper.

Department of Mathematics,
Kansas State University
and
Department of Mathematics,
University of Denver

BIBLIOGRAPHY

[1] Amrien, W. O., 'Localizability for Particles of Mass Zero', *Helvetia Physica Acta* **42** (1969), 149–90.

[2] Bell, J. S., 'On the Hidden Variable Problem in Quantum Mechanics', *Reviews of Modern Physics* **38** (1966), 447–52.

[3] Birkhoff, G., 'Lattice Theory', *American Mathematical Society Colloquium* Publ. **25** (1967).

[4] Birkhoff, G. and von Neumann, J., 'The Logic of Quantum Mechanics', *Annals of Mathematics* **37** (1936), 823–43.

[5] Bodiou, G., *Theorie Dialectique des Probabilities*, Gauthier-Villars, Paris, 1964.

[6] Bohm, D. and Bub, J., 'A Proposed Solution of the Measurement Problem in Quantum Mechanics by Hidden Variables', *Reviews of Modern Physics* **38**, (1966) 453–69.

[7] Bohm, D. and Bub, J., 'A Refutation of the Proof by Jauch and Piron that Hidden Variables can be Excluded in Quantum Mechanics', *Reviews of Modern Physics* **38** (1966), 470–5.

[8] Bub, J., 'Hidden Variables and the Copenhagen Interpretation – a Reconcilliation', *The British Journal for the Philosophy of Science* **19** (1968), 185–210.

[9] Bub, J., 'What is a Hidden Variable Theory of Quantum Phenomena?', *International Journal of Theoretical Physics* **2** (1969), 101-23.

[10] Catlin, D., 'Implicative Pairs in Orthomodular Lattices', *Carribbean Journal of Science and Mathematics* **1** (1969), 69–79.

[11] Davies, E. B., 'On the Repeated Measurements of Continuous Observables in Quantum Mechanics', *Journal of Functional Analysis* **6** (1970), 318–46.

[12] Davies, E. B. and Lewis, J. T., 'An Operational Approach to Quantum Probability', *Communications in Mathematical Physics* **17** (1970), 239–60.

[13] Dirac, P. A. M., *The Principles of Quantum Mechanics*, Oxford University Press, London, 1958.

[14] Eckmann, J. P. and Zabey, Ph., 'Impossibility of Quantum Mechanics in a Hilbert Space over a Finite Field', *Helvetia Physica Acta* **42** (1969), 420–4.

[15] Edwards, C. M., 'The Operational Approach to Algebraic Quantum Theory, I', *Communications in Mathematical Physics* **16** (1970), 207–30.

[16] Edwards, C. M., 'Classes of Operations in Quantum Theory', *Communications in Mathematical Physics* **20** (1971), 26–56.

[17] Einstein, A, Rosen, N., and Podolsky, B., 'Can Quantum-Mechanical Description of Physical Reality be Considered Complete?', *Physical Review* **47** (1935), 777–80.

[18] Emch, G. and Piron, C., 'Symmetry on Quantum Theory', *Journal of Mathematical Physics* **4** (1963), 469–73.

[19] Fay, Gy., 'Transitivity of Implication in Orthomodular Lattices', *Acta Scientiarum Mathematicarum* **28** (1967), 267–70.

[20] Finkelstein, D., *'The Logic of Quantum Physics'*, *Transactions New York Academy of Science* Ser. 2, **28** (1962/63) 621–37.

[21] Gleason, A. M., 'Measures on Closed Subspaces of a Hilbert Space', *Journal of Rational Mechanical Analysis* **6** (1957), 885–93.

[22] Greechie, R. J., 'Orthomodular Lattices Admitting no States', *Journal of Combinatorial Theory* **10** (1971), 119–32.

[23] Greechie, R. J., 'An Orthomodular Poset With a Full Set of States not Embeddable in Hilbert Space', *Carribbean Journal of Science and Mathematics* **1** (1969), 15–26.

[24] Greechie, R. J., and Gudder, S. P., 'Is Quantum Logic a Logic?', *Helvetia Physica Acta* **44** (1971), 238–40.

[25] Greechie, R. J. and Miller, F. R., 'On Structures Related to States on an Empirical Logic: I. Weights on Finite Spaces', *K.S.U. Department of Mathematics Technical Report* No. 14, April (1970).

[26] Gudder, S. P., 'On Hidden Variable Theories', *Journal of Mathematical Physics* **11** (1970), 431–6.

[27] Gudder, S. P., *Spectral Methods for a Generalized Probability Theory*, Transl. *Amer. Math. Soc.* **119** (1965), 428–42.

[28] Gudder, S. P., 'Hilbert Space, Independence, and Generalized Probability', *Journal Math. Anal. Appl.* **20** (1967), 48–61.

[29] Gudder, S. P., 'Dispersion-Free States and the Existence of Hidden Variables', *Proceeding of the American Mathematical Society* **19** (1968) 319–24.

[30] Gudder, S. P., 'Uniqueness and Existence Properties of Bounded Observables', *Pacific Journal of Mathematics* **19** (1964), 81–93; **19** (1966) 588–9.

[31] Gudder, S. P., 'Coordinate and Momentum Observables in Axiomatic Quantum Mechanics', *Journal of Mathematical Physics* **8** (1967), 1848–58.

[32] Gudder, S. P., 'Systems of Observables in Axiomatic Quantum Mechanics', *Journal of Mathematical Physics* **8** (1967), 2109–13.

[33] Gudder, S. P. and Boyce, S., 'A Comparison of the Mackey and Segal Models for Quantum Mechanics', *International Journal of Theoretical Physics* **3** (1970), 7–21.

[34] Gudder, S. P. and Piron, C., 'Observables and the Field in Quantum Mechanics', *Journal of Mathematical Physics* **12** (1971), 1583–8.

[35] Guenin, M., 'Axiomatic Formulations of Quantum Theories', *Journal of Mathematical Physics* **7** (1966), 271-82.

[36] Gunson, J., 'On the Algebraic Structure of Quantum Mechanics', *Communications in Mathematical Physics* **6** (1967), 262–85.

[37] Haag, R. and Kastler, D., 'An Algebraic Approach to Quantum Field Theory', *Journal of Mathematical Physics* **5** (1964), 848–61.

[38] Jauch, J. M., *Foundations of Quantum Mechanics*, Addison-Wesley, Reading, Mass., 1968.

[39] Jauch, J. M., 'The Problem of Measurement in Quantum Mechanics', *Helvetia Physica Acta* **37** (1964), 293–316.

[40] Jauch, J. M., 'Systems of Observables in Quantum Mechanics', *Helvetia Physica Acta* **33** (1960), 711–26.

[41] Jauch, J. M. and Misra, B., 'Supersymmetry and Essential Observables', *Helvetia Physica Acta* **34** (1961), 699–710.

[42] Jauch, J. M. and Piron, C., 'On the Structure of Quantal Proposition Systems', *Helvetia Physica Acta* **42** (1969), 842–8.

[43] Jauch, J. M., 'What is Quantum-Logic?', *Quanta*, University of Chicago Press, Chicago and London, 1970.

[44] Jauch, J. M., 'Can Hidden Variables be Excluded from Quantum Mechanics?', *Helvetia Physica Acta* 36 (1963), 827–37.

[45] Jauch, J. M., 'Generalized Localizability', *Helvetia Physica Acta* 40 (1967), 559–70.

[46] Jauch, J. M., 'Hidden Variables Revisited', *Reviews of Modern Physics* 40 (1968), 228–9.

[47] Jordan, P., von Neumann, J., and Wigner, E., 'On an Algebraic Generalization of the Quantum Mechanical Formalism', *Annals of Mathematics* 35 (1934), 29–64.

[48] Kolmogorov, A. N., *Foundations of the Theory of Probability*, Chelsea, New York, 1956.

[49] Kochen, S. and Specker, E. P., 'The Problem of Hidden Variables is Quantum Mechanics, *Journal of Mathematics and Mechanics* 17 (1967), 331–48.

[50] Lowdenslager, D. B., 'On Postulates for General Quantum Mechanics', *Proceedings of the American Mathematical Society* 8 (1957) 88–91.

[51] Ludwig, G., 'Versuch einer axiomatischen Grundlegung der Quantemechanik und allgemeinerer physikalischer Theorien', *Zeitschrift für Physik* 181 (1964), 233–60.

[52] Ludwig, G., 'Attempt of an Axiomatic Foundation of Quantum Mechanics and More General Theories, II, III', *Communications in Mathematical Physics* 4 (1967), 331–48; 9 (1968), 1–12.

[53] Łukasiewicz, J., *Aristotelic Syllogistic from the Standpoint of Modern Formal Logic*, Oxford University Press, London, 1957.

[54] Mackey, G. W., 'Quantum Mechanics and Hilbert Space', *American Mathematical Monthly* 64 (1957), 45–57.

[55] Mackey, G. W., *Mathematical Foundations of Quantum Mechanics*, W. A. Benjamin, New York, 1963.

[56] MacLaren, M. D., 'Notes on Axioms for Quantum Mechanics', *Atomic Energy Commission Research and Development Report ANL*-7065, Argonne National Laboratory (1965).

[57] MacLaren, M. D., 'Nearly Modular Orthocomplemented Lattices', Mathematical Note No. 358, Mathematics Research Laborary, Boeing Scientific Research Laboratories, (1964).

[58] Maczynsky, M. J., 'A Remark on Mackey's Axiom System for Quantum Mechanics', *Académie Polonaise des Sciences Bulletin* 15 (1967), 583–7.

[59] Mielnik, B., 'Geometry of Quantum States', *Communications in Mathematical Physics* 9 (1968) 55–80.

[60] Mielnik, B., 'Theory of Filters', *Communications in Mathematical Physics* 15 (1969), 1–46.

[61] Misra, B., 'When can Hidden Variables be Excluded in Quantum Mechanics?', *Il Nuovo Cimento* 47 (1967), 841–59.

[62] Plyman, R. J., 'A Modification of Piron's Axioms', *Helvetia Physica Acta* 41 (1968), 69–74.

[63] Plymen, R. J., '*C**-Algebras and Mackey's Axioms', *Communications in Mathematical Physics* 8 (1968), 132–46.

[64] Piron, C., 'Axiomatique' Quantique, *Helvetia Physica Acta* 37 (1964), 439–68.

[65] Piron, C., 'Survey of General Quantum Physics', *University of Denver Report* (1970).

[66] Pontrjagin, L., *Topological Groups*, Gordon and Breach, Inc., New York, 1966.

[67] Pool, J. C. T., 'Baer *-semigroups and the Logic of Quantum Mechanics', *Communications in Mathematical Physics* 9 (1968), 118–41.

[68] Pool, J. C. T., 'Semimodularity and the Logic of Quantum Mechanics', *Communications in Mathematical Physics* **9** (1968), 212–28.

[69] Ramsey, A., 'A Theorem on Two Commuting Observables', *Journal of Mathematics and Mechanics* **15** (1966), 227–34.

[70] Reichenbach, H., *Philosophical Foundations of Quantum Mechanics*, California Press, 1941.

[71] Segal, I., 'Postulates of General Quantum Mechanics', *Annals of Mathematics* **48** (1947), 930–48.

[72] Segal, I., *Mathematical Problems of Relativistic Physics*, American Mathematical Society, Providence, R.I., 1963.

[73] Sherman, S., 'On Segal's Postulates for General Quantum Mechanics', *Annals of Mathematics* **64** (1956), 593–601.

[74] Sherman, S., 'Non-Negative Observables are Squares', *Proceedings of the American Mathematical Society* **2** (1951), 31–3.

[75] Skolem, Th., 'Untersuchungen über die Axiome des Klassenkalküls und über Produktions-und Summations Probleme, welche gewisse Klassen von Aussagen Betreffen', *Videnskapsselskapets Skrifter, I Mat.-Nat. Klasse* (1919) No. 3, p. 37.

[76] Sourian, J. M., 'Quantification Geometrique', *Communications in Mathematical Physics* **1** (1966), 374–98.

[77] Streater, R. F. and Wightman, A. S., *PCT, Spin and Statistics and All That*, W. A. Benjamin, New York, 1964.

[78] Suppes, P., 'Probability Concepts in Quantum Mechanics', *Philosophy of Science* **22** (1961), 378–89.

[79] Suppes, P., 'The Probabilistic Argument for a Non-Classical Logic of Quantum Mechanics', *Philosophy of Science* **33** (1966), 14–21.

[80] Varadarajan, V. S., *Geometry of Quantum Theory*, Vols. I, II, Van Nostrand, Princeton, N. J., 1968, 1970.

[81] von Neumann, J., *Mathematical Foundations of Quantum Mechanics*, Princeton University Press, 1955.

[82] Zierler, N., 'Axioms for Non-Relativistic Quantum Mechanics', *Pacific Journal of Mathematics* **11** (1961), 1151–69.

[83] Zierler, N., 'Order Properties of Bounded Observables', *Proceedings of the American Mathematical Society* **10** (1963), 346–51.

[84] Zierler, N. and Schlessinger, M., 'Boolean Embeddings of Orthomodular Sets and Quantum Logic', *Duke Mathematical Journal* **32** (1965), 251–62.

C. A. HOOKER

METAPHYSICS AND MODERN PHYSICS

A Prolegomenon to the Understanding of Quantum Theory

I. INTRODUCTION

Everyone who has studied the quantum theory knows that it is a 'problem' theory – ever since its inception the debate concerning its interpretation and finality has raged. To-day the debate still rages, the flames fanned no less by those of all persuasions who claim that there is no 'problem' with quantum theory as by their opponents. There is, unhappily, neither agreement as to precisely what the problem really is not yet as to what the solution is (of course!). The 40 year history of quantum theory has made one thing clear: no solution to the 'problem' is easy, nothing simpleminded works – members of all persuasions, including those who claim to have removed the problems, are forced to go through elaborate argument in order to make their positions seem workable.

I have a particular point of view on the nature and origin of the problems of quantum theory which I want to develop in this essay. I have already sketched out some of its basic ideas in a preliminary fashion elsewhere.[1] This essay is still very much in the nature of a preliminary sketch – partly because of my ignorance, and partly because of work still to be carried out (my point of view leads to a research programme) and partly because there are many questions which are still simply unsettled as yet (or on which we do not even have clear ideas). I cannot claim to have developed the view sufficiently to substantiate all of the conjectures which I will make here concerning it, much less claim to have solved all the interpretive problems raised by quantum theory. But it is a relatively unique approach, certainly I believe it deserves emphasis and a fair 'run for its money' – it has many components, though, which are old themes in the history of science. Naturally, I shall choose to formulate problems in a manner which favours this point of view. The view turns out to offer, not only insight into the interpretive difficulties which quantum theory raises, but into the natures and strengths of the various 'solutions' which have been adopted.[2]

Hooker (ed.), *Contemporary Research in the Foundations and Philosophy of Quantum Theory*, 174–304. *All rights reserved*
Copyright © 1973 by D. Reidel Publishing Company, Dordrecht-Holland

Aside. I shall use the following abbreviations: QM for nonrelativistic quantum theory (von Neumann formulation); RQT for relativistic quantum theory; CPM for classical particle mechanics; CSM for classical statistical mechanics; CFT for classical field theory; QFT for quantum field theory and CFFT, QFFT for the free field versions of these latter two theories respectively.

Let me begin by posing the question "What, fundamentally, is the problem over QM(RQM)?" The literature provides many different responses – "There is a measurement problem", "The difficulty of giving a quantum mechanical account of macroscopic objects", "How to understand the full impact of Heisenberg's uncertainty relations" and so on – but I am looking for a more general point of view.

To begin with, the problem is not with the mathematical formalism *per se* of QM at any rate; Hilbert space is a relatively well understood and presumably consistent mathematical structure.[3] Nor is the problem to be found in the practical use of that formalism. Physicists are quite agreed on the application of QM to their problems and are able to derive the relevant statistics in an unambiguous fashion. They are also quite clear on how to determine experimental statistics from their experimental results. The statistics prove to be remarkably accurate.[4]

The problem, then, must be to do with the interpretation, or 'meaning' of QM. And the usual treatment of the problem focuses, indeed, on the use of the physical-descriptive language in the quantum domain. The most refined philosophy of this kind was developed by Bohr. Bohr succeeded, very nearly, in preserving total purity for his approach by restricting himself entirely to a discussion of the meaning of physical-descriptive concepts and physical descriptions – speaking in contemporary jargon, we can say that he proposed a purely semantical thesis concerning physical-descriptive ('empirical') languages.[5] Einstein was at the opposite extreme – he wanted to know what there was in the world. And most discussions have mingled the two levels of inquiry (which are so intimately related to one another through 'semantic-ascent'[6]), now speaking of what exists and now about when words have meaning. It is on the ontological formulation that I wish to focus primarily for, as I shall explain, I believe that this is the key also to understanding properly the linguistic structures involved – though inevitably I shall remain within the domain of words.

Our question then is 'What is there in the quantum domain" or, in more

philosophical jargon, "What is the ontology of quantum physics?" From
this point of view I am able to supply a relatively precise formulation of
the 'quantum problem' : The formal structure and agreed empirical significance (the Born statistical interpretation) of QM suggest a strong analogy between it and CPM, especially CSM; however, *there appears to be no
quantum ontology specifiable which is essentially classical and which stands
to the QM formalism in the same relation as the ontology of CPM and CSM
stands, respectively, to CPM and CSM.* That is, there appear to be no
models of quantum theory of a certain kind.

II. PROGRAMMATIC INTERLUDE

The following discussion will hinge on a particular view of the formal
structure of theories of mathematical physics, so I shall first outline that
structure (in abstracto) as I see it (Section III). Next (Section IV) I shall
explain some of the key ideas in the above formulation – e.g. "essentially
classical" and the classical relation between theory and ontology. This
will carry us deep into the logical structure of CPM and allow a more
precise formulation of the problem. There I shall examine (Section V)
various levels at which an answer can be given to the question "Why not?"
in response to the above assertion concerning the absence of classical
models for QM. One of these is my own point of view which I wish to
present to the reader. Thereafter we will be occupied by it alone (Sections
VI–XIII).

III. ABSTRACT FOR A THEORY OF MATHEMATICAL PHYSICS

Later on I shall to have something to say about the general structures of
classical and quantum theories. To this end I shall now characterize the
components of a theory of mathematical physics in a general, rather
abstract way. Later on I shall be able to compare structures by pointing
out how these various components relate in the various theories.

The most obvious feature of a physical theory to anyone reading a textbook is the fundamental dynamical equation of the theory, the equation
that governs the time evolution of the physical systems in the domain of
the theory. In CPM it is the Euler-Lagrange equations, in QM the Schrödinger equation (or equivalent), in RQM and QFT the Klein-Gordon or

or Dirac equations or the various quantum field equations. What I want to focus on is not these equations, which do not offer the deepest insight into a theory, but the set of functions which are (actual and possible) solutions to this equation, for a given theory. This I shall call the *function set* $\mathcal{S}_T$ of the theory or, where the set forms a space or may be imbedded in a space, the function space $\mathcal{S}_T$ of the theory. All linear theories, which include CFFT, QM and QFFT have a function space since, by the linearity of the fundamental dynamical equation, if ϕ_1 and ϕ_2 are two solutions then $\alpha\phi_1 + \beta\phi_2$ is as well, $\alpha, \beta \in F$ and F is an algebraic field (in fact, always the real or the complex number field). Whether the space has additional structure, and what that structure is, depends upon the theory in question. (For example, in the cases of CFFT, QM and QFFT the function space is – or is postulated to be – a real or complex, separable Hilbert space.) The members of $\mathcal{S}_T$ are in every case functions, primarily, of the space and time variables $x_i (i = 1, 2, 3)$, t.[7]

For reasons which will later become clear, a *state set* $\mathcal{S}_S$, not necessarily identical with $\mathcal{S}_T$, will also be introduced. Like the notion of *observable* (and *time* and *space*) that of *state* will be taken to be a primitive notion here. States have this logical function in a theory: they yield maximal information concerning a system's properties, i.e. state descriptions are complete. Pseudo state descriptions for which this is not true can only be regarded as would-be state descriptions which reflect our ignorance. (Of course the *criteria* of completeness will have to be carefully specified!) In general we would expect that $\mathcal{S}_T$ and $\mathcal{S}_S$ would be intimately related (after all, solving the dynamical equation is supposed to say *something* about the system!) but I shall resist the temptation to specify any rigid relationship in advance.

The point of not studying just the equation of dynamical evolution is that its structure is determined in essentially the same way in every case: it is a one-parameter group of transformations that leave invariant a certain structure $\mathcal{S}_D$ say. Thus we may regard the structure of $\mathcal{S}_D$ as essentially determining the character of the dynamical evolution. For any given theory what $\mathcal{S}_D$ is will be well determined, it has so far in fact been either $\mathcal{S}_T$ or $\mathcal{S}_S$. (In CPM these transformations are canonical contact transformations and it is the Jacobi invariants that express the fact that these transformations leave the state space of CPM metrically unaltered; in QM the transformations are unitary in complex Hilbert space, i.e. they represent generalized rotations in that space which leave the metrical structure again invariant.[8])

But while these considerations determine the structure (the algebraic form) of the dynamical equation, the specific 'content' of the equation is not thereby determined. To explain. In the cases of CPM and QM (and also under certain conditions in CFT and QFT) the foregoing structure determines an algebraic form for the group elements in terms of the infinitesimal generator of the dynamical group, this infinitesimal generator is known in both theories as the Hamiltonian. Because of its specific role the Hamiltonian is constrained with respect to its algebraic structure, but not uniquely so (even when formulated in terms of the 'preferred' coordinates, i.e. those identified as the generators of the space-time symmetries – positions, momenta and angular momenta).[9] The specific algebraic form chosen for the Hamiltonian then determines the specific dynamics of the theory. The specific dynamics has a great deal to tell us about the specific experimental predictions of a theory but is of relatively little relevance in itself to those theoretical structures important for the basic understanding of the theory – for this latter purpose it is the structure of $\mathscr{S}_D$ and the space-time symmetries which are the key features. Since the temporal evolution transformations leave the Hamiltonian invariant we are further able to specify the structure of the dynamical evolution group of a specific theory (in fact we are able then to determine it exactly) – in each case the Hamiltonian appears as a bilinear form over $\mathscr{S}_D$ and hence the relevant groups are the symplectic groups.[10]

Equally important is the set of *observables* of the theory, those physical magnitudes, represented by certain mathematical entities, which (so we hope) together exhaustively characterize all the relevant physical properties of the physical systems in the domain of the theory. How the observables relate to one another depends very much on what the mathematical representatives are, but in general we would want the following kind of relationship to hold: the mathematical entity representing the observable $\mathcal{O}' = f(\mathcal{O})$, where f is some suitable function (a borel function), gives as its values just the $f(O)$, where the O are the values given by the mathematical entity representing the observable $\mathcal{O}$. Moreover, at least where two observables $\mathcal{O}_1$, $\mathcal{O}_2$ can have precise simultaneous values (can be jointly well-defined) we can define sums and products of observables according to:

$$\mathcal{O}_1 + \mathcal{O}_2 = f(\mathcal{O}) + g(\mathcal{O}) = (f + g)(\mathcal{O})$$
$$\alpha \mathcal{O}_1 = \alpha f(\mathcal{O}) = (\alpha f)(\mathcal{O})$$
$$\mathcal{O}_1 \mathcal{O}_2 = f(\mathcal{O}) g(\mathcal{O}) = (fg)(\mathcal{O}).$$

Upon this basis one can erect the *partial algebra* of observables $PA_{\mathcal{O}}$.

The relations between the observables $\mathcal{O}$ and the state space elements s_i is determined by the logical role that states, by definition, play: a particular state s_i determines precise values for a maximal consistent set of observables. Only in this way can states determine complete physical descriptions of physical systems. This logical function can be made clearer if we introduce a minimal propositional structure. Consider the the proposition $P_{\mathcal{O}, E} = $ The observable $\mathcal{O}$ has a value in (borel) set E, $\subseteq R$ (R is the real line). Clearly, if our foregoing restriction obtains, $P_{\mathcal{O}, E} = P_{f(\mathcal{O}), f(E)}$. Define on the set $P_{\mathcal{O}}$ of propositions the relation $\leqslant$ by

$$P_{\mathcal{O}, E} \leqslant P_{\mathcal{O}', E} \quad \text{iff} \quad P_{\mathcal{O}, E} \to P_{\mathcal{O}', E'},$$

where '$\to$' means implies. Then $\leqslant$ is a partial ordering relation on $P_{\mathcal{O}}$.[11] Label the resulting partially ordered set (POset) of propositions $P_{\mathcal{O}}^o$. Further structure to $P_{\mathcal{O}}^o$ must await the analysis of specific theories.

We can now introduce propositions of the form $P_{s_i} = $ The system is in state $s_i \in \mathscr{S}_s$. These propositions then determine truth values for maximal consistent sets of propositions in $P_{\mathcal{O}}^o$. Note that the maximal consistent sets thus determined need not include all of $PA_{\mathcal{O}}^o$ since only a partial algebra has been defined.[12] In general the maximal consistent sets can be determined as the sets of all functions of one of a special subset of the observables (the maximal observables).[13] Such a maximal consistent set (indeed, any consistent set) forms a complete algebra (or simply, an algebra) which is a subalgebra of $PA_{\mathcal{O}}$, since all sums and products are well defined (as above). In fact an alternative definition of compatibility is this: two observables are compatible just when they generate a (sub) algebra. The subalgebra so formed will exhaust $PA_{\mathcal{O}}$ if and only if all observables are compatible, i.e. if and only if $PA_{\mathcal{O}}$ is itself an algebra. In propositional terms the foregoing is re-expressible as: A complete compatible set of observables determines a maximal boolean subalgebra of the POset $P_{\mathcal{O}}^o$. (Two observables are compatible when they generate a boolean subalgebra of $P_{\mathcal{O}}^o$, actually a subalgebra of those determined by every complete compatible set of observables of which they are both members.) A proposition P_{s_i} then determines an ultrafilter on some maximal boolean subalgebra formed by the maximal consistent sets of observables.[14] In the case where all observables are compatible P_{θ}^o is a boolean algebra and the P_{s_i} determine ultrafilters on this algebra.

Of course, we have not said that determining values for a maximal consistent set of observables is *sufficient* for state descriptions being complete, though it obviously is necessary. Whether or not it is sufficient depends upon whether or not one has a purely formal criterion of completeness in mind. If one has then the sufficiency condition seems reasonable. If, on the other hand, there are other (presumably physical and/or metaphysical) considerations that play a role in determining what counts as a complete description then it could turn out to be the case that the maximal consistent descriptions permitted by a theory were nonetheless not complete. In fact, I adopt this second approach and I shall later claim that completeness fails for QM descriptions, for a variety of possible approaches, in just this latter sense.

An important connection between PA_{θ} and P_{θ}^o comes through the idempotent observables, I_{θ}, of PA_{θ} which take only the values 0,1. The I_{θ} form a partial boolean algebra $P\mathscr{B}_I$. The I_{θ} pick out sets of elements $S(I_{\theta})$ in $\mathscr{S}_S$ as follows:

$$S(I_{\theta}) = \{s_i : s_i \in \mathscr{S}_s \quad \text{and} \quad P_{I_{\theta},\{1\}} \quad \text{for} \quad s_i\}.$$

Thus the I_{θ} determine a partial boolean valuation on the subsets of $\mathscr{S}_S$. To each proposition $P_{I_{\theta},\{1\}}$ there corresponds exactly the proposition "The system state is one of the set $S(I_{\theta})$" and intuitively this corresponds to "Either P_{s_1} or P_{s_2} or ..." where the disjunction exhausts the $s_i \in S(I_{\theta})$. (The exact specification of this disjunction in terms of the structure of $\mathscr{S}_S$ has yet to be specified and will be carried out for each theory in turn as we proceed.)

Now the observables and the function set must also be related to one another in the following general way: an element of the function set determines, partially or completely, the values of the observables. If this were not so the connection between the properties of the system and its behavior in time would be severed thus removing the foundation for physical description and experimental testing. Unfortunately it is impossible to be much more precise than this about the relationship. Quantum theory already loosens the tie between the two to the point where, as we shall see, an element of the function set in general determines only a probability distribution for the values of given observables (although it compensates somewhat by restricting the ranges of values which observables may assume in a relatively precise fashion) and I cannot think of a solid objection to the proposal that an element of the function set determine only a probability distribution over probability distributions over value ranges of observables. We can apparently demand nothing stronger than some probabilistic connection between the two.

Finally there is a set of probability measures $\mathscr{S}_\mu$ on some domain, say $\mathscr{D}_\mu$. Since these measures are intended to characterize ensembles of physical systems and since what characterizes individual systems are states it is natural to take $\mathscr{S}_S$ as $\mathscr{D}_\mu$. (Since the elements of $\mathscr{S}_S$ determine truth values for maximal consistent sets in $PA_\mathcal{O}$ and $P_\mathcal{O}^o$ the probability measures can also be assigned $PA_\mathcal{O}$ or $P_\mathcal{O}^o$ as domain, though how this is done depends upon the structures of $PA_\mathcal{O}$ and $P_\mathcal{O}^o$. Moreover, since the function set is related statistically to the observable values we must ultimately have at least a statistical connection between the set of probability measures and the elements of $\mathscr{S}_T$.)

The set $\mathscr{S}_\mu$ we shall assume to be a strongly convex set, i.e. every convex linear combination of the members of $\mathscr{S}_\mu$,

$$ \mathfrak{s}'_\mu = \sum_i a_i \mathfrak{s}_{\mu_i}, \quad \mathfrak{s}_{\mu_i} \in \mathscr{S}_\mu, \quad 0 \leqslant a_i \leqslant 1, \quad \sum_i a_i = 1, $$

is also in $\mathscr{S}_\mu$. In general the set $\mathscr{S}_\mu$ will contain extreme measures or points. This notion is arrived at as follows. I shall label a probability measure μ a mixture of measures μ_i just in case μ can be expressed as a convex, linear combination of the μ_i in the foregoing manner. Measures not so expressible, i.e. for which $\mu = \sum a_i \mu_i \rightarrow \mu_i = \mu$, all i (or $a_i = 0$, all $i \neq j$, $a_j = 1$ for some j), I shall call pure and these, if any, are the extreme points of the set $\mathscr{S}_\mu$. In case all the measures of $\mathscr{S}_\mu$ can be expressed as pure measures or mixtures of them I shall say $\mathscr{S}_\mu$ is atomic. In CSM for example, $\mathscr{S}_\mu$ is atomic, the pure measures being also dispersion-free and 'concentrated' at the points of phase space, while in QM $\mathscr{S}_\mu$ is atomic but the pure measures, which are projections on the one-dimensional subspaces of a Hilbert space, are not dispersion-free. In the central cases of concern in this essay it will also transpire that the set $\mathscr{S}_\mu$ is complete, i.e. that the set of probability measures on $\mathscr{D}_\mu$ which the theory admits also exhausts all the probability measures on $\mathscr{D}_\mu$.[15] Establishing completeness is in general non-trivial; in the case of CSM it follows from theorems in abstract functional analysis, in the case of QM the proof is known as Gleason's Theorem.[16]

Using the $\mu \in \mathscr{S}_\mu$ we may construct a new partial algebra of observables, the statistical partial algebra $SPA_{\mathcal{O}}$, as follows: Define equivalence classes of observables $[\mathcal{O}]$ by

$$\mathcal{O}' \in [\mathcal{O}] =_{\mathrm{df}} \mu(P_{\mathcal{O}', E}) = \mu(P_{\mathcal{O}, E}), \quad \text{all} \quad \mu, E,$$

where $\mu(P_{\mathcal{O}, E})$ is the probability that $\mathcal{O}$ takes a value in $E, \subseteq R$; now define compatibility between observables as

$$\mathcal{O}C\mathcal{O}' =_{\mathrm{df}} \text{ there exists } f, f' \text{ and } \mathcal{O}'' \text{ such that } \mathcal{O} \in [f(\mathcal{O}'')] \text{ and } \mathcal{O}' \in [f'(\mathcal{O}'')],$$

where $f, f': R \to R$ are borel functions on R, $\mathcal{O}''$ is an observable, and then construct sums and products of compatible observables as before. Since we deal here only with an algebra of *statistical* equivalence classes, whereas in $PA_{\mathcal{O}}$ we dealt with *logical* equivalence, there is no a priori necessity that $SPA_{\mathcal{O}}$ should correspond to $PA_{\mathcal{O}}$. Corresponding to $SPA_{\mathcal{O}}$ there is a *PO*set of propositions $SP_{\mathcal{O}}^o$ formed in a similar fashion to $P_{\mathcal{O}}^o$ but the $SP_{\mathcal{O}, E}$ standing for the proposition "Any $\mathcal{O}' \in [\mathcal{O}]$ takes a value in E". Again there is no a priori reason why $P_{\mathcal{O}}^o$ should closely resembles $SP_{\mathcal{O}}^o$.[17]

These then are the five basic structures in a theory of mathematical physics – the function set $\mathscr{S}_T$, the state set $\mathscr{S}_S$, the partial algebra of observables $PA_{\mathcal{O}}$, and its associated *PO*set $P_{\mathcal{O}}^o$, the set of probability measures $\mathscr{S}_\mu$ and the statistical partial algebra $SPA_{\mathcal{O}}$ and its associated *PO*set $SP_{\mathcal{O}}$, constructed from $PA_{\mathcal{O}}$ and $\mathscr{S}_\mu$. What determines the character of a theory is how these elements are related to one another.

IV. THE CHARACTER OF CPM AND CSM

Elsewhere [52] I have attempted to develop the characterization of the *spirit* or *essence* of the ontologising of classical physics in some detail. For our purposes here it suffices to point out the two outstanding features: *definiteness* and *exhaustiveness*.

Definiteness. It is now an historical commonplace to note that one of the great changes forming part of the Galilean-to-Newtonian revolution in physics was the turn to the 'mathematization' of nature, to the demand that all theoretically significant physical magnitudes be continuous, real valued functions. The result of this demand is that the classical conception of the world was of a 'sharp', precisely defined world, where every individual has only mathematically precisely specified properties.[18]

Exhaustiveness. The classical physical ontology was what physical theory said there was. But it was also supposed to be exactly what there actually was. All of the physical properties of matter were considered captured by classical physics and that in such a fashion that a *state description* exhausted all of relevance that was to be said.[19] In the classical scheme, offering an exhaustive description falls into two parts: (i) specifying the masses and charges of each fundamental constituent, (ii) specifying the kinematics (geometry, location and velocities) of each fundamental constituent. This suffices for generating the full set of exhaustive state descriptions for the entire lifetime of any one of these fundamental constituents or for any complex system constructed from them or which they constitute, held in isolation.[20]

It ought to be remarked that the question of what constitutes an exhaustive characterization of the physical aspects of the world is not a *mere* empirical question (though it may ultimately prove to be an entirely empirical matter if some advocates of 'quantum logic' are after all correct). Once we admit space and time and motion to the physical world we are *logically* bound to admit that an exhaustive characterization must include all of the *logically constructible* geometrical and kinematical properties (thus shape, size, position, velocities – angular and linear – acceleration and so on). But whilst motion itself is logically sufficient to distinguish matter from mere extension, contra Descartes, we require some additional properties to yield up its distinctive features. These – mass and charge – are also (luckily?) just those required to yield up also the dynamics. And now, with the introduction of these, we are logically required, for exhaustiveness, to specify exactly all properties logically constructible from these and the foregoing kinematical properties (for example, momentum and energy). In this sense, what counts as an exhaustive description is determined as much by the logical structure of the physical descriptive language as by the way the world is.[21]

So much then for the general character of the classical outlook. I note in passing that QM retains the strong sense of the mathematization of the world and the classical catalogue of physical properties. On the other hand it displays a fundamental element of discontinuity and a certain exclusiveness among the mathematical representatives of the classical properties. This is the backdrop against which an understanding of the exhaustiveness or completeness of the quantum mechanical description of the world must occur or, more fundamentally, the context in which we must decide how, and indeed whether at all, we are to specify the quantum ontology. But before all this can be achieved we must gain some insight into the relations between CPM (and CSM) and classical physical ontology.

The world of CPM and CSM is conceived of as composed of ideal point masses from collections of which sizable, massy bodies are constructed. I shall immediately exploit Newton's proof that a uniform spherical massy body behaves gravitationally exactly like an ideal punctiform one, to pass on to the standard 'billiard ball' conception of the fundamental constituents of the world, for ontological finesse is not my objective just now.[22] Each such billiard ball – I shall call them atoms hereafter – had a definite mass, shape size, position and motion. Complexes of atoms could themselves form complexes and so on up.

The most powerful tool for representing the states of systems (of any complexity) proved to be *phase space*, i.e. a space of $3N$ dimensions for N individual systems (these systems may be atomic or complex, what counts is that they are being treated as individuals for the purposes of theoretical analysis), each constituent system contributing three dimensions corresponding to its location on three orthogonal coordinate axes and three dimensions corresponding to its component momenta measured along the three coordinate directions. This phase space is so constructed that the specification of a location in it exhaustively describes the physical properties of the N-component system (at the moment of its location there). Phase space is the *state space* of CPM and the set of its locations the state set $\mathscr{S}_s$ (cf. Section III).[23]

The trajectory of a system in its phase space parametrized by time is its dynamical evolution. This evolution is determined by the dynamical group $U_{t,t'}$ for the system.[8] In the standard formulation we find instead the infinitesimal generator of this group, the Hamiltonian, which determines a differential equation giving the possible system trajectories. Equivalently (or generally so), one may commence from a Lagrangian and an extremal principal (Hamilton's Principal) to arrive in turn at the corresponding differential equations of motion (in this case the $3N$ second order Euler-Lagrange equations which are equivalent to the $6N$ Hamilton's equations given by the Hamiltonian). The Hamiltonian generates canonical contact transformations, transforming the state specification at one time to that of another along a trajectory.[24] The specific dynamics is determined by the Hamiltonian form for N particles as $\sum^{N}(p_i^2/2m_i)+$ $+ V(q_1, ... q_{3N})$. Such canonical transformations leave the structure of phase space invariant, hence $\mathscr{S}_D$ is identical with phase space, i.e. with $\mathscr{S}_S$, here as anticipated. Note that canonical transformations in phase space are not orthogonal transformations (i.e. real generalized rotations) but include translations as well – the trajectories do not in general lie on the 'unit sphere' in phase space. The set of canonical transformations leaves the Hamiltonian invariant, as remarked earlier, and has the structure of the symplectic group – this group determines the dynamical symmetries of the theory.[25] Since the dynamical equations of motion are at most of second order in time and the known physical forces are functions at most only of geometry and momentum, the specification of a location in phase space, which fixes the instantaneous spacial locations and momenta, completely determines the ensuing dynamics. Since all other mechanical properties of the system can be obtained as functions of the positions and momenta, state specifications are complete.

To prepare the way for what follows we need a characterization of the structure of phase space. Now if one were interested just in the mathematical foundations of CPM from a purely mathematical point of view it would be natural to characterize the phase space as a topological vector space, or perhaps even some more abstract manifold.[26] But although such a characterization is mathematically useful, it does not properly illuminate the true logical structure of the theory. In fact, although a $6N$-dimensional Euclidean space is commonly thought of as the model for phase space this is misleading just to the extent that it is also the model for a real $6N$-dimensional vector space. For it is the *set-theoretic* structure of phase space that is the key logical-structural feature of it and not at all any vectorial structure which its most common model shares. Consider now the structure of all borel subsets of the $6N$-dimensional real space. This is well known to have the structure of a distributive, orthocomplemented lattice or, what comes to the same thing, of a boolean algebra.[27]

Hereafter, to draw attention to this set-theoretic structure, I shall denote the space by the symbol S^{6N} and the associated boolean algebra by $\mathscr{B}_S^{CPM}$.

The observables of CPM are usually taken to be represented by real valued borel functions on the phase space S^{6N}. Consider an arbitrary system location in phase space given by $\langle q_1, q_2, \ldots q_{3N}; p_1, p_2, \ldots p_{3N}\rangle$, abbreviated to $\langle q, p\rangle$ for convenience, then

$$f_{q_i}: \langle q, p\rangle \to q_i$$

$$f_{p_i}: \langle q, p\rangle \to p_i$$

$$f_p: \langle q, p\rangle \to \left(\sum_1^{3N} p_i^2\right)^{1/2}$$

$$f_E: \langle q, p\rangle \to \sum_1^{3N} (p_i^2/2m_i) + V(q_1, q_2, \ldots q_{3N})$$

give respectively the position of the ith component of the system, the momentum of the ith component, the total momentum and total energy.

Generalizing after the fashion of quantum theorists we may say that all real valued borel functions over phase space represent observables (adding quickly that most will not, indeed could not, practically speaking, be directly measured – fortunately such measurements are not necessary, for it suffices to determine the p_i and q_i to determine the values of all such functions).

The set of observables in this case is well known to have the structure of a complete, commutative algebra $A_{CPM,\mathcal{O}}$[28]. It is a complete algebra, rather than a partial algebra, because every pair of observables may be jointly precisely defined. Thus we obtain a stronger algebraic structure for CPM than the partial algebra $PA_\mathcal{O}$ of Section III. Since a state determines a location $\langle q, p\rangle$ in phase space and since all observables are functions of the q_i, p_i it follows that a state determines unique values for all observables. In this case the maximal consistent set of observables exhausts $A_{CPM,\mathcal{O}}$. Notice that this strengthening is possible only because all observables are representable as functions of the same basic observables, the q_i and p_i, and that ultimately – when the dynamical evolution is determined – these in turn are functions of only one observable (external parameter would be a better description), the time t.

Consider briefly the subset of these generalized observables given by the functions

$$f_{c_s} : \langle q, p \rangle \to \begin{cases} 1 \text{ iff } \langle q, p \rangle \in S \\ 0 \text{ otherwise} \end{cases},$$

where S is a borel subset of the phase space S^{6N}. I shall call these observables the characteristic observables for S, or the observables C_S for short. Notice the following correspondence: C_S has the value 1 if and only if the location of the system is in S. The f_{c_s} are the idempotents of the algebra of observables $A_{\text{CPM}, \varnothing}$ ($f_{c_s} f_{c_s} = f_{c_s}$) and it is well known that the idempotents of an algebra form a boolean algebra – hence the f_{c_s} form a boolean algebra $B_{f_C}^{\text{CPM}}$, isomorphic to $\mathscr{B}_S^{\text{CPM}}$.[29]

I note in passing that this set of observables does not, on the face of it, capture everything that would ordinarily count as observable in CPM. Thus $\dot{p}_i = dp/dt$, for example, would ordinarily count as an observable and yet it is not of the form $f(q, p)$. This is only a superficial appearance, however, because of the following three facts: (i) the Hamiltonian $H = = H(q, p)$ is an observable, (ii) the poisson bracket of any two observables is an observable, (here the poisson bracket is defined by

$$\{f, g\}_{q, p} = \sum_1^{3N} \left(\frac{\partial f}{\partial q_i} \cdot \frac{\partial g}{\partial p_i} - \frac{\partial g}{\partial q_i} \cdot \frac{\partial f}{\partial p_i} \right)$$

and clearly again furnishes a function of the q, p), and (iii) it follows from the Hamiltonian dynamical equations that $\dot{q}_i = \partial H/\partial p_i$, $\dot{p}_i = -\partial H/\partial q_i$, thus in each actual case the $\dot{q}_i$, $\dot{p}_i$ can also be expressed as functions of the q, p and hence count also as observables in the foregoing sense. (Incidentally, if we consider the algebra of all sums and products of observables, with the product defined as $\{\mathcal{O}_1, \mathcal{O}_2\}_{q,p}$ we obtain the lie algebra $LA_\varnothing^{\text{CPM}}$ of observables for CPM, with $A_\varnothing^{\text{CPM}}$ as the enveloping algebra for $LA_\varnothing^{\text{CPM}}$ – cf. e.g. Rosen [111]). On the other hand there is a distinction which the above formulation does not explicitly observe. In each case the q_i, p_i are ultimately given as functions only of time t. For *all* observables then their values must be specified, ultimately, at a particular time. (Since in CPM this amounts to specifying a particular location in phase space the foregoing formulation is not faulty, but it does obscure this basic shift in dependence from the location $\langle q, p \rangle$ in phase space to the time t.) Thus we can make a distinction between what I shall call the 'static' observables – the observables at a fixed time t – and the full dynamic observables. Notice that the algebra A_0^{CPM} can also be regarded as a dynamic sequence of 'static' algebras of real numbers $A_0^{\mathrm{CPM}}(t)$, one for each time t and all structurally isomorphic to each other.

Finally I want to draw attention to a logical feature of classical state descriptions of some significance. The phase space description requires simultaneous precise assignments of positions and momenta, i.e. essentially velocities, 'conjugate' to the positions. Now the strict definition of a velocity is that of a *finite* distance divided by a *finite* time. In particular, the Weierstrauss theory of limits demands this

$$\left(\dot{q}(t_0) = \frac{dq}{dt}\bigg|_{t_0+} = \mathrm{Lim}_{\delta t \to 0} \frac{q(t_0 + \delta t) - q(t_0)}{\delta t} \right)$$

and is formally understood to be precisely the following assertion: For every $\varepsilon > 0$ there exists a $\delta > 0$, ε, $\delta \in R$, such that

$$\left| \frac{q(t_0 + \delta t) - q(t_0)}{\delta t} - \dot{q}(t_0) \right| < \varepsilon \text{ for } \delta t < \delta.$$

Thus a velocity cannot, strictly, be specified at a point location but only in some (δ) neighbourhood of the point. (Essentially this point was already recognized by Zeno over two millenia ago and formed an important part of the bases of his famous paradoxes.[30]) Of course, there is no mathematical inconsistency in associating with each point q a function $\dot{q}$ via the Weierstrauss *algorithm* but, conceptually, the logic of the situation is clear: descriptions restricted to precise positions strictly preclude velocities and descriptions restricted to precise velocities strictly preclude precise position descriptions. Indeed, once we understand the logical structure of the situation in this case we can see that it applies to *every* pair of canonically conjugate variables, for example to angles and angular momenta and so on, for the logical relations between the definitions of all such pairs are the same. The practical impact of these relationships is considerably lessened in CPM because the δ neighbourhoods can be made arbitrarily small (though this does not alter the logical situation!) but, as the argument briefly referred to in Section VII (and given in detail in Hooker [61]) so clearly shows, it is this same logical structure which is at the root of the powerful indefiniteness arguments for QM and which are reflected in the non-vanishing commutators of QM. Those arguments, like so much else in QM, already find their roots in classical physics.[31]

At this point I want to expand my attention to CSM as well. This is possible because CSM rests upon CPM in a fundamental way. The *grounding* state space of CSM is the phase space of CPM, i.e. probability measures concern ensembles of CPM states. Moreover, the dynamics of CSM is also derivable from that of CPM.[32]

We must consider now, therefore, the set of probability or statistical measures on a phase space (more precisely, on the borel subsets of phase space). A probability measure μ is interpreted as giving the probabilities that a system will be found located in various regions of phase space. Although this formulation does not commit me to any specific interpretation of the notion of probability, I shall nonetheless insist on at least this: probability or statistical measures are attributed to, or said to *characterize*, *ensembles* of systems; such characterizations are to be experimentally tested by the preparation of an ensemble of systems whose various long run frequencies are estimates of the probabilities concerned (and good estimates, i.e. accurate within certain limits, with a given probability (!) which depends in general upon those limits).[33] Since in CSM the system is held in fact to have a precise location in phase space, probability measures are interpreted as expressions of ignorance.

Probability measures on phase space are defined exactly on the borel subsets of phase space and certain mathematical sophistries irrelevant to this discussion aside, this is exactly the structure S^{6N}. A probability measure μ is then a normed, countably addative, real valued, set function on the σ-ring $\mathscr{R}$ of borel subsets of S^{6N} satisfying $\mu(S^{6N}) = 1$; $\mu(\phi) = 0$, ϕ the null set; $0 \leqslant \mu(E) \leqslant 1$, $E \in R$; if $E_i \cap E_j = \phi$, all i, j $i \neq j$ then $\mu(\bigcup^{\infty} E_i) = \sum^{\infty} \mu(E_i)$. It is a theorem that in S^{6N} *the extreme measures are exactly the dispersion-free measures and that these latter are in turn exactly the measures* $\mu_{\{\langle q, p \rangle\}}$ *concentrated at the points* $\langle q, p \rangle$ *of phase space and hence in 1-1 correspondence to the points of phase space.*[34] Thus if the probability measure $\mu_{\{\langle q, p \rangle\}}$ characterizes an ensemble K of systems every member of K has the value 1 for the observable $C_{\{\langle q, p \rangle\}}$. On the other hand these probability measures are somewhat singular in character. For the 'physically realistic' measures employed in CSM for actual physical problems assign probability zero to any finite or countably infinite set of points in phase space, they therefore cannot be constructed as countable convex combinations of the extreme measures. The method of resolving measures into convex combinations of other measures is through the use of the characteristic functions χ_E of the borel subsets E,

$$\chi_E : \langle q, p \rangle \to 1, \quad \langle q, p \rangle \in E$$
$$\to 0, \quad \text{otherwise}$$

Notice that the χ_E are not themselves probability measures, except that in the special cases of the $\chi_{\{\langle q, p \rangle\}}$ corresponding to the extreme or pure discrete measures $\mu_{\{\langle q, p \rangle\}}$ concentrated at single points there is a formal equivalence between the two; the χ_E are, however, exactly the characteristic observables f_{C_E}. If μ is a probability measure on S^{6N} and $\bigcup^N E_i = S^{6N}$ (E_i a borel set, $E_i \cap E_j = \phi$ all i, j, N finite or countably infinite) then $\mu = \sum^N \mu_{E_i} \chi_{E_i}$, where μ_{E_i} is a probability measure coinciding with μ on E_i. In this fashion we see that any continuous probability measure may be resolved into other such measures in infinitely many different ways. Also it is clear that by this construction we cannot build up the continuous measures from the discrete measures (i.e. from those measures concentrated on a finite or countably infinite number of points of S^{6N}). The discrete measures, on the other hand, clearly form an atomic set with the dispersion-free measures as atoms. It is, however, possible to extend the above construction to uncountable convex combinations, or integrals, of extreme probability measures and in this way capture even the continuous probability measures. Both the atomicity and the completeness of the set of probability measures on phase space introduce by CSM is thereby demonstrated.[35] This latter set is the set $\mathscr{S}_\mu^{\text{CPM}}, = \mathscr{S}_\mu$ of Section III, for CPM-CSM.

We immediately note the following: *because of the 1-1 correspondence between the pure or extreme measures of $\mathscr{S}_\mu^{\text{CSM}}$ and the points of phase space the statistical algebra of observables, $SPA_{\mathcal{O}}^{\text{CPM}}$, is isomorphic to $PA_{\mathcal{O}}^{\text{CPM}}$.* For we have

$$\mathcal{O}' \in [\mathcal{O}] \quad \text{iff} \quad \mu(P_{\mathcal{O}', E}) = \mu(P_{\mathcal{O}, E}), \quad \text{all} \quad \mu, E$$

and in particular that

$$\mathcal{O}' \in [\mathcal{O}] \quad \text{iff} \quad \mu_{\{\langle q, p \rangle\}}(P_{\mathcal{O}', \{r\}}) = \mu_{\{\langle p, q \rangle\}}(P_{\mathcal{O}, \{r\}}), \quad \text{all} \quad \mu, \{r\}.$$

It follows immediately from the foregoing that $\mathcal{O}' \in [\mathcal{O}]$ if and only if $\mathcal{O}$ and

$\mathcal{O}'$ take the same values at all points of phase space and hence if and only if $\mathcal{O}$, $\mathcal{O}'$ are the same function.[36]

In the case of CSM the members of $\mathscr{S}_{\mu}^{CPM}$ are usually described as the *states* – the *statistical* states – of the theory. This use of the word "state" is apt to be confused with its use as referring to elements of $\mathscr{S}_S$ and this confusion leads very easily to serious difficulties in interpreting physical theories. If not the confusion itself, then certainly a way of talking which tends to generate the same errors, is extremely widespread in the discussions of the interpretation of QM. For this reason I wish to distinguish sharply between the elements of $\mathscr{S}_S$ and those of $\mathscr{S}_{\mu}$. I have already said that the elements of $\mathscr{S}_S$ *describe*, or *determine*, the state of a *single* physical system and that the elements of $\mathscr{S}_{\mu}$ *characterize* an *ensemble* of such systems. But to emphasize the point even more clearly I shall abstain entirely from any talk whatever of statistical states in what follows. Talk of such states would be relatively more harmless if all statistical theories stood in in the same relation to a grounding non-statistical theory as CSM does to CPM. But there is no such grounding theory in the case of QM.[37]

Finally we come to inquire about $\mathscr{S}_T$ in CPM – though one might have expected this to be our first topic of discussion. Here we strike a peculiarity – at least I hope that the sense of peculiarity will become clear before the end is reached – it is not clear what is being asked for! A priori we might expect the theory to yield a differential equation whose independent variables were the x_i, t which could be solved for the solving functions $\phi(x_i, t)$; but the traditional Lagrangian-Hamiltonian formalism does not ask for the Lagrangian or Hamiltonian as a solution to the Euler-Lagrange equations, rather we obtain a specific solution to these equations only *after* we supply such a function. Moreover there is not just one such equation but $3N$ of them (for an N-component system) and these are solved for some of the 'independent' variables (the q_i) in terms of another (t)!

Of these initial reactions we may dispense immediately with the $3N$ equations in favor of 1. This equation is to determine the evolution of the location vector $\phi(q_1, q_2, \ldots q_{3N}, p_1, \ldots p_{3N})$ in phase space. To do this we simply recall that we may represent $\phi(q, p)$ as

$$\phi(q_i, p) = \sum_i^{3N} q_i \mathbf{I}_{q_i} + p_i \mathbf{I}_{p_i}$$

where I_{q_i}, I_{p_i} are unit vectors along the q_i, p_i axes respectively, whence

$$\dot{\phi}\left(=\frac{d\phi}{dt}\right) = \sum_i^{3N} \dot{q}_i I_{q_i} + \dot{p}_i I_{p_i},$$

$\dot{q}_i, \dot{p}_i$ being given by the solutions of the Euler-Lagrange equations. Moreover, reflection strongly suggests that we choose the ϕ's as the members of $\mathscr{S}_T$ rather than any Lagrangian or Hamiltonian functions.[38] Thus our set $\mathscr{S}_T^{CPM}$ is the set of all CPM-admissible vector valued functions over S^{6N} of 1 real variable (time), where to be a CPM-admissible function just means to be the solution of the Euler-Lagrange equations for some admissible Lagrangian. $\mathscr{S}_T^{CPM}$ is a subset of the set S_T of all vector-valued functions of 1 real variable over R^{6N}.

Note immediately that it is logically impossible to obtain a 1-1 correspondence between any subset of $\mathscr{S}_T^{CPM}$ and the points of phase space S^{6N}. This is simply because, so long as $p_i \neq 0$ (for some i), the system cannot have $q_i =$ constant. On the other hand each $\phi \in \mathscr{S}_T^{CPM}$ corresponds uniquely to the subset $S_\phi \in S^{6N}$ of points that lie on the trajectory which it determines. In general the mappings $\phi: T \rightarrow S^{6N}$, T time, will be many-one since a system may pass through the same state on several different occasions (though it cannot be in several distinct states on the one occasion). We may easily convert this into a one-one map, however, if we add a 'phase variable' which is monotic increasing along the system trajectory as an additional *observable*, for then distinct recurrences of the same state will be labelled with distinct phase values. Strictly this means an additional dimension to phase space, making it S^{6N+1}, though in what follows I shall ignore this unimportant complication because we could always eliminate it in favour of the usual parametrization of the curves in S^{6N}. Effectively, this device simply introduces time as an observable (since the phase variable must be some function of time and which function it is is essentially arbitrary). Note too that since the $\phi \in \mathscr{S}_T^{CPM}$ are functions only of time it makes no sense to ask whether temporal evolution can be construed as a transformation *in* $\mathscr{S}_T^{CPM}$ – each element of $\mathscr{S}_T^{CPM}$ determines a complete temporal evolution for a system. For the same reason there is no set of *functions* derivable from $\mathscr{S}_T^{CPM}$ in which dynamical evolution could be reconstructed.

The set $\mathscr{S}_T^{CPM}$ does not form a linear space. The sum of two phase space trajectories is not necessarily, or in general, an admissible phase space trajectory. (Note, however, that S_T forms a linear vector space.)

At this point we may wonder (with an eye on QM) whether the observables of CPM cannot be reconstrued as operators in some way. Now phase space S^{6N} has a purely set theoretic structure and there is no way that I can see to reconstrue the observables as operators on it. On the other hand the function space S_T does have a vector space character which we can take advantage of for this purpose. Define 'projection' operators in S_T as follows:

$$\tilde{P}_{q_i}\phi = q_i\mathbf{I}_{q_i}, \quad \tilde{P}_{p_i}\phi = p_i\mathbf{I}_{p_i}$$

which project ϕ on the ith position and momentum coordinate axes respectively. The values of the basic observables q_i, p_i are given by the 'eigenvalues' of these projection operators. Other observables will be *functions* of these projection operator eigenvalues. *Only in this sense* can they be regarded as themselves operators; though, formally, we may define, for each observable $\mathcal{O}(q, p)$, an operator $\tilde{\mathcal{O}}$ such that $\tilde{\mathcal{O}}\phi = \mathcal{O}(q, p)\phi$. Then we may write, in a symbolic fashion, $\tilde{\mathcal{O}} = \mathcal{O}(\tilde{P}_{q_i}, \tilde{P}_{p_i})$ as the formal counterpart here to the spectral theorem in QM. Clearly, the algebra of these operators is identical with $A_{\mathcal{O}}^{\mathrm{CPM}}$ if we define $(\tilde{\mathcal{O}}_1 + \tilde{\mathcal{O}}_2)\phi = (\mathcal{O}_1 + \mathcal{O}_2)(q, p)\phi$, $(\alpha\tilde{\mathcal{O}}_1\tilde{\mathcal{O}}_2)\phi = (\alpha\mathcal{O}_1\mathcal{O}_2)(q, p)\phi$, $\alpha \in F$. (Note that this reconstrual of the observables of CPM involves two dimensions of additional complexity. In the first place the operators will be in general non-linear on the space $\mathscr{S}_T^{\mathrm{CPM}}$ and so it will be difficult to obtain any significant general structure from them. But in the second place we are still mostly ignorant of the structure of the space $\mathscr{S}_T^{\mathrm{CPM}}$ appearing in CPM. Since these spaces will also be non-linear spaces the structure of operators on these spaces will be that much more difficult to characterize. We remain, then, doubly ignorant of the details of this reconstrual – which can only appear as extremely artificial in its simple, abstract form above.)[39]

The probability measures over $\mathscr{S}_T^{\mathrm{CPM}}$ are independent of time. Precisely because the $\phi \in \mathscr{S}_T^{\mathrm{CPM}}$ give the complete time evolution of systems the measures μ over $\mathscr{S}_T^{\mathrm{CPM}}$ do not alter in time. This is in contrast to the situation in phase space S^{6N} where the probability measures μ must undergo a temporal evolution precisely to give the same measures to sets of trajectories (i.e. systems) at different times. But this latter condition is automatically realized by constant measures over $\mathscr{S}_T^{CPM}$.

Whatever our knowledge of $\mathscr{S}_T^{CPM}$, we can be sure of the following important structure: even though a linear combination of members of $\mathscr{S}_T^{CPM}$ is again a member of $\mathscr{S}_T^{CPM}$, *no physical sense attaches to such combinations in $\mathscr{S}_T^{CPM}$ – there is no superposition principal operative in $\mathscr{S}_T^{CPM}$.* Indeed, such combinations could not represent systems compounded out of the component systems corresponding to the component trajectories since (i) the former trajectory would be represented in a higher dimensional phase space than that of the component trajectories and hence $\mathscr{S}_T^{CPM}$ would be a set of different functions than those associated with the component trajectories and (ii) the component systems will in general interact with one another (this is determined by their respective trajectories) producing a trajectory in general distinct from any linear combination of the trajectories of the components even when represented in the higher dimensional phase space. For it is possible to represent a trajectory in S^{6N} in the phase space of S^{6N+6N}, the location vector ϕ simply having zeros for the coefficients of $3N$ of the position terms and $3N$ of the momentum terms. For two trajectories in S^{6N} we have

$$\phi_1 = \sum_1^{3N} q_i^1 \mathbf{I}_{q_i} + p_i^1 \mathbf{I}_{p_i} + \sum_{3N+1}^{6N} O \cdot \mathbf{I}_{q_i} + O \cdot \mathbf{I}_{p_i}$$

$$\Phi_2 = \sum_1^{3N} O \cdot \mathbf{I}_{q_i} + O \cdot \mathbf{I}_{p_i} + \sum_{3N+1}^{6N} q_i^2 \cdot \mathbf{I}_{q_i} + p_i^2 \mathbf{I}_{p_i}$$

and the *interaction-free* trajectory in S^{6N+6N} is given by $\phi_1 + \phi_2$. If the two systems interact the trajectory in S^{6N+6N} cannot in general be given as any simple combination of ϕ_1 and ϕ_2. Even in the interaction-*free* case there is no genuine superposition of system states – the two systems remain distinguishable, the total energy, momentum etc. is just the sums of the component system totals. Indeed, no physical sense attaches to an expression of the form $\alpha\phi_1 + \beta\phi_2$, α, $\beta \neq 1$. Moreover, there is no physical sense attached to the compounding of probabilities of component trajectories to obtain the probability of the compound trajectory in any fashion reflecting a genuine superposition of states (as, e.g., in QM). In no sense is there a superposition principal operative in S^{6N+6N} either.[40] This absence of a superposition principal is deeply related to the ontology of CPM

(CSM) – atomic individuals spatio-temporally exclude one another, they do not superpose (cf. Section VII below).

The four major abstract components of CPM-CSM have been briefly described: the function set $\mathscr{S}_T^{\text{CPM}}$ of vector-valued functions of time over phase space S^{6N} describing the admissible trajectories of systems whose states are represented by locations in the phase space; the algebra $A_{\text{CPM},\mathscr{O}}$ of observables, i.e. of real valued functions of phase space; the set $\mathscr{S}_\mu^{\text{CPM}}$ of probability measures on phase space characterizing ensembles of systems; and, of course, phase space S^{6N} itself which represents the states of physical systems. To complete this account it is only necessary to consider the specific relations among these components. These have already been spelled out in the preceding; to summarize: phase space is the domain of $A_{\text{CPM},\mathscr{O}}$ and $\mathscr{S}_\mu^{\text{CPM}}$ (i.e. of members of these sets) and the range of $\mathscr{S}_T^{\text{CPM}}$. Fixing a location in phase space determines the values of the $\mathscr{O} \in A_{\text{CPM},\mathscr{O}}$, and conversely fixing values for the members of appropriate subsets of $A_{\text{CPM},\mathscr{O}}$ (in particular the f_{q_i}, f_{p_i}, but other subsets can also suffice) will determine a point in S^{6N}; also, because it happens that the mapping $\phi : T \to S^{6N}$ (strictly S^{6N+1}, to include phase) is one-one (the theory is *deterministic*[41]), fixing a point in S also determines a point $t \in T$ and a value for ϕ, conversely a value for ϕ (including phase) determines a point in $S \times T$ and hence a point in S^{6N}.[42] Notice that $S^{6N} \neq$ $\neq \mathscr{S}_T^{\text{CPM}} \neq \mathscr{S}_\mu^{\text{CPM}}$. Graphically:

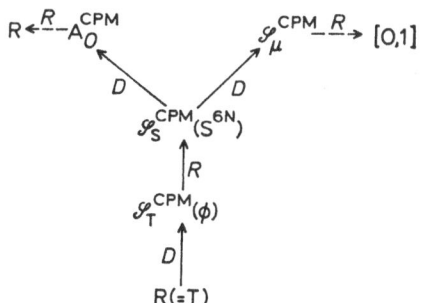

Although intuitively the connection between the characteristic probability measures and the observables on the phase space is clear enough, this connection is brought out clearly and elegantly when the logical structure of the corresponding propositions is considered.

We begin by considering propositions of the form:

The value of observable $\mathcal{O}$ lies in $E (\subseteq R)$

where E is a borel subset of the real line and corresponding to the physical magnitude $\mathcal{O}$ there is a real valued borel function on phase space $\mathcal{O}(\langle q, p \rangle)$. As before denote this proposition by $P_{\mathcal{O}, E}^{CPM}$. In this case we have:

$$P_{\mathcal{O}, E}^{CPM} \quad \text{and} \quad P_{\mathcal{O}, E'}^{CPM} \quad \text{iff} \quad P_{\mathcal{O}, E \cap E'}^{CPM}$$
$$P_{\mathcal{O}, E}^{CPM} \quad \text{or} \quad P_{\mathcal{O}, E'}^{CPM} \quad \text{iff} \quad P_{\mathcal{O}, E \cup E'}^{CPM}$$
$$\text{not} - P_{\mathcal{O}, E}^{CPM} \quad \text{iff} \quad P_{\mathcal{O}, R-E}^{CPM}$$

So that the $P_{\mathcal{O}, E}^{CPM}$ form a boolean algebra $\mathscr{B}_{\mathcal{O}}^{CPM}$. Moreover, every proposition $P_{\mathcal{O}, E}^{CPM}$ selects a borel subset $\mathcal{O}(E)$ of phase space,

$$\mathcal{O}(E) = \{ \langle q, p \rangle : \mathcal{O}(\langle q, p \rangle) \in E \}.$$

Thus the $P_{\mathcal{O}, E}^{CPM}$ stand in one-one correspondence with a subfield of the field of subsets of S^{6N} and hence $\mathscr{B}_{\mathcal{O}}^{CPM}$ is isomorphic to a subalgebra of $\mathscr{B}_{S}^{CPM}$ (the boolean algebra of such subsets). The same is true for any other observable $\mathcal{O}'$. Consider now the singleton sets $\{\langle q, p \rangle\}$. These are atoms of $\mathscr{B}_{S}$ and hence determine ultrafilters on $\mathscr{B}_{S}^{CPM}$ (i.e. they determine maximal consistent sets in $\mathscr{B}_{S}^{CPM}$).[43] They therefore also determine ultrafilters on the boolean algebras $\mathscr{B}_{\mathcal{O}}^{CPM}$, $\mathscr{B}_{\mathcal{O}'}^{CPM}$, ..., the atoms of $\mathscr{B}_{\mathcal{O}}^{CPM}$ being the propositions $P_{\mathcal{O}, \{r\}}^{CPM}$, where $\{r\}$ is the singleton set of r. In particular, precise values for $\mathcal{O}$, $\mathcal{O}'$... are thereby determined.

The product of two such boolean algebras, $\mathscr{B}_{\mathcal{O}}^{CPM} \otimes \mathscr{B}_{\mathcal{O}'}^{CPM}$, is the boolean algebra pertaining to the joint assertions concerning the observables $\mathcal{O}$, $\mathcal{O}'$ (the propositions take the form: the values of $\mathcal{O}$, $\mathcal{O}'$ fall in the sets E, E' respectively; these are equivalent to $P_{\mathcal{O}, E}^{CPM}$ and $P_{\mathcal{O}', E}^{CPM}$), and we may continue in this fashion for all of the observables of $A_{CPM, \mathcal{O}}$. Let $\mathscr{B}_{\mathcal{O}}^{CPM}$ stand for the boolean algebra $\mathscr{B}_{\mathcal{O}}^{CPM} \otimes \mathscr{B}_{\mathcal{O}}^{CPM} \otimes ...$ where we exhaust all of the observables. The atoms of $\mathscr{B}_{\mathcal{O}}^{CPM}$ are the conjunctions of atoms one each from $\mathscr{B}_{\mathcal{O}}^{CPM}$, $\mathscr{B}_{\mathcal{O}'}^{CPM}$... respectively. Clearly the $\{\langle q, p \rangle\}$ determine ultrafilters on $\mathscr{B}_{\mathcal{O}}^{CPM}$ and stand in one-one correspondence with the atoms of $\mathscr{B}_{\mathcal{O}}^{CPM}$. Thus $\mathscr{B}_{\mathcal{O}}^{CPM} \approx \mathscr{B}_{S}^{CPM}$. This is the formal expression of the fact that specification of a location in phase space determines a complete description in CPM.

We can also view this remark from another perspective. Consider propositions of the form:

The location of the system lies in the region S'

where S' may be any borel subset of phase space.
Denote this proposition by $P_{S'}^{\text{CPM}}$. Clearly the $P_{S'}^{\text{CPM}}$ form an algebra $\mathscr{B}_P^{\text{CPM}}$ isomorphic to $\mathscr{B}_{S'}^{\text{CPM}}$.[44] Moreover we have

$$P_{S'}^{\text{CPM}} \quad \text{iff} \quad P_{\mathscr{O}, \mathscr{O}(S')}^{\text{CPM}} \quad \text{and} \quad P_{\mathscr{O}', \mathscr{O}'(S')}^{\text{CPM}} \quad \text{and} \dots$$

in particular, for the sentences $P_{\{\langle q, p \rangle\}}^{\text{CPM}}$ we have $P_{\{\langle q, p \rangle\}}^{\text{CPM}}$ if and only if $P_{\mathscr{O}, \mathscr{O}(\{\langle q, p \rangle\})}^{\text{CPM}}$ and ... But $\mathscr{O}(\{\langle q, p \rangle\}) = \mathscr{O}(\langle q, p \rangle)$ for this special case and the proposition $P_{\mathscr{O}, \mathscr{O}(\{\langle q, p \rangle\})}^{\text{CPM}}$ informs us that the observable $\mathscr{O}$ takes on *the* value $\mathscr{O}(\langle q, p \rangle)$. Thus a proposition of the form $P_{\{\langle q, p \rangle\}}^{\text{CPM}}$, specifying a point location for the system in phase space, simultaneously determines precise values for all observables. Thus the $P_{\{\langle q, p \rangle\}}^{\text{CPM}}$ determine maximal, consistent sets of propositions, or ultrafilters, on the logic $\mathscr{B}_0^{\text{CPM}}$ and are equivalent to the atoms of $\mathscr{B}_0^{\text{CPM}}$. (Note that because of this correspondence the observables and probability measures of CPM could equally well be defined, respectively, as real valued functions on, and real valued measures on, the propositions $P_{\{\langle q, p \rangle\}}^{\text{CPM}}$.) To complete the correspondence with the algebra $A_{\text{CPM}, \mathscr{O}}$ consider the subset of propositions having the special form $P_{Cs, \{1\}}^{\text{CPM}}$. Then $P_{Cs, \{1\}}^{\text{CPM}}$ is true if and only if the system has its location in S. Thus the $P_{Cs, \{1\}}^{\text{CPM}}$ correspond one-one with the borel subsets in phase space and so form a boolean algebra $\mathscr{B}_C^{\text{CPM}} \approx \mathscr{B}_{fc}^{\text{CPM}} \approx \mathscr{B}_S^{\text{CPM}}$.

The probability measures on S^{6N} can equally well be regarded as measures on $\mathscr{B}_0^{\text{CPM}}$ – this much follows from the fundamental Stone representation theorem. What also follows and is of great importance is the rule for conditional probabilities: Prob. $(p_1 \mid p_2) = $ Prob. $(p_1 . p_2)/$Prob. $(p_2) =$ = Prob. $(p_2 . p_1)/$Prob. $(p_2) = \mu(S_{p_1} \cap S_{p_2})/\mu(S_{p_2})$, where p_1, p_2 are two propositions, S_{p_1}, S_{p_2} the subsets of phase space to which they correspond and μ the probability measure on phase space we are employing. This result may be found in Løs [80]. Its importance centers on the fact that just on this score QM diverges crucially from CPM (cf. Bub in this volume, and my own extended discussion of the most famous case where it occurs, the Einstein-Podolsky-Rosen experiment, especially as shown by Furry's treatment of it, in [52], cf. [36]).

To complete the catalogue of categorial propositions consider finally propositions of the form:

> The location of the system in $\mathscr{S}_T^{\text{CPM}}$ lies in the region S''

and denote this proposition by $T_{S''}$. $T_{S''}$ determines a region of phase space S' at each time t and $T_{\{\phi\}}$ determines a unique location in phase space for each time t; $S'(t) = \{\langle q,p \rangle : \phi \in S'' \text{ and } \phi(t) = \langle q,p \rangle\}$. The $T_{S''}$ hence form a boolean algebra (this is already evident from the boolean character of the algebra of subsets $S'' \subseteq \mathscr{S}_T^{\text{CPM}}$) and if we extend the $P_{S'}^{\text{CPM}}$ to record the time explicitly for which those assertions are made, we have $T_{S''}$ iff $P_{S'_1, t_1}^{\text{CPM}}$ and $P_{S'_1, t_2}^{\text{CPM}}$ and ... Since the $P_{S', t}^{\text{CPM}}$ form a boolean algebra $\mathscr{B}_P^{\text{CPM}}$ isomorphic to $\mathscr{B}_{S'}^{\text{CPM}}$ at each time t, the $T_{S''}$ determine automorphisms of $\mathscr{B}_{S'}^{\text{CPM}}$ parameterized by time. This is the reflection in the logical structure of dynamical evolution.[45]

We come now to consider the probabilistic propositions associated with CPM-CSM (but more especially with CSM). I shall offer only a brief sketch of these propositional structures since in fact relatively little seems to be known about them. To begin with we should like to consider probabilistic propositions corresponding to the categorical proposition of the form $P_{\emptyset, E}^{\text{CPM}}$. The probabilistic propositional form has to be chosen with some care. For example, there is no sense in CPM attached to a proposition of the form "The probability that in state s the observable $\emptyset$ has a value lying in E", which is the form quantum mechanical probability assertions might suggest, since in CPM a specific state determines an exact value for the observables. On the other hand, taking our cue from that dangerous tendency to talk of the probability measures of CSM as states also, there is sense to considering propositions of the following form:

> The probability is r that in an ensemble characterized by probability measure μ the observable $\emptyset$ takes a value in borel set E

where $\mu \in \mathscr{S}_\mu^{\text{CPM}}$, $E \subseteq R$ and $0 \leqslant r \leqslant 1$. These probabilities r are calculated as follows: μ and $\emptyset$ induce on R a probability measure $\mu_\emptyset$ defined by $\mu_\emptyset(E) = = \mu(\emptyset^{-1}(E))$, where $\emptyset^{-1}(E) = \{\langle q,p \rangle : \emptyset(\langle q,p \rangle) \in E\}$; $\mu_\emptyset$ then gives the requisite probabilities. Denote these propositions by $M_P^{\text{CPM}}(\mu, \emptyset, E, r)$. I shall begin (and end) by considering the special case for which $r = 1$. For fixed μ and $\mathcal{O}$ this set of propositions can be made to be formally equivalent to a boolean algebra $\mathscr{B}_{M,\mu,\mathcal{O}}^{CPM}$. For $M_P^{CPM}(\mu, \mathcal{O}, E_1 \cap E_2, 1)$ iff $M_P^{CPM}(\mu, \mathcal{O}, E_1, 1)$ and $M_P^{CPM}(\mu, \mathcal{O}, E_2, 1)$; and if we now *define* '*or*' and '*not*' so that $M_P^{CPM}(\mu, \mathcal{O}, E_1 \cup E_2, 1)$ iff $M_P^{CPM}(\mu, \mathcal{O}, E_1, 1)$ *or* $M_P^{CPM}(\mu, \mathcal{O}, E_2, 1)$, *not* – $M_P^{CPM}(\mu \mathcal{O}, E, 1)$ iff $M_P^{CPM}(\mu, \mathcal{O}, R-E, 1)$ then we obtain a structure isomorphic to a boolean algebra. Notice, however, that it has been necessary to introduce definitionally what are in fact non-standard connectives, since for the standard connectives the "only if" clauses in the above would not be valid in either case (*or*, *not*). An interesting property of these non-standard connectives is that $M_P^{CPM}(\mu, \mathcal{O}, E, 1)$ *or not* $-M_P^{CPM}(\mu, \mathcal{O}, E, 1)$ is always true (it is just $M_P^{CPM}(\mu, \mathcal{O}, R, 1)$) but neither of the component propositions need be true – for example, consistently with the above we may have $M_P^{CPM}(\mu, \mathcal{O}, E, 1)=r$, *not* $-M_P^{CPM}(\mu, \mathcal{O}, E, 1)=1-r$, $r \neq 1,0$. (It will not escape the reader's attention that *or, not* are thus strikingly like those connectives quantum logicians introduce – cf. below.) Thus although we obtain the correspondence $P_{\mathcal{O},E}^{CPM} \leftrightarrow M_P^{CPM}(\mu, \mathcal{O}, E, 1)$ and hence eventually a formal isomorphism between $\mathscr{B}_S^{CPM}$ and $\mathscr{B}_{M,\mu,\mathcal{O}}^{CPM} \otimes \mathscr{B}_{M,\mu,\mathcal{O}'}^{CPM} \otimes \ldots$ we should not, contrary to what intuition may suggest, identify the two sets of propositions. Indeed, this correspondence does not even preserve truth values, though it does map tautologies into tautologies and contradictions into contradictions.[46] Probability 1 is not quite the same as truth.

I shall not pursue further here the calculus of these propositions when either μ is not held fixed, or $\mathcal{O}$ is not held fixed or $r \neq 1$. Despite some obvious facts which one can write down,[47] remarkably little seems to be known about the structure of the calculi for any of these cases (i.e. even for the cases where two of the three variables are held fixed). The investigation of these calculi, not only for CPM-CSM but for QM as well, is one of the more urgent tasks in the investigation of the logical structure of physical theories.[48]

I shall pass on instead to consider very briefly another set of probability propositions. These take the form:

> The probability is r that a member system from an ensemble characterized by a probability measure μ is located in a borel subset S' of phase space

and I shall denote them by $M_S^{CPM}(\mu, S, r)$. Again I shall consider only the special case $r=1$ and μ fixed. It is easily seen that with the connectives as

defined in the preceding discussion this set can be made to be formally
equivalent to a boolean algebra $\mathscr{B}^{CPM}_{M, \mu}$, where now we have the correspondences

$$M^{CPM}_S(\mu, S, r) \leftrightarrow M^{CPM}_P(\mu, \mathcal{O}, \mathcal{O}(S), r) \text{ and } \ldots$$

$$M^{CPM}_P(\mu, \mathcal{O}', \mathcal{O}'(S), r) \text{ and } \ldots$$

$$M^{CPM}_S(\mu, S, 1) \leftrightarrow P^{CPM}_S$$

where it is to be noted that the first set of correspondences are in fact
exact logical equivalences and hold for all values of r (not only $r = 1$)
while the second set of equivalences are only the weak formal equivalences
introduced earlier between the $P^{CPM}_{\mathcal{O}, E}$ and the $M^{CPM}_P(\mu, \mathcal{O}, E, 1)$. Again I
refrain from considering any other cases because so little is known about
them.

I shall illustrate the difficulties in specifying a logical structure for
probabilistic propositions by considering the special subsets of the fore-
going propositions given by the $M^{CPM}_P(\mu_S, \mathcal{O}, E, r)$ and the $M^{CPM}_S(\mu_S, S',$
$r)$ where the μ_S are probability measures such that $\mu_S(S^{6N} - S) = 0$.
This case is worth investigating because of the contrast it makes with
the role played by the corresponding probability measures in QM. Only
the values $r = 0, 1$ are considered. We have that $M^{CPM}_P(\mu_S, \mathcal{O}, E, 1)$ if and
only if $\mathcal{O}^{-1}(E) \supseteq S$ and $M^{CPM}_S(\mu_S, S', 1)$ if and only if $S' \supseteq S$. The smallest
sets E, S' such that $M^{CPM}_P(\mu_S, \mathcal{O}, E, 1)$ and $M^{CPM}_S(\mu_S, S', 1)$ are true then
pick out S exactly. Though for fixed $\mu_S, \mathcal{O}, r = 1$ and varying E or S'
these propositions may again be made formally into boolean algebras, we
can also get a small distance, in this special case, with fixed $\mathcal{O}, E$ and S',
$r = 1$ and varying μ_S. We can obtain, for example, $M^{CPM}_S(\mu_S, S'', 1)$ and
$M^{CPM}_S(\mu_{S'}, S'', 1) \to M^{CPM}_S(\mu_{S \cap S'}, S'', 1)$, $M^{CPM}_S(\mu_{S \cup S'}, S'', 1) \to M^{CPM}_S(\mu_S,$
$S'', 1)$ or $M^{CPM}_S(\mu_{S1}, S'', 1)$, though the reverse arrows do not hold. More-
over, I know of no useful way to define not $- M^{CPM}_S(\mu_S, S', 1)$ in terms of
$M^{CPM}_S(\mu_{\bar{S}}, S', 1)$ for any choice of $\bar{S}$ that could be specified. Thus even in
this special case we are a long way from discovering the logical structure
of the propositions concerned. Similar remarks hold exactly for the
$M^{CPM}_P(\mu_S, \mathcal{O}, E, 1)$ propositions.

Finally we could consider propositions $\mathcal{M}^{CPM}_S(\mu, S, r)$ which concern
probability distributions over $\mathcal{S}^{CPM}_T$ rather than over phase space. Nothing

new can be added about the structure of such propositions, except that
we can write down the following exact correspondence:

$$\mathscr{M}_S^{CPM}(\mu, S, r) \leftrightarrow M_S^{CPM}(\mu_1, t_1, S_1, r) \text{ and } M_S^{CPM}(\mu_2, t_2, S_2, r)$$

$$\text{and} \dots,$$

where the S_i in phase space are calculated from the sets in $\mathscr{S}_T^{CPM}$ and the
probabilities μ_i over phase space are calculated in a similar fashion. (The
temporal sequence of such probabilities is determined by the Louiville
equation $d\mu/dt = \{\mu, H\}$.) These are all the remarks which I shall make
about the propositional structure of CPM-CSM, though I hope that the
variety of propositions and our relative ignorance of the probabilistic
propositions has been made clear.

Summing up, we may represent the structure of CPM-CSM as follows:

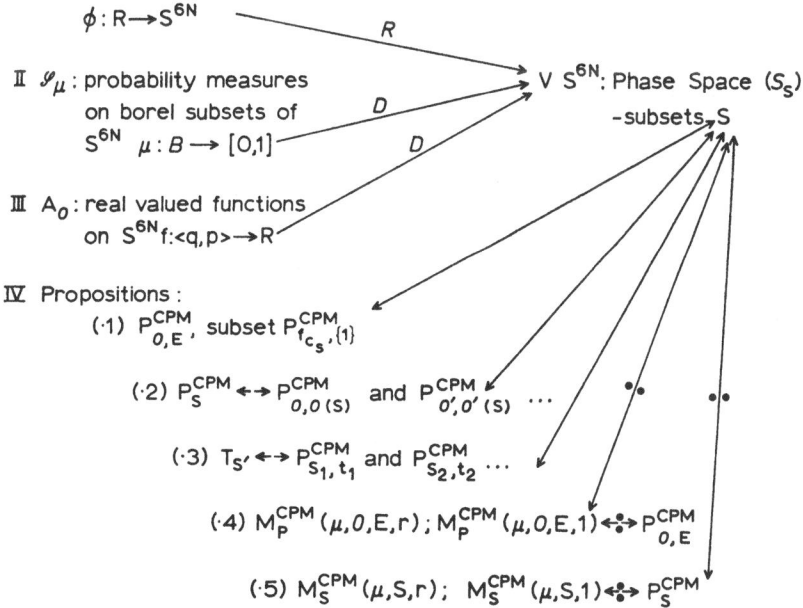

This diagram should sufficiently emphasize the importance of the *set-
theoretic* structure of phase space to CPM-CSM. Only once is any vector
character to phase space introduced and that, significantly enough (cf.

below), has to do with the introduction of the function space $\mathscr{S}_T^{\text{CPM}}$.

A final remark will complete this sketch of CPM-CSM. Phase space is a state space of CPM. For this to work successfully, however, several conditions must be met. In the first place, specification of a precise location in phase space must be sufficient to determine all of the physical properties of the system. This, we have seen, actually happens (a succinct statement of which is that the $P_{\{\langle q, p \rangle\}}^{\text{CPM}}$ determine ultrafilters on the appropriate logic, i.e. on $\mathscr{B}_S^{\text{CPM}}$). Secondly, there must be a certain coherence between the statistical measures on phase space and the values of the observables. I expressed this loosely by saying that the states of CPM were grounding states for CSM. More precisely, we want it to make sense to say that the $P_{\emptyset; \{k\}}^{\text{CPM}}$ propositions, where $\{k\}$ is the singleton set of k, describe precisely individual systems and that probability measures can be interpreted as characterising the distributions of observable values in an ensemble of such precisely describable systems. This will be possible only if (i) the resolutions of the probability measures into extremal probability measures is unique and (ii) these extremal measures are measures concentrated at the points of phase space. The first of these clauses captures what is required for interpreting the probability measures as characterising distributions of values in an ensemble of definite individuals (i.e. individuals each of which has precise values for all its observables). Labelling, after the fashion in QM, ensembles characterised by an extremal probability measure pure ensembles, this clause is equivalent to saying that sub-ensembles of pure ensembles are themselves pure and characterised by the same probability measure. The second clause then affirms that the precise states of the individuals are indeed given by specifying point locations in phase space.[49] Finally, note that CPM is a deterministic theory, i.e. any two systems having the same location in phase space at any one time have the same location in phase space at all other times.[50] This requires that extremal statistical measures dynamically evolve only into extremal statistical measures. It also requires that an ensemble of pairs of interacting systems which commences pure dynamically evolves again into a pure ensemble (i.e. each of the two sub-ensembles consisting of one member each of the interacting pairs starts out as a pure ensemble and finishes as a pure ensemble after the interaction has ceased).[51] Whether these things happen or not is already decided by our choice of the representation of CPM (and of course they do). Talk of a CPM state of a system and the CSM interpretation of probabilities thereby receive their necessary, correlative foundations.

So much then for the formal structure of CPM-CSM; we understand almost all of it rather thoroughly. Now the point of introducing this analysis of CPM-CSM was to say something about the way in which the classical ontology relates to classical theory. Recall then that the ontology of CPM (CSM) consists of atomic individuals. Here I use the generic sense of the term *atomic* – the individuals are regarded as spacio-temporally localized, impenetrable and (relatively) indestructible substances. Complex objects can be analyzed ultimately into collections – it would almost be more illuminating to say *sets* – of non-complex or irreducible atomic objects related spacio-temporally to one another in a particular way. Certainly the configuration and motion of any such complex is specified by giving a *set* of spacial locations and a set of velocities at those locations. The set theoretical structure of phase space at once expresses the spacial localizability and individuality of the basic objects (atoms). Especially is the individuality of the atoms expressed through their relative indestructibility and their impenetrability (they exclude one another from the same spacio-temporal locations). The structural transposition of this metaphysics is the fact that no superposition principal can be operative for phase space or function space. The *boolean* structure of the resulting propositions sets arises directly from, and is a logical transposition of, this set-theoretic structure. Thus I conclude that *the set-theoretic structure of CPM (CSM) is mirrored exactly by the specific kind of individuality of the objects in the classical ontology and the logic of their state descriptions is given exactly by the logic of propositions determined by this set-theoretic structure.* The nature of the atomic ontology is to be explored in some detail shortly (Section VII) and its unique suitability to CPM will I hope become even clearer as the essay proceeds.

V. RATIONALES FOR QUANTUM THEORY

A rationale for quantum theory I shall take to be an attempt to show in a relatively penetrating way how to understand the quantum world (or, semantically ascending, how to understand the quantum theoretic account of the world) in such a fashion that problems, puzzles, difficulties no longer arise and those structures peculiar to the quantum theory find a natural and clearly understood place in the whole. Rationales for quantum theory fall into two classes. Class 1: those rationales which accept quantum theory as it is, as having its own internal completeness and integrity. Class 2: those which regard the quantum theory as an unfinished theory of a certain kind, standing in certain historical traditions and still in need of undergoing further evolution.

Needless to say, virtually all of the 'main current' attempts at understanding quantum theory are of class 1. They all have this much in common: we are asked to adjust to suit some aspect either of the world or of our knowledge of the world in the light of the structures peculiar to quantum theory.

(α) Thus, for example, there are those who would have us adjust our conception of physical reality to suit the peculiarities of quantum theory. At one time or another, for example, Bohm, Heisenberg and Margenau all favoured this approach, suggesting that reality was no longer to be conceived in the classical tradition as being everywhere always mathematically definite, rather, properties were not to be actually possessed by quantum objects until 'forced' to do so by an act of measurement, such properties were 'unrealized', 'latent', 'potential' (see [11, 47, 84]). What is interesting is that all three of these thinkers have now essentially changed their points of view. Bohm turned to the development of 'hidden variable' theories and lately has turned increasingly to the exploration of a much more radical conceptual structure for quantum theory – I think there is significance in this. Heisenberg has turned to concentrate on the development of non-linear 'unified' field theories – I think there is significance in this. Margenau, by contrast, has increasingly emphasized what was part of his original view of quantum theory – its nature as a statistical theory – and increasingly de-emphasized any discussion of a corresponding nature to physical reality – I see significance in this too. For this latent-property approach to quantum theory must inevitably prove less than satisfying. *Any theory whatever*, so far as I can see, could have its problems 'solved' by this approach – simply because the concept of an 'actualization' of a 'potential' is so vague and intrinsically not open to direct investigation of its structure. Margenau and his followers have in reaction turned increasingly Positivist, retaining only the experimental results and the theoretical machinery which predicts the statistics of those results in what they count as the significant component of a theory – or so the recent writings of his group strike me.[52] Bohm and Heisenberg, however, have reacted to this dissatisfying position by attempting to produce improved theories which if successful, would return us to a conception of reality which was everywhere definite and everywhere open to clear-cut investigation, while Heisenberg has turned away from particle theories altogether. But, as will become clear in the sequel, I see a much greater significance in the *particular* directions to which their efforts have been recently turned.

(β) Then there are those who would have us adjust the epistemology of quantum theory to suit. The view was expounded very clearly by Schrödinger[53], who was expressing (though not necessarily agreeing with!) the general views of the day. In its extreme form it involves renouncing Realism entirely, moving to an essentially Positivist-Phenomenalist view on which the only reality are measurement results. To this view belongs the 'disturbance of the system upon measurement' doctrine originated by Heisenberg and which was intended to account for the probabilistic features of quantum theory by representing them as an expression of ignorance. Von Neumann added to this stream of thought with his (mathematically beautifully clear but conceptually peculiar) discussion of measurement, in particular of the so-called 'reduction of the wave packet'. This discussion has lead to the view that the consciousness of 'the observer' plays a peculiar role in the theory and this has in turn supported the doctrine that the ψ function is fundamentally a representation of our ('the observers') state of knowledge (though it may also be used to arrive at a doctrine of the class α above in which the conscious act of observing effects a unique, substantial change in the state of the observed system – cf. Wigner's article in this volume). The 'disturbance' doctrine is known to be inadequate so long as some broad physical principles, such as energy conservation, are retained.[54] The general epistemological approach, however, need not thereby be undermined. Indeed, it is so 'powerful' that it could never be decisively undermined, for once again any theory whatever can be cast in this light – even Realists about theories[55] concede that, whatever else a theory does, it tells us what we know of the systems in its domain and if one wishes to divorce the representation of knowledge of the world from a true representation of the world *severely enough* then likely any theory could be counted among the former group.

There are also positions half-way between types α and β – those who would retain a pure particle ontology as a quantum ontology and regard the 'wave aspects' of quantum theory as describing only our knowledge (or rather lack of it) of the particle states. Such seems to have been Popper's position (see [101], though his latest pronouncements are problematic in this regard, cf. [102, 103] and Bub's article in this volume).

(γ) Finally there are those who argue for a logical and/or conceptual quantum 'revolution'. In this approach we are to accept a new logical and semantical structure to language such that within this new structure quantum theory is shown to be a clear, non-problematic theory. Such, at heart, was Bohr's approach – he argued for a particular semantical analysis of physical descriptive language which, if accepted, would have had the effect that quantum mechanical descriptions were logically consistent, exhaustive and complete. (However reality was for Bohr very much like a Kantian thing-in-itself.)[56] Such is also the approach of the quantum logicians. These people would also have us accept that the logical structure of quantum theory is non-standard (non-boolean) and once again that quantum theory stated in this new structure is consistent, exhaustive and complete. (But whereas Bohr was essentially Kantian in his conception of the relations between human thought and perception on the one side and the world on the other, the essence of the quantum logic approach is to reflect the non-standard logical structure of quantum theory directly out into the world – to argue that the world itself has this non-standard logical structure – in a thoroughly Realist fashion.)[57] Although such new semantical-logical insights as these theories propose are exciting discoveries (*in any case* – i.e. whether or not we actually adopt this approach to QM) and though there seems to be some ability to argue clearly about the validity of such insights, these approaches, like those preceding them, still stand desperately in need of criteria for their acceptability – they also have the alarming property of apparently being able to remove the problems in any physical theory whatever. (Cf. also my remarks on the resulting ontology in Section XI.)

The only well-known approach within class 2 is that of the 'hidden variable' theorists. In an important sense this approach also wishes to accept the present structure of quantum theory as is – but it seeks to show that that structure may be 'supplemented' in such a fashion that we are returned to an essentially classical conception of physical reality and an essentially classical statistical theory. There are two reasons why the approach is not 'main current', why it has not attracted a large following – though the debate has been vigorous and wide spread and there is a substantial literature on the subject.[58] The trivial reason is simply that most physicists have, largely through unphilosophic non-reflection, simply not been interested in these alternatives. The non-trivial reason is because there exist very powerful theorems to the effect that interesting varieties of hidden variable theories would be incompatible with the accepted structure of existing quantum theory. (For a clear exposition see the article by Bub in this volume.) I shall not pursue this line of approach either.

Let us be clear about what is happening in each of these cases. There are three major components to the structure of a physical theory: the mathematical structure, the conceptual structure and the ontology type.[59] The first of these has already been analysed in some detail. By the second and third categories I mean the set of fundamental descriptive concepts of a theory (e.g. space, time, mass, etc.) and the *kind* of things which are taken to exist (e.g. atoms and a void, or fields and no void etc.). The importance of these latter two categories will become clearer in the sequel. For the moment I wish only to assert that every good scientific theory exhibits a coherence among these. How deep this coherence runs we will see in what follows (though even here the investigation is barely begun), what it is important to appreciate here is that all of these views of QM seek to recapture the coherence among these three components generally felt to be exhibited by CPM and CFT: The doctrine α by eschewing all talk of fields, waves, interferences etc. and then by tailoring the ontology to suit the remaining essentially particle language (i.e. conceptual structure) and the probabilistic interpretation of the mathematical structure in its terms, and so on down to Bohr (γ) who tried to alter the conceptual structure so as to admit (in a Kantian sense) the familiar mixed classical ontology of particles and fields once again consistently with the mathematical structure, and the quantum logicians who would retain only the particle concepts but alter the logical relations among them so as to accommodate the mathematical structure – with what consequences for ontology and with what success we shall later see.

There are, of course, other views on QM. For example, there are various non-linear theories that seek to approximate QM under certain conditions – hence an interpretation of QM is to be developed as an approximation to a true account of Reality.[60] There are similar approaches based upon even more radical departures from QM. I shall be mentioning some of of these later on. What they all have in common is that they propose to discard QM, not to interpret it. Since my own view lies somewhat in this direction (though not of its essence) I shall not attempt to discuss this potentially infinite class of alternatives at this point, rather discussion is postponed until presentation of my own view can colour it (!).

Finally, there is my own view. So that the reader's patience and understanding be tried no further I shall devote the next section to outlining it (crudely) before proceeding with the analysis.

VI. SUMMARY OF DOCTRINE AND PROGRAMMATIC INTERLUDE (II)

I shall first propose three theses to the reader and argue for them.

(1)   The ontologies appropriate to particle and field theories are radically different in character.

(Sub-Thesis: The entire history of Western Science has been dominated by the exploration of, and opposition between, these two onto- logies.)

(2)   The conceptual structures of physical description appropriate to particle and field theories are correlatively radically diffe- rent.

(3)   The mathematical structure of CFT is significantly different from that of CPM.

(Sub-Thesis: Each kind of theory (particle, field) will in general share a mathematical structure common to theories of that kind and the distinctive to that kind of theory.)

The argument for these theses will support a fourth:

(4)   There is a profound and initimate connection between the mathematical structure of a theory, its conceptual structure for physical description and its basic ontology.

The argument for Theses (1), (2), (3), will be conducted respectively in Sections VII, VIII, X. Section IX is devoted to an analysis of the mathematical structure of CFT; since the similar analysis of CPM (CSM) has already been completed, the ground work is then laid for the comparative study of Section X. If I am correct about the depth of the disparities between particle and field theories then an understanding of their differences would be crucial to an understanding of the foundations of physics. For I need hardly argue the first, historical sub-thesis, it suffices to point out that the ancestors of the modern terms *particle, field* are the ancient terms *atom* (generic sense), *plenum*. These latter terms, and theories formulated in terms of them, have a venerable history in the scientific tradition – one thinks of the great pre-Socratic debates (about the reality of change) with Democritus on one side (the particle side) and Parmenides on the other,[61] of Descartes and Huygens versus Newton, of Dalton versus Faraday, Maxwell, of Einstein (relativity, unified field theory) and ...QM? To the extent that we do not understand clearly the fundamental differrences between these two kinds of theories, to just that extent we do not understand at all deeply the character and evolution of Western science.[62]

In Sections VII, VIII, X (+IV, IX) I explore the fundamental characters of these two theories. It will become clear in the course of the analysis that we might expect any two theories of these respective types ot differ in the same characteristic ways, provided we are dealing with essentially classical versions of the two ontologies.

But all of this constitutes background from which to tackle QM; what of it? Again I propose and argue for a thesis:

(5)   QM is a fusion of the characteristic mathematical structures of a field theory and a particle theory.

(Sub-Thesis: The overall abstract structure of QM is that of a field theory.

Sub-Thesis: The dynamical structure of QM is that of a particle theory.)

The two sub-theses give the detailed content to the thesis, although they are a little too black-and-white. These theses are based upon an analysis of the mathematical structure of QM carried out in Section XI.

Now I began this paper by posing the question "What is the basic ontology of QM?". I did not pose the correlative linguistic question "What is the appropriate conceptual structure for quantum descriptions?" because of the danger of being directed into the traditional discussions of the meaningfulness of such concepts as position and momentum. We can now see, Thesis (4), that the original question leads directly to the question "What is the basic mathematical structure of QM?" and if Thesis (5) is correct as an answer to this question we can see how dangerous to insight it would be to be diverted by the linguistic question into traditional discussions, for though by Thesis (4) there is a linguistic correlate to Thesis (5), Thesis (5) strongly suggests

(6)   There is no strictly coherent conceptual structure for QM couched in terms of the traditional conceptual structures for particle and field theories, neither is there any coherent ontology for QM specifiable in terms of the classical particle and field ontologies

as an answer to the linguistic and ontological questions. This answer is developed in Section XII.

This answer would be trite if all that was intended was that QM was neither CPM nor CFT but a new theory. This is not all that Thesis (6) asserts. What is asserted is that the interpretational difficulties arise in QM out of a deep-seated attempt to fuse the mathematical forms of CPM and CFT. It follows that the approach to an understanding of QM must be through an understanding of the structures of CPM and CFT and of the way they interact in QM – only then are we likely to gain some significant idea of what to say about QM.

Actually my own (tentative) view is that the fusion is such that it is better to say nothing, declare QM a bastard offspring and pass on to consider more viable alternatives. These I discuss in Section XIII. A premise of this "Einsteinian" attitude is that we have, basically, only the classical ontologies to work with, only these are clearly definable. This, of course, may not really be true – for example, the ideas discussed in Section V, α might possibly be formulated more precisely as an alternative conception of ontology. Moreover, Thesis (5) naturally favours field theories as the superseding view to QM; moreover, field theories do display a discreteness of the right structural sort (those generated through periodicity because of boundary conditions). But, though I do not believe we understand it at all deeply, there is a prima facie case for quantized field theories and quantization is not an element of either classical conception.

VII. ATOM AND PLENUM I: METAPHYSICS

The basic ontologies of the two great theoretical traditions in western science are radically different. This difference is not perhaps more apparent to everyone only because elements of both views are built into ordinary language. In any event it is remarkable that relatively so little attention seems to have been paid to the importance of the contrast – for they differ in ways of direct relevance to contemporary science. In this section I shall summarize very briefly the salient features of the two ontologies – the reader is directed to [61] for the detailed analysis and for many features of interest not mentioned here.

In the atomic schema the fundamental individuals are the atoms (here I use the generic sense of the term), a plurality of relatively indestructible substances each localized in space. All other individuals are reducible to spatio-temporal structures of these fundamental individuals (I shall refer to them as derivative individuals).

Likewise, there are fundamental and derivative properties. The intrinsic properties of the fundamental individuals are the fundamental properties and these are atomic or quantized (i.e., there is only a finite number of values for their magnitudes) just in case there are only a finite number of kinds of fundamental individuals. All other properties (i.e. all properties of derivative individuals) are reducible to spatio-temporal structures of these fundamental properties.

All change is reducible to spatio-temporal rearrangements of fundamental individuals. All laws (in particular, the laws of change) are reducible to fundamental laws concerning spatio-temporal relations among fundamental individuals, and the fundamental properties.

The situation with the plenum schema is in maximal contrast to this. There is only one fundamental individual, the plenum, and it is present everywhere. All individuals consist of spatio-temporal distributions of properties. There is no fundamental/derivative contrast inherent in the schema. (Scientific experience may, however, lead us to introduce a distinction between some complexes of properties as being more 'fundamental' than others – just in the sense determined by their respective roles in theories.)

Similarly, there is no intrinsic distinction between fundamental and reducible properties, all properties are, prior to actual scientific experience, on an equal ontological footing with one another.

All change is basically qualitative change, i.e. change in properties. Thus even motion is to be viewed as the systematic disappearance and re-instantiation of clusters of properties in such a fashion that the cluster sequence forms a spatio-temporal trajectory. The laws of nature will concern fundamentally the relations amongst properties and property-complexes and again, prior to any roles which they might play in actual scientific theories, all such relations are on an equal footing.

There is, then, the deepest possible contrast between the two kinds of schemes, between the two kinds of metaphysics. The foregoing features are the really fundamental ones from which everything stems, though their brief statement perhaps belies their importance and the profound differences between them. One could go on developing the contrast between them at considerable length, but I shall develop only three more features of it which are pertinent to the discussion of quantum theory.

The first of these has to do with the sense of globalness or wholeness in the two schemas. The reduction principles of the atomic schema are founded on the fundamental localizability of the atoms and this in turn leads to local laws, in the sense that the behaviour of a collection or complex of atomic individuals is built up from the actions of each individual separately on the others. (Of course these actions are, in actual physics, usually viewed as local in the sense of being mediated by a *field*, but this is just to introduce the alternative ontology, if we confine ourselves to the atomic schema there can only be direct actions of one individual on another – whether by action at a distance or by contact). This means that the state of any physical system can be analysed into its constituent substances and the physical interactions among them, changes in state are determined then entirely by the properties of the individuals, in particular the individual positions and velocities (for the relations between them are then fully determined by these properties and the intrinsic properties of the individuals). In this fashion one produces a description of a physical system just when one has described the local behaviour of each fundamental individual in it.

By contrast there is no such reduction principle operative in the plenum schema. There is no natural object whose state can be specified except the entire plenum itself. The state of the plenum is specified at an instant of time just when the spatial distribution of property values is everywhere specified at that instant. Changes in the plenum (even the simulated motion of an individual) are specified just by specifying the temporal sequence of such global plenum states. And indeed in field theories we always obtain equations involving the spatial derivatives of the field as well as the time derivative, thus requiring spatially global solutions for the field state.[63] These solutions are 'wholes' they cannot, unlike the situation in the atomic schema, be analysed into local parts. (Without this wholeness – which expresses the absence of a grounding in the fundamental ontology for localizable individuals – the superposition principle for fields would not be possible.)

The second feature concerns the notion of continuity and discontinuity. We already have the following results: there is a fundamental element of of discontinuity in the atomic schema, namely the substantial discontinuity introduced because the fundamental individuals are spatially bounded substances and the fundamental property discontinuity introduced because each kind of fundamental individual has only one value for each of its intrinsic properties. Were space and time themselves quantized there would be no element of continuity whatever in the atomic schema (for then all change would likewise be discontinuous and so would every reducible property). In fact the assumed continuity of space and time in classical science provides the only entrance of continuity into the atomic schema (and then only at the level of derivative individuals and changes in derivative individuals). We might say that the 'natural', certainly the homogeneous, development of the atomic schema is one in which space and time are quantized and discontinuity reigns supreme.

In contrast, the 'natural', certainly the homogeneous, situation with the plenum schema is one of total continuity. A detailed examination leads to arguments with the following conclusion: if space, time are continua, then physical states are well defined if and only if every change (whether in properties or in spatiotemporal location) is also continuous.[64]

On the other hand, so long as we consider only essentially static (steady state) situations the plenum schema is capable of displaying a form of discreteness properly so-called. This arises from periodicities in the field introduced by boundary conditions. For example, standing waves in a rigid box can have only certain, pure discrete wavelength values. These discretenesses are closely related to those arising in QM (though I think the relation is not well understood). They are to be distinguished from discretenesses not properly so-called that arise in non-linear field theories where one attempts to ape the behaviour of particles via singularities. Moreover, the whole connection of such discretenesses with the structure of space-time (cf. above and below) seems not at all well understood yet. (These matters are also discussed further in Section XIII below.)

For the situation is complicated by the following important fact: if we do introduce a quantized or discontinuous space and time to the plenum schema (combinations of a continuous space with a discontinuous time or *vice-versa* turn out not to be very attractive) then we do introduce a fundamental element of discontinuity into that schema. If space is quantized then the spatial distribution of properties is also quantized (though not of course the values of those properties). Similarly, a quantized space and time leads to discrete or quantized changes. In this case, *and only in this case*, the characteristics of the plenum schema approach somewhat those of the atomic schema. Since the 'marriage' of the these two schema lies at the heart of quantum theory, these considerations certainly suggest that under the conditions of a quantized space and time may be found a way to erect a conceptually coherent quantum structure in which all states are precisely defined and all processes well understood.

Thirdly, there is the question of the criteria of identity in the two schemas. We can begin to see the fundamental contrasts between the atomic and plenum schemas over this issue by considering the question of the possible relations among individuals in the two schemas.

In the atomic schema the possible relations among fundamental individuals is clear cut – it is one of total exclusion. In fact fundamental individuals obey the following exclusion principle: E.P.: no two fundamental individuals can occupy the same spatio-temporal location. Only of derivative individuals do we allow interpenetration (and only then through interpenetration of their respective spatio-temporal structures, not of their substantial components[65]).

The situation is quite different in the plenum case. There is nothing intrinsic to the idea of a quality complex which prevents the mutual interpenetration, or superposition, of such quality complexes. To have this happen would be to have no more than the following happen: systematic changes in the pattern of quality instantiations occur which are such that two quality complexes approach one another from differing spatial locations and merge into one another (perhaps later re-emerging as two, perhaps not). Such merging is a commonplace among such entities as, e.g., waves in a fluid, and energy or charge contours etc. in fields (gravitational, electromagnetic, nuclear) associated with several different sources – but these are plenum phenomena. Again, of course, we shall discover that some quality complexes do not superpose (namely, just those referring to the usual physical individuals) but there is nothing inherent in the notion of such complexes which prevents their superposition, it lies only in the physical relations which experience teaches us hold among them. Now the concept of superposition, for that matter the concept of being wave-like, is central to the peculiarities of the quantum mechanical treatment of matter.

It behoves us, therefore, to look at the fundamental principles of identity within the two schemas. Within the atomic schema, the fundamental identities are those of its fundamental individuals and these are determined just by the spatio-temporal trajectories of those individuals. (Qualitative likeness and difference is irrelevant here since all fundamental individuals of the same type have exactly the same qualities.) Fundamental individuals obey the following identity principle:

I_1:   A continuous spatio-temporal trajectory constituting the successive locations of a substance also constitutes the identity of the individual whose trajectory it is.

In addition, there is the question of the identities of the derivative individuals. Of these we may make two distinct judgements: those concerning substantial identity and those concerning morphic identity (i.e., identity of an individual *qua* structure of a certain sort). Clearly questions concerning substantial identity are reducible to those concerning the identity of the fundamental individuals constituting the derivative individual in question. But on occasion we judge an individual to be the same (derivative) individual just in case it continues to possess the same structure (e.g., cathedrals). In this case it is the continuous spatio-temporal trajectory of a spatio-temporal structure which constitutes the identity of the derivative individual in question. In all cases we can see that questions of identity are effectively reducible entirely to spatio-temporal considerations.

By contrast, the fundamental identity in the plenum schema is that of qualitative sameness or difference. All other identities are reducible to these or spatio-temporal terms. Qualitative identity is in maximal contrast to substantive identity since the same quality may be instantiated at many

different spatial locations at a single time whereas this is not possible for
the fundamental individuals of the atomic schema. The identities of all
individuals, i.e., of all spatio-temporal quality complexes, in the plenum
schema has two aspects, that pertaining to the sameness or difference of
qualities which are instantiated in the quality complexes and that pertaining to the spatio-temporal distributions of the qualities in the complexes.
All quality complexes obey the following identity schema:

$I_2(1)$:   Two quality complexes are identical (are instances of the same
              quality complex) if and only if their component qualities are,
              pair-wise, instances of the same qualities and their spatio-
              temporal structures are identical.

To construct individuals within this schema, we need to distinguish between quality complexes which are structurally and qualitatively indistinguishable but have distinct locations. Thus we have the following
identity principle for plenum individuals (*p*-individuals):

$I_2(2)$:   A continuous spatio-temporal trajectory for a quality complex
              which is everywhere the same ($I_2(1)$) along the trajectory con-
              stitutes the identity of the *p*-individual whose trajectory it is.

Finally, and deeply related to the preceding, there are distinct algebraic
structures which the two schemas find natural to employ. Lying at the
heart of the atomic schema is a *set-theoretic* structure. There is an important reason for this. Fundamental individuals can be neither created nor
destroyed in the true atomic schema. One counts them, and counts groups
of them, using the ordinary laws of arithmetic. The algebraic structure of
groups of fundamental individuals is therefore that of a power set algebra.
The same applies to derivative individuals, essentially because derivative
individuals are reducible to spatio-temporal complexes of fundamental
individuals.[66] This set-theoretic structure, together with the spatial local-
isability of the fundamental individuals, provides the natural foundation
for the phase-space representation of the state of a collection of fundamental individuals in classical physics.

In the plenum schema, however, there are no such natural basic individuals to form the basis for counting. For property complexes are in general
neither precisely spatio-temporally definable, nor do they preserve their
identity by obeying an exclusion principle – rather they superpose – nor

do they have in general a definite spatio-temporal trajectory to form the basis for re-identification – rather they appear and disappear. All of the necessary conditions for applying an arithmetic schema to them are therefore absent. The schema that does apply is an essentially *vectorial* one – for vectors may be superposed and be transformed into combinations of other vectors. The vectorial schema has distinctive algebraic structures associated with it that differ markedly from the set-theoretic structure at the basis of arithmetic. But these ideas are shortly to be explored.

To summarise: There are differences between the two ontologies that go to the very roots of the two conceptions of the nature of the world. We must expect these differing conceptions to have a profound influence on the characteristics of the theories that embody them. The significance of the set-theoretic character of CPM-CSM we have already seen, the structure forms the entire foundation for the theory – algebraically, logically and in the interpretation of probability. I shall turn shortly to consider CFT, but before that some remarks on the correlative conceptual structures for physical description are in order.

VIII. ATOM AND PLENUM II: CONCEPTUAL STRUCTURE

It will not surprise anyone that where our conception of Reality differs descriptive conceptual structure follows suit – this much is conveyed by speaking our *conception* of Reality. It does not always happen, however, that where the nature of Reality differs our conceptions follow suit – whether that happens depends upon our adaptability, perceptiveness and imagination. For example, it is quite striking that whereas we have a well developed conceptual apparatus for use with an atomic conception of nature we have only a poor, fragmentary conceptual apparatus for offering physical descriptions from within the plenum schema. Psychological explanations could doubtless be developed – for survival and social purposes the macro world is most easily perceived in terms of localised individuals in motion, this idea is reinforced by scientific successes with such models etc. – but they are not my concern. Rather do I wish to point out that there are two such structures, differing radically from another, and intimately related to the two ontologies.

The power and ubiquity of atomic schema concepts need hardly be emphasised. The last section (Section VII) was full of them, they were even pressed into service (usually negatively) in characterising the plenum ontology! The very subject/object grammatical structure of our natural languages attest to the fundamental permeation of our conceptual patterns by the atomic conceptual schema. To this schema belongs concepts such as *individual, interaction* (=action between), *cause, collection* (= group, complex), *counting* and so on. The structure of the descriptive conceptual scheme is most economically stated here by pointing to the analysis of CPM-CSM in Section IV. (Abstract principles could be generated but this would be a lengthy business not to the point here, and the ultimate precision and usefulness of such an enterprise is still in doubt in my mind.) State descriptions are given by specifying catalogues of properties, one predicated of each of a number of individuals. There is a logical structure to such descriptions which determines how properties relate to one another, when a description is exhaustive (complete) and so on.

The plenum schema shares with the atomic schema the categories of predication, space and time. Thus one speaks of properies, locations and times and logical constructions out of these (spatio-temporal property distributions, gradients of these etc.). Even so, the uses to which the two schemas put these concepts is quite different. Thus in the atomic schema we speak of *velocity* and *acceleration* (at a spatio-temporal point), of sets of *locations*, velocities etc., of *distances, torques* etc. Whereas the key notions in the plenum schema are continuous distribution properties: *momentum* and *energy density, energy-momentum* and *inertia tensors, currents*, and various *volume integrals* of these quantities.

We have, however, only a fragment of the distinctive part of the plenum schema. From physics one thinks of such concepts as *wave, field* (in the sense of continuous distribution of property values), *superposition, interference*. Important though they are (especially that of superposition) these concepts still do not bite deep enough. For they still operate within an 'atomic' subject/object structure pressed into service for use in this strictly incompatible context. We need concepts that reflect the basic global, non-individual character of plenum states. And there are some clues as to what the useful concepts might be.

Consider the case of a fluid surface with waves upon it. It is futile (pragmatic purposes aside) to attempt to regard waves as individuals. It is almost equally futile even to regard them as processes, where there is still the 'atomic' connotation of individual processes, for the fluid surface is a whole whose state (property value distributions) evolves coherently. Nonetheless the idea of a *process* clearly has a place in this conceptual scheme – it seems to be in fact the counterpart to that of the individual in the atomic scheme for the purposes of analysis. Thus just as complex states are conceived as being analyzable into relations among basic individuals, so complex processes may be analyzed into related simpler processes, and if the evolving field state is regarded as a complex process then it may be analyzable into related simpler processes. (This may be the abstract description, for example, of fourier analysis. Note that the possibility of fourier analysis is related intimately to the superposition principle.) There is also a question of a 'pure event' language which has been investigated by several philosophers (though not with the specific goals and sensitivities operative here).[67] Certainly this language has the possibility of offering a description which is neutral between the two, though whether it can be adapted to do positive justice to the characteristics of the plenum schema is an open question. The appropriate plenum discourse is still under investigation. (A student of mine, Mr. P. English, is investigating conceptual structures that do not presuppose a base of localizable individuals but the work is in too early a stage to report on further here.)

We already have convincing evidence, I believe, for the view that the descriptive conceptual schemes appropriate to the atomic and plenum ontologies are as radically distinct as are the two ontologies. The impact this thesis has for understanding the significance of the 'informal' language of physical description attached to a physical theory will not be lost on the reader. Science (with some scientists excepted) has paid relatively little attention to this language and almost universally failed to appreciate its deep, if subtle, connection to the formulation of physical theory. Some notable exceptions are Mach, Einstein and Bohr, and more recently Bohm has returned to emphasize the point.[68] (To take an example Bohm cites: The notion of *signal* is central to the informal exposition of relativity and yet can be given no precise meaning within a quantum theory – since it requires a simultaneously precise velocity and trajectory. Surely this 'informal' fact is of importance in understanding why relativity theory and quantum theory are not readily joined.)

The relevance of these conceptual differences to the interpretation of QM need hardly be emphasized. The most common textbook informal language has the form of a bewildering, not-quite-systematic sliding from particle to plenum concepts (particles and waves) as situations seem to demand. But before anything more is said about QM we must gain some idea of the alternative mathematical structure.

IX. THE STRUCTURE OF FIELD THEORIES

We have now good reason to think that the field-plenum view of the world is radically different from the atomic-particle view, and different in ways of relevance to quantum theory. It behooves us therefore to investigate the mathematical structure of field theories, both for its own sake and for the light it should throw upon QM.

It seems obvious to begin the investigation of CFT by examining the phase space, or at least the state space of field theory. It is surprising to discover, then, that this kind of formulation is noticeably absent from the standard presentation of CFT.[69] This discovery calls for a more cautious approach. I shall commence by following the standard presentations and attempt to abstract what I need from them – the reader should be warned, however, that many of the questions to be asked are simply unanswered (at least to the knowledge of the author) or in some cases the entire approach undeveloped, so that I shall not succeed in obtaining here even as complete a characterization of CFT as was possible for CPM.

It is customary to begin the exposition by emphasizing the closeness of the *dynamics* of CFT to that of CPM. In CPM one derives the dynamic equations from an extremal principal and an associated N-particle Lagrangian[70]:

$$\delta \int_t^{t'} L \, dt = 0$$

$$L = L(q_1, \dots q_{3N}, \quad \dot{q}_1, \dots \dot{q}_{3N})$$

with the resulting Euler-Lagrange equations

$$\frac{\delta L}{\delta q_i} - \frac{d}{dt}\left(\frac{\delta L}{\delta \dot{q}_i}\right) = 0, \quad i = 1, 2 \dots 3N. \tag{D_1}$$

In CFT we begin with the extremal principal

$$\delta \int_{t}^{t'} \left(\int_{V} \mathscr{L} dx \right) dt = 0$$

where $dx = dx_1 dx_2 dx_3$, V is a spacial volume, $\mathscr{L}$ the Langrangian density,

$$\mathscr{L} = \mathscr{L}(\phi_1, \nabla\phi_1, \dot{\phi}_1; \phi_2, \nabla\phi_2, \dot{\phi}_2; \ldots)$$

$$\nabla\phi = \left(\frac{\partial\phi}{\partial x_1}, \frac{\partial\phi}{\partial x_2}, \frac{\partial\phi}{\partial x_3} \right)$$

to obtain as the corresponding Euler-Lagrange equations

$$\frac{\partial\mathscr{L}}{\partial\phi_i} - \sum_{k=1}^{3} \frac{d}{dx_k} \left(\frac{\partial\mathscr{L}}{\partial\left(\frac{\partial\phi_i}{\partial x_k}\right)} \right) - \frac{d}{dt}\left(\frac{\partial\mathscr{L}}{\partial\dot{\phi}_i}\right) = 0, \quad i = 1, 2, \ldots \text{(D}_2\text{)}$$

This equation bears a clear resemblance to (D$_1$) with each of the fields ϕ_i playing the role of the q_i. But let us examine the comparison a little further.

Equation (D$_1$) is a *total* differential equation for q_i as a *function of t alone*. Equation (D$_2$) is a *partial* differential equation for ϕ_i *as a function of t and the x_k*. This difference is not accidental or minor, it has to do with the fact that in the CPM formalism fields are viewed as systems with infinite degrees of freedom – such systems are associated in a deep going way with partial, rather than total, differential equations. This point of view on the mechanical nature of fields can be developed as follows. Suppose space to be broken up into cells, labelled by an index s. The value of the field ϕ in the cell s, $\phi(s)$, is then chosen as the generalized coordinate q_s and the Lagrangian $L = \sum_s \delta x_s L(s)$, where $L(s)$ is the value of the Lagrangian in the cell s of volume δx_s and can be written as a function of the q_s, $L(s) = L(q_{s_1}, q_{s_2}, \ldots, \dot{q}_{s_1}, \dot{q}_{s_2}, \ldots)$ (since the spacial derivatives of ϕ appear here as differences). The $\phi(s) = q_s$ are now solely functions of time. The extremal principal then has the same form as in CPM. When the transition is made to the limit of cells of infinitesimally small volume the field formalism is regained. But in this limiting case we clearly need infinitely many q_s's, one for each of the non-denumerable infinity of spacial points, so that the field system corresponds to a mechanical system with infinite degrees of freedom (with infinitely many components, one for each spacial location).[71] Note that in this development we have infinitely many generalized coordinates for each field. The equations (D_2) are regarded as an infinite system of equations for each field separately, namely an equation for each spacial location. Here the fundamental *globalness* of the fields is made manifest. In sharp contrast to CPM, here the notion of a field at a point location is, strictly speaking, logically untenable – at most only fields over indefinitely small, but finite, volumes are well defined. In all actual physical applications we deal with fields defined over the entire spacial universe under consideration.

We have now two distinct analogies to CPM dynamics, the one expressed by $q_i \leftrightarrow \phi(s_i)$ and the other by $q_i \leftrightarrow \Phi_i(x)$. The former analogy is the traditional one to emphasize – it hinges on a specific (and not strictly exactly accurate) *dynamical* analogy. But in view of the *logical* role played by the Φ_i as the determiners of single field states (the field is global) it is the latter analogy which I shall emphasize – it is a *structural-logical* analogy.[72]

The formal or abstract dynamical analogy with CPM can be extended in the standard classical fashion to the Hamiltonian formalism under suitable conditions.[70] The canonical momenta are defined by $\pi_i = \partial \mathscr{L}/\partial \dot{\phi}_i$ and the Hamiltonian density $\mathscr{H} = \sum \pi_i \dot{\phi}_i - \mathscr{L}$. In analogous fashion to CPM, we obtain the Hamiltonian form of the equations of motion

$$\dot{\phi}_i = \frac{\delta H}{\delta \pi_i}, \quad \dot{\pi}_i = -\frac{\delta H}{\delta \phi_i}, \quad H = \int_V \mathscr{H} \, dx.$$

In this context the Poisson formalism may be introduced with $\{\Psi, \chi\}$ understood as

$$\{\Psi, \chi\}_{\phi_i, \pi_i} = \int \int \sum_i \left(\frac{\delta \Psi}{\delta \phi_i} \cdot \frac{\delta \chi}{\delta \pi_i} - \frac{\delta \chi}{\delta \phi_i} \cdot \frac{\delta \Psi}{\delta \pi_i} \right) dx \, dt$$

where $\delta \psi / \delta \phi$ is the functional differentiation of ψ with respect to ϕ.[73] We then obtain the dynamical equation of evolution for observables as

$$\frac{d\mathcal{O}}{dt} = \{\mathcal{O}, H\}_{\phi_i, \pi_i}.$$

Most of the other correspondences carry across, in particular the development of Hamilton's characteristic equation and characteristic function. Moreover the reader is reminded that the reciprocal indeterminacy relations holding between precise definitions of canonically conjugate quantities analysed for CPM will also continue to apply here, in particular the ϕ, π cannot, strictly speaking, be simultaneously precisely specified. Hamiltonian formalism also forms the basis for the canonical quantization of CFT, in analogy with the quantization of CPM.[74]

Note that in (D_2) the space and time *parameters* play similar roles, and roles different from the ϕ_i, whereas in (D_1) only the time parameter had this peculiar function. In (D_2) the x_k, t are not themselves subject to a dynamical equation of evolution, in (D_1) the q_i, which also refer (in general) to spacial location, were subject to dynamical evolution and only t was an independent variable. Moreover the symmetry between the x_k and t in (D_2) suggests the possibility of a covariant relativistic formulation of (D_2) in a fashion not readily apparent for (D_1), namely

$$+ \frac{\partial \mathscr{L}}{\partial \phi_i} - \sum_{\mu=0}^{3} \frac{\mathrm{d}}{\mathrm{d}x_\mu} \left(\frac{\partial \mathscr{L}}{\partial \left(\frac{\partial \phi_i}{\partial x_\mu} \right)} \right) = 0 \quad (x_0 \quad ct)$$

and will be relativisticly invariant provided $\mathscr{L}$ and the ϕ_i are world scalars. Thus CFT allows a straightforward relativistic formulation in a fashion not accessible to CPM.[75] Clearly the functions ϕ_i which are the solutions to the various field equations arising from particular choices of Lagrangian form the elements of the corresponding function spaces of CFT. The free field equations of physics are all linear. (These are the real and complex scalar and vector fields and the Dirac or spinor field. The real vector field contains the physically important case of the electromagnetic field.) In this case the corresponding function spaces are linear vector spaces.[76] Indeed, it is always possible to introduce a suitable scalar product and norm such that these free field function spaces become real or complex Hilbert spaces.[77] In the interacting field cases the dynamical equations typically become non-linear. The function sets will not then form linear spaces. Unfortunately, once again relatively little seems known about the structure, and very often about even the membership, of these function sets. Whether any of them could be understood as forming non-linear spaces is unknown to me at this time.[78] I shall give the function spaces of CFT the symbol $\mathscr{S}_T^{\text{CFT}}$ and those particular spaces which are linear I shall denote by $_L\mathscr{S}_T^{\text{CFT}}$.

Since the functions ϕ_i are taken to be the fields, the spaces $\mathscr{S}_T^{\text{CFT}}$ are the natural *state spaces* of CFT. Thus, following the terminology of Section III, $\mathscr{S}_S^{\text{CFT}} = \mathscr{S}_T^{\text{CFT}}$. Specifying a particular field ϕ fixes the value of the field intensity (for example the electromagnetic four-potential) at each space-time location, thereby specifying the field state at all times and places.

Theories of the linear CFFT sort then obey a *Principle of Superposition*, for if ϕ_1, ϕ_2 represent two permissible states of the field then so also does $\alpha_1\phi_1 + \alpha_2\phi_2$, α_1, $\alpha_2 \in F$ (the real or complex number field). But this principal has a converse, the *Principle of Resolution Ambiguity*, for if ϕ is the actual state of the field in given circumstances then equally well can the state be regarded as the sum of any set of fields ϕ_i such that $\sum_i \alpha_i \phi_i = \phi$. In a general there will be infinitely many such possible resolutions of a field. (Any situation involving linear vector spaces supports similar principles – in particular, the various vector quantities of CPM.) Since fields are taken realistically in CFT we can raise the question of which fields are really present in any given situation. From a physical point of view this question has no clear answer. *The reason for this has to do with the fundamental kind of ontology supported by field theories (cf. Section VII), only the resulting structure can be viewed as physically significant because of the basic construction of physical features as property complexes.* There is no physical basis for viewing any of the infinitely possible component fields of a field as having a preferred status. (Notice that similar questions do *not* arise in CPM for the CPM *states* – cf. Section IV – they can arise only for some other vectorial quantities characterizing states in CPM, such as forces, accelerations, and so on. In these latter cases the resolution question is again in some doubt, though on most philosophical positions, one would be committed, I believe, to holding that the vector resolution was factually arbitrary or undecided. I have argued elsewhere, however, that we can view CPM so that each vectorial quantity is uniquely resolvable into real (actually existing) components – but there are still some subtle questions surrounding this issue which by and large have gone ignored.[79])

Now let us turn our attention to the observables of CFT. It is not obvious, ab initio, how to go about defining observables at all for CFT. Fields are continuous entities over space-time so that one has no localized objects such as particles with well known properties from which to commence. What we do have is a dynamical analogy with CPM, physical intuition and an important theorem due to Noether. Noether's theorem states that to every continuous transformation of the space-time variables such that $\delta \int_{t}^{t'} (\int_{V} \mathscr{L} dx) dt = 0$ is preserved and for which we are given the transformations of the fields, there corresponds a function of the fields which represents a conserved quantity. A closely similar theorem applies in CPM from which the present one draws its inspiration and in CPM the key conserved quantities are energy, linear momentum and angular momentum corresponding to invariance of the action under time displacements, spacial translations and spacial rotations respectively. Since these are the basic coordinate transformations they are also the first applied in CFT. In addition, we know in the case of the electromagnetic field that certain other functions of the field are also observable (we know this because, for example, we build up the field Lagrangian so as to yield Maxwell's equations whose fields are observable) and that the field equations are invariant under certain transformations called gauge transformations. Thus we apply also this transformation and interpret the corresponding invariant as the total charge. This essentially exhausts the linear transformations at our disposal and hence this method of generating (conserved) observable quantities (or at any rate quantities which have an analogous place in the dynamical formalism to well known physical magnitudes in CPM). The quantities obtained in this way are tensors and functions of both the field and its derivatives.[80]

It is not clear (to me) whether *other observables* may be constructed by this method, but certainly other *conserved quantities* may be so constructed. A more thorough investigation reveals that the derivation of a conserved quantity in the manner just indicated requires two distinct assumptions – (1) that the transformations leave the Lagrangian invariant, (2) that they leave the corresponding Euler-Lagrange equations invariant – which are neither singly nor jointly sufficient for defining a conserved quantity. It can be shown that there are many other conserved quantities besides those mentioned above which may be defined.[81] Whether any of these correspond to physically interpretable observables is not determined thereby. However the view which I shall shortly adopt concerning the set of observables in CFT will entail that all these conserved quantities are themselves observables.

The Noether construction then does not really help to answer the question of what to count as an observable in CFT. The only hint of how to proceed further is to exploit even more extensively the dynamical analogy with CPM. In CPM, the reader will recall, all the observables were given as functions of the q_i, p_i (or we may reduce these to functions of the q_i alone). According to the dynamical analogy $q_i \leftrightarrow \phi(s_i)$ and so we begin by considering all functions of the $\phi(s_i)$. As we allow $s_i \to 0$, however, we must pass to considering spacial derivatives of the field functions. Thus the analogy, though not strictly valid at the last, suggests that we take as the set of observables the set of all functions Ψ on the fields (which include derivatives of all orders). (If the Hamiltonian formalism is the basis for the analogy we would choose functions on the ϕ, π instead. This will make no essential difference, since the π are functions of the ϕ.) The tacit generalization here beyond strict analogy with CPM, namely to introducing time derivatives of ϕ as well, is necessary because, unlike the situation with CPM, the space and time variables occur symmetrically in CFT and we must include both space and time derivatives if we are to capture all the observable functions (indeed, if we are to capture even those generated by Noether's theorem). (The symmetry between the space and time variables in CFT means that we deal basically with observable functions which constitute field densities or intensities, currents and continuity equations. Only in CPM, where we have spacially localizable individuals and hence where the space and time variables enter the theory asymmetrically do we deal with collections of magnitudes which describe individual states at each moment of time. Only the integral or global properties of the field have this latter feature, for they are properties of the entire field individual.)

Now we must consider the structure of the set of such functions. A preliminary observation is in order. The functions are not linear on the fields. (The observable densities generated by the Noether theorem are, for example, bilinear in the ϕ's.) Thus $\mathcal{O}(\phi_1 + \phi_2) \neq \mathcal{O}(\phi_1) + \mathcal{O}(\phi_2)$ in general. Nevertheless if addition, multiplication and multiplication by a scalar are now defined for observables in the same manner as they were in CPM we clearly obtain an algebraic structure $A_{\mathcal{O}}^{\mathrm{CFT}}$ which is again a complete algebra. Where the fields are complex, however, there is no guarantee that these observables will be real valued, in fact they quite obviously will not all be real valued. We have then two choices, to attempt to attach some empirical significance to the complex values occuring or limit the observables to the class of real valued functions of the ϕ (ϕ,π). I shall take the second course. In this case we obtain a subalgebra $A_{R\emptyset}^{CFT}$ of $A_{\emptyset}^{CFT}$. Clearly $A_{R\emptyset}^{CFT} \approx A_{\emptyset}^{CPM}$, since here the set of observables is just the set of all functions of the ϕ (ϕ,π). Moreover, as with CPM, the Lie algebra of all observables closed under the Poisson bracket will form a subalgebra of $A_{R\emptyset}^{CFT}$, with $A_{R\emptyset}^{CFT}$ as the enveloping algebra. The ϕ (or ϕ,π), or (I suppose) their real components (for ϕ, π complex), become the fundamental observables and determine the values of all other observables.[82]

Notice that one cannot say "...determine the *values* of the other observables", since observables as I have defined them are functions of the x_i,t through the ϕ,π. Determining the ϕ, π *functions of x_i,t* determines the observables as *functions of x_i,t*, but only fixing a particular location x_i', t' will fix a numerical value for the observables. Thus it seems preferrable to say that the observables are actually the functionals on the fields, the functional $\mathcal{O}$ mapping, at each point x_i', t' of space-time, the field into the value of $\mathcal{O}$ at x_i', t'. It will turn out in fact that this is the only sense one can attach to the notion of observable if a correlative propositional structure is to be carried through. Thus I shall take observables to be the real-valued functionals on the fields (or field, momenta pairs) which map functions of the fields into their values at each point x_i', t' of space-time. For completeness sake I record the fact that the idempotent observables are the functionals $\tilde{I}_S$ which map any $\phi \in S \subseteq \mathcal{S}_T^{CFT}$ into 1 at all x_i,t and map every other $\phi \in \mathcal{S}_T^{CFT}$ into 0 at all x_i,t. (It would be possible to further decompose the $\tilde{I}_S$ into functionals mapping $\phi \in S$ into 1 only for $x_i,t \in V$ a space-time region, but this will not be pursued.)

A state is, recall, a particular field $\phi(x_i,t)$ and thus indeed determines the values of all observables. (Immediately below I shall consider the 3-dimensional picture of evolving states in $\mathcal{S}_{ST}^{CFT}$ rather than the static 4-dimensional space-time picture. In this case a state is a particular spacial field $\phi(x_i) = \phi(x_i,t')$ for some fixed t' and the specification of such a state clearly determines the values of all observables at the fixed time t'.)

Now let us consider briefly the representation of interacting fields. There are two kinds of interaction to be considered, 'self-interactions' and coupled-field-interactions. In the first case an additional term is introduced to the free field equation 'coupling' the 'free' field to itself. (For example, instead of the Klein-Gordon equation $(\Box^2 - M^2)\phi(x_i, t) = 0$ we have $(\Box^2 - M^2)\phi + \alpha\phi^n = 0$, n real. For $n \neq 1$ these equations are non-linear.[83]) In the second case coupling terms are added to the Lagrangian of N fields $\phi_i (i = 1, \ldots N)$ to produce coupled field equations. Since the coupling terms also have to be relativisticly invariant their forms are determined as world scalars. Again one obtains in almost every case non-linear coupled field equations.[84] In the first case the effect of the self-coupling is to alter the function space $\mathscr{S}_T^{\mathrm{CFT}}$ to a more complex non-linear set (or perhaps space). In the second case there corresponds to each field ϕ_i its own function space $\mathscr{S}_{T,i}^{\mathrm{CFT}}$ and it will in general also alter from a linear space to a non-linear set (or perhaps space) when interactions are introduced. In this latter case, however, the elements of the function spaces are then also coupled to one another and we may (in abstracto) define a tensor produce space $\otimes \mathscr{S}_{T,i}^{\mathrm{CFT}}$ (the elements may be looked upon, though not rigorously, as N-tuples of field functions, one each from the $\mathscr{S}_{T,i}^{\mathrm{CFT}}$) in which the composite solution may be expressed. Observables for the composite system will then be defined as functionals on $\otimes \mathscr{S}_{T,i}^{\mathrm{CFT}}$. (These comments will hold equally well if we use instead the derived spacial function spaces, to be discussed shortly, where in each case we may construct $\otimes \mathscr{S}_{ST,i}^{\mathrm{CFT}}$.)[85]

Finally we turn our attention to the probability measures of CFT. Again one finds very little discussion of these in the literature and none at all in the standard presentations of CFT. At least part of the reason for this lacuna is surely the fact that one cannot realize ensembles in any simple fashion. In CPM one had spatio-temporally localized individuals governed by local laws; one could therefore realize both spatial ensembles and temporal ensembles, i.e. one could arrange to reproduce a given particle state either at the same time in other spatial locations, or successively at the same location. Strictly speaking neither device is applicable to the global field. Since each field occupies the whole of space one obviously cannot produce spatial ensembles. (Recurring spatial parts of fields do not form such an ensemble but simply a single field with periodic properties.) Equally, a truly global field requires at least the propagation time from boundary to boundary for its full character to be revealed and in general this is infinite – temporal truncation is simply not possible, and if it could be done and an alleged temporal ensemble produced such a temporally recurrent field would be viewed as a single field whose boundary conditions again introduced periodicity. But while literal ensembles are not possible it is always possible to introduce the quite different Gibbs ensemble and view the actual field configuration as one among infinitely many possible field configurations satisfying certain conditions (eg. having a certain total energy). In direct analogy with these ensembles for CPM, it is the field state functions ϕ which determine membership in an ensemble and so it is natural to take the probability measures as representing or characterizing the states of these ensembles of fields. That is it seems natural to define probability measures over the borel subsets of $\mathscr{S}_T^{\text{CFT}}$.[86] We are then able to speak about the probability that the state of the field is given by some particular ϕ_i. It follows that the structure of the set $\mathscr{S}_\mu^{\text{CFT}}$ here is exactly similar to that in CPM also, in particular it is strongly convex, atomic with the measures $\mu_{\{\phi\}}$ as the extreme points. Moreover, it follows that as for CPM the characteristic functions on $\mathscr{S}_T^{\text{CFT}}$ again coincide with the idempotent observables there and are formally equivalent to the extreme probability measures for that special case. Notice that $\mathscr{S}_T^{\text{CFT}}$ satisfies trivially the conditions for being a state space stated at the end of Section IV. Resolution into pure probability measures is unique, these measures are concentrated at the points of $\mathscr{S}_T^{\text{CFT}}$ and it is trivial that a pure measure remains pure through time since the $\phi \in \mathscr{S}_T^{\text{CFT}}$ already include all times. (We have already seen that a particular member of $\mathscr{S}_T^{\text{CFT}}$ determines maximally precise state specifications at all x_i, t.)

At least for CFFT the probability measures will be defined over real or complex Hilbert spaces (for the $_L\mathscr{S}_T^{\text{CFT}}$ are such spaces in these cases). But since probability measures are defined only on the *sub-sets* of the space the structure of $_L\mathscr{S}_T^{\text{CFT}}$ as a *space* is *irrelevant* (even though it is not irrelevant for the specification of states). If $\mu_{\{\phi\}}$ is the probability measure concentrated at ϕ and we write $\phi = \sum_i \alpha_i \phi_i$, $\phi_i \neq \phi$, then it will not be posto regard $\mu_{\{\phi\}}$ as a function of the $\mu_{\{\phi_i\}}$ since any such function will inform us that the actual state of the field will be ϕ_i and so not ϕ at all. This argument hinges, of course, on the fact that the ϕ are *taken realisticly as representing actual physical states*.

If $\mathscr{S}_T^{\text{CFT}}$ is a space it would be possible to represent any field as a combination of other, suitably selected fields. Is this fundamental superposability of fields such that we can view any field as really composed of such constituent fields? The answer to this question hinges on the behavior of the observables and probability measures on $\mathscr{S}_T^{\text{CFT}}$. The situation with respect to the probability measures, we have just seen, is not very promising. What would be required to properly exploit the space structure of $\mathscr{S}_T^{\text{CFT}}$ would be a way of introducing probability measures defined on the subspaces of $\mathscr{S}_T^{\text{CFT}}$. Unfortunately (it would suit my purposes to be otherwise) I can see no way of doing this and this is essentially because fields in this case are real, so that the α_i represent measures of the actual component field strengths present in a given field and cannot be interpreted within CFT, that I can see, as probabilities. (Otherwise, we might exploit the operator structure to observables in CFT now to be outlined to introduce probability measures also in a fashion similar to that of QM.)

Some progress can, however, be made with respect to the observables of CFT and this has an independent interest. The observables can certainly be regarded as functionals of the component fields of a field. Thus if $\phi = \sum_i \alpha_i \phi_i$ we may substitute this expression into $\mathcal{O}(\phi)$ to obtain $\mathcal{O} = \mathcal{O}(\phi_i)$. On the other hand it is not in general possible to obtain the $\mathcal{O}(\phi)$ as a linear combination of the $\mathcal{O}(\phi_i)$. However we may exploit the vector space character of $\mathscr{S}_T^{\text{CFT}}$ (when it has one) by reconstruing the observables as operators on this space (a formally smaller jump here than in CPM). Suppose that the ϕ_i form a basis for $_L\mathscr{S}_T^{\text{CFT}}$. Then any $\phi \in {_L\mathscr{S}_T^{\text{CFT}}}$ may be expressed as $\phi = \sum_i \alpha_i \phi_i$. We define the projection operators as usual as $\tilde{P}_{\phi_i}\phi = \alpha_i \phi_i$. Denote by $\tilde{\tilde{I}}$ the 'unit' functional taking ϕ into its values at each x_i, t. We have $\tilde{\tilde{I}}\,\phi = \sum_i \alpha_i \tilde{\tilde{I}} \phi_i$. Now define the 'functional valued projection operator' $\tilde{\tilde{P}}_{\phi_i}$ by

$$\tilde{\tilde{P}}_{\phi_i}\phi = \alpha_i \tilde{\tilde{I}} \phi_i.$$

Then we may 'decompose' $\tilde{\tilde{I}}$ as follows:

$$\tilde{\tilde{I}}\phi = \sum_i \tilde{\tilde{P}}_{\phi_i}\phi = \left(\sum_i \tilde{\tilde{P}}_{\phi_i}\right)\phi.$$

For any functional $\tilde{\mathcal{O}}(\phi)$ we may write, formally,

$$\tilde{\mathcal{O}}\phi = \mathcal{O}\left(\sum_i \tilde{\tilde{P}}_{\phi_i}\phi\right) = \left[\mathcal{O}\left(\sum_i \tilde{\tilde{P}}_{\phi_i}\right)\right]\phi = \mathcal{O}\left(\tilde{\tilde{P}}_{\phi_1}, \tilde{\tilde{P}}_{\phi_2}\cdots\right)\phi$$

so that each observable becomes a functional valued operator on $_L\mathscr{S}_T^{\text{CFT}}$ expressible as a function of the functional valued projection operators on $_L\mathscr{S}_T^{\text{CFT}}$ (an extremely crude counter part of the spectral theorem in QM).

Thus, at least in respect of the observables, it would be possible to argue that the component fields ϕ_i are the basic fields from which everything else is constructed. Of course, such bases will not be unique! – the super-position principal is operative in full force here.

At this point we may make a move logically not open to us in CPM. Concider an element $\phi \in \mathscr{S}_T^{CFT}$ for some Lagrangian and introduce the set of purely spacial functions $\phi(x_i) = \phi(x_i, t')$ for various, fixed t'. Let $\mathscr{S}_{ST}^{CFT}$ be the set of all such purely spacial functions generated by all members of $\mathscr{S}_T^{CFT}$. Call $\mathscr{S}_{ST}^{CFT}$ the derived spacial function set or space. (This move is logically not open to us in CPM because there the elements of $\mathscr{S}_T^{CPM}$ are functions *only* of time.) The field dynamics can now be recast within $\mathscr{S}_{ST}^{CFT}$. The field state undergoes a dynamical evolution in $\mathscr{S}_{ST}^{CFT}$ determined by a 1-parameter group of transformations on $\mathscr{S}_{ST}^{CFT}$. This evolution is espe-cially simple in the cases of the Hilbert spaces $_L\mathscr{S}_T^{CFT}$ of CFFT, the dyna-mical evolution is in each case a unitary transformation on the space. In the cases where $\mathscr{S}_T^{CFT}$ is non-linear, $\mathscr{S}_{ST}^{CFT}$ can in general be expected to be non-linear as well and dynamical evolution in those sets, though it will still be determined by a dynamical group, will not be representable as a linear transformation on the set. In such cases little seems known about either the spaces or the dynamical groups.

Observables become real valued functionals or operators on this space and their values functions of time. (Alternatively, at least in the linear cases, we can switch to the 'Heisenberg picture' and have the observables undergo the dynamical evolution and the state remain unaltered.) Proba-bility measures may be defined over the space and will likewise undergo a dynamical evolution. We have the relation $\mu_t(E) = \mu_0(U_{t,t'}^{-1}E)$, where E is a borel subset of $\mathscr{S}_{ST}^{CFT}$. Again the dynamical evolution of such proba-bility measures remains undiscussed in the standard literature. (However, Hopf [66] and Rosen [110] have studied the evolution of such probability measures for fields governed by a dynamical equation of the form $\partial \phi / \partial t = F(\phi)$, where F is an operator containing only spacial derivatives. Though in their studies they had hydrodynamics primarily in mind, the Dirac equation can be written in this form so that their work applies at least in this case.[87]) On the whole the entire subject (derived spaces, observables and probability measures) remains even more completely open when it comes to the non-linear equations.

To date nothing has been said of any correlate to phase space in the theory. It is of great importance to see that *no such correlate arises naturally within CFT*. Just as in CFT phase space played the natural central role of the structure, for it was state space and observables and probability measures were defined over it as domain, so in CFT the function space plays this same logical role – $\mathscr{S}_T^{CFT}$ is the state space of CFT and observables and probability measures are defined over it as domain. The natural structure of CFT is then expressed graphically by:

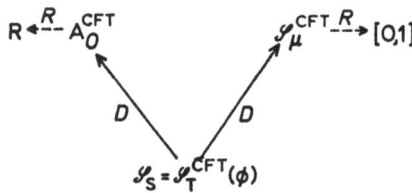

But just as, with some difficulty, we were able to introduce a function space to CPM and, with considerably more difficulty and artificiality, reconstrue observables and probability measures as taking this space as domain, so also in CFT we can attempt to reverse the procedure and force the introduction of phase space S^{CFT}. Following the initial dynamical analogy with CPM, S^{CFT} will be taken to be an infinite dimensional space having an axis for each distinct spacial location, for each distinct field ϕ. Specifying a location in S^{CFT} will fix the values of ϕ at every spatial location, hence determining $\phi(x_i) = \phi(x_i, t')$ and the dynamical equations will then give the trajectories of ϕ in S^{CFT} as a function of time. If we wish to avoid this marring of the symmetry between the space and time variables an extra dimension can be introduced so that a point in the new phase space determines an entire field $\phi(x_i, t)$. Such spaces are, however, patently artificial constructions. In the first place not all locations in S_T^{CFT} are physically permissible or accessible, contra CPM, because one cannot alter arbitrarily the relations between the field values at various spatial or spacio-temporal, locations and still satisfy the demand that the fields be continuous functions of the space and time variables. (For there will be locations in S_T^{CFT} such that the value of the field at x_i, t differs by a finite amount from its value at $x_i + \Delta x_i, t$, no matter how small Δx_i is made. This is essentially because the various locations are each treated independently of the others in this phase space.) These inaccessible locations must be ruled out over and above any limitations on the trajectories within

S^{CFT} imposed by the dynamical equation of evolution. Moreover, observables are not, contra CPM, real valued functions on S^{CFT}, nor are probability measures defined over its borel subsets. (Indeed any such probability measure would not make physical sense in view of the fact that not all points of S^{CFT} are physically admissible.) The concept of phase space has neither a natural place in CFT, nor is it even clear that we can properly reconstruct CFT as a theory over S^{CFT}, nor can it in any case be given a logical role in CFT analogous to that played by S^{6N} in CPM. (Nor is this situation appreciably altered if we switch to S^{CFT} (ϕ, π) and introduce an additional uncountable infinity of dimensions for the π at each x_i, t.) The concept of a phase space for CFT will therefore be pursued no further. The natural structure of CFT as presented above is also the only acceptable structure. (The reader will again observe that what forces on us the peculiar nature of a putative phase space and also leads to its difficulties is just the global nature of the fields.)

To complete this sketch of CFT it is needful to introduce the corresponding propositional structures. Some care must be shown in the specification of propositional forms, however, because of the global nature of the fields. I shall consider first propositions of the form

The observable $\mathcal{O}$ has a value lying in a borel set E of R (the real line) at the location x_i, t.

These propositions, which I shall denote by $P^{\text{CFT}}_{\mathcal{O}, E}(x_i, t)$, are the correlates of the $P^{\text{CPM}}_{\mathcal{O}, E}$. For fixed x_i, t they have a boolean structure due to the boolean character of the algebra of borel sets E on R. Since the propositions for distinct x_i, t values are logically independent from one another, they may be conjoined and disjoined in the usual boolean fashion. Thus the entire set of propositions $P^o_{\mathcal{O}}(x_i, t)$ forms a boolean algebra $\mathcal{B}^{\text{CFT}}_{\mathcal{O}}$. (Note that we could consider the requirement of continuity for the ϕ, π as imposing a *logical* restriction on the relations among the $P^{\text{CFT}}_{\mathcal{O}, E}(x_i, t)$. In this case, for example, for suitable E and E' the conjunction $P^{\text{CFT}}_{\mathcal{O}, E}(x_i, t)$ and $P^{\text{CFT}}_{\mathcal{O}, E'}(x_i + + \Delta x_i, t + \Delta t)$ is an always false proposition. I have not explored this possibility in detail but it is certain that such a restriction would break down the boolean structure to the set of propositions and very likely that it would reduce it to just a POset $\mathcal{P}^o_{\mathcal{O}}(x_i, t)$.)

There is no analogue to the $P^{\text{CPM}}_{s_t}$, $P^{\text{CPM}}_{s_t, t}$ and $M^{\text{CPM}}_S(\mu, S, r)$ propositions, all of them hinge on the existence of a phase space. There is, however, a direct analogue of the T_S^{CPM} propositions, namely the propositions

> The system is characterized by a field ϕ in a region S of $\mathscr{S}_T^{CFT}$.

Denote these propositions by T_S^{CFT}. They form a boolean algebra because of the corresponding set-theoretic structure of the $S \subseteq \mathscr{S}_T^{CFT}$. These propositions determine ranges of values for all of the observables, for let

$$\mathcal{O}(S)(x_i, t) \subseteq R = \{x \in R : \mathcal{O}(\phi(x_i, t)) = x \text{ for } \phi \in S\}$$

Then we have

$$T_S \equiv P_{\mathcal{O}, \mathcal{O}(S)(x_i, t)}^{CFT}(x_i, t) \text{ and } P_{\mathcal{O}', \mathcal{O}'(S)(x_i, t)}^{CFT}(x_i, t) \text{ and } \ldots$$

Notice that in this case the temporal relationship between the T_S^{CFT} and the $P_{\mathcal{O}, E}^{CFT}(x_i, t)$, as well as the correlative spacial relationship, falls out automatically.

Finally there are the probabilistic propositions corresponding to

$$M_P^{CPM}(\mu, \mathcal{O}, E, r) \text{ and } \mathcal{M}_S^{CPM}(\mu, S, r)$$

The analogue of $M_P^{CPM}(\mu, \mathcal{O}, E, r)$ is $\mathcal{M}_P^{CFT}(\mu, \mathcal{O}, E, r, x_i, t)$:

> The probability is r in an ensemble of fields characterized by probability measure μ that the value of the observable $\mathcal{O}$ at the location x_i, t lies in the borel subset E,

and the analogue of $\mathcal{M}_S^{CPM}(\mu, S, r)$ is $\mathcal{M}_S^{CFT}(\mu, S, r)$:

> The probability is r that in an ensemble of fields characterized by probability measure μ the field state lies in the region S of $\mathscr{S}_T^{CFT}$.

Nothing essentially new can be said about the structure of these propositional sets over above what was said in Section IV when CPM-CSM was discussed, except to add that the propositions for varying x_i, t are here treated as logically independent of one another, possible restrictions flowing from the demand for continuity for the fields set aside, and that we have, as before

$$\mathcal{M}_S^{CFT}(\mu, S, r) \equiv M_P^{CFT}(\mu, \mathcal{O}_1, \mathcal{O}_1(S)(x_i, t), r_1, x_i, t) \text{ and }$$
$$M_P^{CFT}(\mu, \mathcal{O}_1, \mathcal{O}_1(S)(x_i', t'), r_1', x_i', t')$$
$$\ldots \text{ and } M_P^{CFT}(\mu, \mathcal{O}_2, \mathcal{O}_2(S)(x_i, t), r_2, x_i, t)$$
$$\text{and} \ldots$$

Finally note that we obtain an analogous proposition set to the $\mathcal{M}_S^{\text{CFT}}$ (μ, S, r), the $\mathcal{M}_{SS}^{\text{CFT}}(\mu, S, r)$, if we replace $\mathscr{S}_T^{\text{CFT}}$ by $\mathscr{S}_{ST}^{\text{CFT}}$ in the definition of $\mathcal{M}_S^{\text{CFT}}(\mu, S, r)$, and then we find

$$\mathcal{M}_S^{\text{CFT}}(\mu, S, r) \equiv \mathcal{M}_{SS}^{\text{CFT}}(\mu_1(t_1), S, r_1, t_1) \text{ and } \dots$$
$$\mathcal{M}_{SS}^{\text{CFT}}(\mu_2(t_2), S, r_2, t_2) \text{ and } \dots$$

I shall comment no further on the structure of these proposition sets since they are not discussed in the literature and I have nothing further to say of them at present which would cast any new light on the structure of CFT.

Thus we have for CFT:

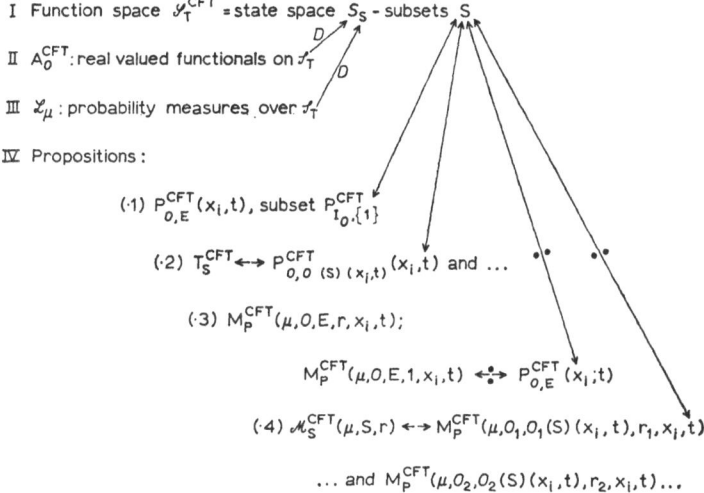

I Function space $\mathscr{S}_T^{\text{CFT}}$ = state space S_S - subsets S

II A_0^{CFT}: real valued functionals on $\mathscr{S}_T$

III $\mathscr{L}_\mu$: probability measures over $\mathscr{S}_T$

IV Propositions :

(·1) $P_{0,E}^{\text{CFT}}(x_i,t)$, subset $P_{I_0,\{1\}}^{\text{CFT}}$

(·2) $T_S^{\text{CFT}} \longleftrightarrow P_{0,0\ (S)\ (x_i,t)}^{\text{CFT}}(x_i,t)$ and ...

(·3) $M_P^{\text{CFT}}(\mu,0,E,r,x_i,t)$;

$M_P^{\text{CFT}}(\mu,0,E,1,x_i,t) \longleftrightarrow P_{0,E}^{\text{CFT}}(x_i;t)$

(·4) $\mathcal{M}_S^{\text{CFT}}(\mu,S,r) \longleftrightarrow M_P^{\text{CFT}}(\mu,0_1,0_1(S)(x_i,t),r_1,x_i,t)$

... and $M_P^{\text{CFT}}(\mu,0_2,0_2(S)(x_i,t),r_2,x_i,t)$ ...

From the foregoing it is perfectly clear how the mathematical structure of field theories reflects, or represents, the peculiar characteristics of the ontology of the plenum view (cf. Section VII). All observables are constructed from functionals or functional valued operators on the fields themselves and are, in general, spacio-temporally continuous. Thus the entities of CFT are, besides the fundamental entity (the field itself), just spacio-temporal distributions of properties determined by field values. The fundamental superposability of fields leads to the non-exclusiveness of such entities, i.e. the property complexes may be superposed to yield new complexes. (This remains true in a fundamental sense even for non-linear fields, for even if such theories prevent the construction of a state *space* because of the appearance of singular regions in their solutions this will mean no more than that we have been able to ape within the field theory the localized behavior of atomic objects; such aping does not speak against the fundamental character or structure of field theories). Dynamical evolution is simply the change of the field distribution with time, neither the field nor parts of it move. This latter is clear because of another feature of fields, their globalness, which we have seen to enter into and influence the mathematical structure of field theories at every turn. When the field equations are solved the solution must be given throughout a spacial volume, the specific solutions accepted being determined by the boundary conditions of that region, thus the field solutions in a region form a whole which cannot be regarded as the sum of spacial parts. *Thus, as with CPM and the atomic ontology, the peculiarities of the plenum ontology find a direct reflection in the mathematical structure of CFT – there is the most intimate connection between the view of the world the theory expresses, the fundamental conceptual apparatus required to express physical descriptions from that point of view and the mathematical structure in which the view finds expression as a theory in exact science.*

There is still much that we do not know about CFT (and about CPM for that matter). In particular there are only elementary beginnings on the development of an adequate process and/or event language that is better suited to the offering of physical descriptions from the plenum point of view. And when we come to the construction of the representation of objects even the mathematical apparatus is lacking, for we might hope to represent macroscopic objects, and even molecular objects, by some form of non-linear field that would express our factual inability to superpose them (i.e. express their relative impenetrability). (A non-linear theory is actually forced on us anyway because of the non-linear character of the gravitational interaction.[88]) But relatively little is known of the mathematical structure of such theories. Not until these constructions are completed, and the correlative propositional structures thoroughly investigated, will we really be in a position to compare with the set-theoric, boolean structure of CPM the logical structure of the plenum point of view and to evaluate in a meaningful fashion the significance of the co-existence of the partial similarities and deep dissimilarities to the atomic structure within this classical theory. In the next section, however, I attempt some initial evaluation of the relations between the two.

X. PLENUM VERSUS ATOM II: CLASSICAL FIELD VERSUS PARTICLE THEORIES

The contrast in mathematical nature and structure between CFT and CPM is quite substantial, despite some similarities. There are several significant differences between field and particle structures but the central one is the absence of any natural phase space in the abstract structure of CFT to play the role of a state space. That structure, which plays an absolutely essential role in CPM, finds no home in CFT (and could not play the same logical role even if it were artificially created) – its function as a state space is taken over by the function space of CFT, conversely, the function spaces of CPM are virtually never discussed, play a some-what different logical role in CPM to that which they play in CFT and little seems to be known of them (very special cases aside, see below), though they are at the center of attention in CFT.

Now this difference, as I have had occasion to emphasize, pivots on the fact that particles are spacially localized systems whereas fields are global entities. (Recall also the discussion of the global and local characteristics respectively of the plena and atom in Section VII.) Thus systems in CPM may be regarded as spacial aggregates of spacially localized systems and their locations may thus be represented by a set of triples. Essentially the 'configuration space' is the sufficient and proper mathematical object on which to build the entire edifice of CPM. Fields cannot be represented in this fashion, their single, whole field solutions over space must be represented at each time. This accounts for the centrality of two quite different fundamental structures in the two mathematical theories. To CPM belongs *set theoretic* structure of *phase space*, expressing in its representation the individuality of the fundamental entities. To CFT belongs the *vector space* structure of *function space*, expressing the fundamental globalness and superposability of the fields. No superposition principle applies in CPM, neither in its phase space nor in its function space, whereas such a principle is fundamental to CFT. Correspondingly, the composition principles of the two theories are quite different. In CPM we construct a *higher dimensional phase space* in which the composite state is still represented by a *point location*; in CFT we construct the *tensor product function space* of solutions to the coupled field equations in which a state is given by a *vector*, or superposition of vectors, in the space.

Moreover, the function spaces of CPM, if we obtain spaces at all, are in general non-linear whereas at least those of CFFT are all linear spaces, indeed are all real or complex Hilbert spaces (or may be made so or embedded in same). And, correlatively, CPM temporal evolution in phase space is not given by an orthogonal transformation on the space, whereas temporal evolution in CFFT can always be represented in the form of such a transformation (unitary for complex spaces). In addition we do not and can not represent temporal evolution *in* the function space of CPM, rather each function contains a temporal history for a system, evolution occurs in phase space; indeed, the construction of the derived spacial function space makes no sense within CPM. By contrast we have temporal evolution in the derived spacial function set or space of CFT.

It is perhaps worthwhile to pause and reflect on the fact that, at least from a formal point of view, the phase space of CPM plays a very special and peculiar role in that theory. Formally, one might have expected to develop the function space $\mathscr{S}_T^{CPM}$, define observables and probability measures on it and proceed, entirely ignoring phase space. In fact we do not do this because the natural representation space for spacially localized systems is their configuration space (i.e. phase space). Moreover, temporal evolution cannot be represented as a dynamical evolution within function space but only on phase space. But in the phase space approach we entirely transform the algebraic setting of a physical problem from its 'natural' formal setting into a special setting in which the description problem is reduced to the specification of a location on a geometrical space, rather than being determined by a characterizing function. This special reduction of the physical description problem is possible just because some physical systems meet the following three stringent conditions: (i) their components are spacially localized, (ii) they have finite degrees of freedom and (iii) the description of their physical characteristics is entirely reducible to the specification of their dynamical configuration (i.e. component positions and momenta) as a function of time. (Thus observables, which are naturally construed as operators on function space, here become transformed into functions on the phase space.)

I labour this point for two reasons. Firstly because there is, I think, a general impression abroad that in any theory of mathematical physics the obvious thing to do is to construe the states of physical systems as elements of a mathematical space which is the counterpart of phase space in CPM.

As against this view I wish to emphasize the specialness and peculiarity of phase space in CPM and bring out the stringent conditions which must be met before a physical theory can be reconstrued from this point of view. Correspondingly, I wish to emphasize that this is not the general character which you expect of theories of mathematical physics and to direct attention to the formulation of theories from within the function space point of view. The second reason for this emphasis stems from a particular result of the foregoing prejudice, namely the currency which phase space has as the believed analogue to Hilbert space in QM. I now regard this approach to QM as a highly misleading one whose end result is interminable fruitless, frustrating speculation and discussion of 'paradoxes'.

In this connection it is interesting to observe that the observables of CFT are functionals on the function space and can be viewed as functional valued operators on that space, whereas the observables of CPM are functions on phase space and can only be represented as operators if a shift is made to the function space and its vector space character is exploited. (I note here that the reason why phase space is generally viewed as a vector space lies in the efficacy of this richer structure for the mathematical representation of dynamical evolution – e.g. Mackey [83], but cf. the digression to follow shortly. Insofar as it obscures the logical structure of phase space, this is unfortunate. Although I have made some remarks here on the proper understanding of the representation of space and time in a physical theory, cf. also Varadarajan [133], I do not yet properly understand the philosophical significance, if any, of this dual character to phase space. When one moves to function space the duality is much less in evidence.)

This latter similarity, and the similarities between the algebras of observables and the structure of the sets of probability measures in the two theories, arises because both theories exploit essentially similar mathematical structures at these points. (Thus, for example, both theories essentially define products of observables in the same algebraic way and in both theories probability measures are construed as measures on borel subsets of the appropriate measure space.) I think, though I am not sure, that specifying these distinctive mathematical formulations and structures is perhaps the most perceptive way of characterizing what is essentially *Classical* in classical physics.

The remaining similarities between the two theories spring not from their

fundamental structure, but from the Lagrangrian-Hamiltonian dynamical formalism both employ. Now this formalism is just a mathematical
construction permitting the derivation in a simple fashion of the infinitesimal generator of the dynamical group, thus it is not so surprising that both
theories can share such an approach to dynamics. More surprising is the
fact that it seems that we do not yet have available a sharp characterization of what precise assumptions are made, what precise mathematical
structure imposed (if any), by all theories adopting this approach to dynamics. (Acquiring this abstract characterization would be of considerable
importance – not only would it deepen our insight into the nature of dynamics and our present physical theories, it might lead to the formulation of
alternative and more powerful approaches to physical dynamics[89].)

Digression. There is one class of cases in which a more striking similarity
may be exhibited between the structures of CPM and CFT and for this reason it is worth discussing. This class, discussed briefly by Mackey[90], and
elaborated in detail by Marsden[91], is also the exception where the function
space of CPM is of importance to the exposition. Mackey begins by defining a *linear* system. Formally, linear systems are systems S where (i) the
"configuration space" $\mathcal{M}$ of S is a vector space, (ii) the potential energy
function $V(\phi, \phi)$ is quadratic in the generalized coordinates of the space,
(iii) the kinetic energy function $T(\phi, \phi)$ does not depend explicitly on the
generalized coordinates. Informally, we notice that these conditions are
met for the simple harmonic oscillator and all other cases turn out to be
essentially generalizations of this, for they are cases where the generalized
coordinates of the system oscillate simple harmonicly, and do so independently of one another (a possibility founded deep in the theory of
generalized fourier analysis). Mackey is then able to show that essentially
the configuration space can be formed into a real Hilbert space under the
norms $(V(\phi, \phi))^{1/2}$, $(T(\phi, \phi))^{1/2}$ and that the direct sum of it and its *-dual
also forms a Hilbert space under the norm $(V(\phi, \phi))^{1/2} + (T(\phi, \phi))^{1/2}$.
He is then able to exploit the fact that in linear systems the kinetic and
potential energy terms are related in a very special way to construct a skew
adjoint operator A on the Hilbert space $\mathcal{M} \otimes \mathcal{M}^*$ that generates in $\mathcal{M} \otimes \mathcal{M}^*$,
$U_{t,t'}: t \rightarrow e^{(t'-t)A}$. (That each of the generalized coordinates oscillate
simple harmonicly is the key to understanding how it is possible to represent dynamical evolution as an orthogonal transformation on the Hilbert
space. For in CPM in general dynamical evolution is not an orthogonal

transformation on phase space, but in this special case since every coordinate oscillates the 'location' vector in the Hilbert space undergoes a generalized rotation.) Importantly, this scheme holds good for systems of infinite degrees of freedom as well for those of finite degrees of freedom – in fact it is applicable to real linear fields. (Mackey treats the electromagnetic field as an example.) Moreover, the real Hilbert space can easily be generalized to a complex Hilbert space preserving everything of importance. (The nearness of the resulting mathematical structure to QM will not be missed by the reader.)

Now we notice following important fact: in this case, unlike the typical cases in CPM, the configuration space is chosen as the space of all functions in the class to which the solution of the corresponding dynamical equations belong, i.e. "configuration space" becomes what I have called the function space of the theory. (That we obtain a function *space* is just what is assured by requiring the systems to be *linear*.) The 'generalized coordinates' become (essentially) basis vectors in the function space. Thus this structure bears an even more striking resemblance to the structure of CFT then it does to CPM (and in fact only the dynamical analogy between CPM and CFT allows one to develop this approach starting out with a discussion of CPM dynamics). Thus this important class of cases, so far from constituting a counter example to my view of the disparity between particle and plenum theories, actually reinforces the claim for that disparity, reemphasizing also the importance of distinguishing between the abstract structure of a theory which expresses the distinctive features of the ontology of the theory and the general dynamical formalism which is incorporated into that structure. *End of digression.*

It was the genius of classical physics to keep these two great metaphysical-conceptual-mathematical schemas for physical theorizing almost entirely segregated from one another. Physicists of previous centuries, by and large, were more conscious of the differences between the two – at least in specific cases – because they were involved in the great post-Newtonian debates concerning the true nature of heat, electricity, magnetism and light (and because they did not have their minds drugged with the dogma that nothing need now be understood except quantum theory and that it was to be understood *sui generis* as superceding all classical ideas). The debates of the previous centuries struck deeply enough to raise penetrating discussions of the foundations of these two approaches, especially of the atomic approach as propounded by Newton and of the reality of the 'lines of force' introduced by Faraday.[92] Classical science was not, however, completely successful at segregating the two approaches. Early on, material objects, viewed atomistically and as ontologically independent of fields, were considered the sources of the physical fields, especially of the electromagnetic and (later) gravitational fields. Conversely, attempts to obtain a purist atomic conception broke down when it became clear that electromagnetic interaction between charged atoms required a time lapse (the action 'propagated' at a finite velocity) and fields were introduced to 'carry' these actions. Now these were rather cautious attempts at marriage between the two schemas – ontologically and with respect to their basic mathematical structures the two theories were kept segregated, only through their dynamical evolutions were the two kinds of entities related to one another. It is a striking witness to the fundamental differences between these two schemas that even these relatively cautious attempts at marriage never really worked. The classical theory of the electron produced infinite self-energies and run-away accelerated states as a consequence of the action between it and its own field. The field functions, on the other side, were prevented from being the continuous, highly differentiable functions usually assumed in CFT because of the presence of material objects. And when the passage was made to relativity theory, which is the field theory par excellence, not only was action at a distance (one of the two basic modes of action of the atomic schema) excluded but the very conception of an extended fundamental atomic individual is excluded (because the rigidity, which is a consequence of the representation in the theory of these objects as materially simple, partless objects is incompatible with the relativistic transformation laws).[93]

There is still much that we do not understand about CPM and CFT. In particular, I feel strongly that the foregoing remarks alone have only just begun to scratch the surface of the problem of providing a sharp penetrating formulation of the distinctive features and consequences of the atomic and plenum conceptions of the world. However, I hope the discussion has made it least at clear that the debate between these two fundamental approaches has a conceptual and mathematical depth and penetration to it which is the proper counterpart to the great empirical debate that has been carried on throughout the history of physics in the western world. Against the background of this historical experience it is remarkable how little attention has been paid to the significance of the ramifications of the differences between these approaches in post-QM physics and philosophical thought. This lack of attention is even more striking in view of the fact that the early founders of quantum theory clearly felt that the understanding of quantum theory was intimately bound up with the atomic/plenum debate. Perhaps it is that respect for authority and social position which has been the bane of scientific thought over the past few decades that is responsible for the waning of the critical scientific tradition. More than likely an equally important influence in the case of physicists has been the peculiarly arrogant disregard for historical tradition and theoretical conceptions of the world which lies at the heart of the Positivist dogma that science is founded only on experimental observation. In any event, let us re-open the debate concerning the nature of QM in the light of the foregoing study of classical physics.

XI. THE STRUCTURE OF QM

A mathematical theory, so the approach claims that I have been stressing, has an intimate connection with the fundamental view of the world assumed by the theory and with the language of physical descriptions appropriate to that view. For QM one thing at least is clear, the changes in conceptual structure introduced are of the deepest kind. The arguments outlined in Section VII showed that the introduction of quantized magnitudes in the presence of continuous space-time leads to an indeterminacy in state descriptions and that this indeterminacy arises out of the fundamental logical structure of our descriptive concepts, it is based on the complementary indeterminacies already noted in CPM and CFT concerning joint precise specifications of pairs of canonically conjugate quantities. Unfortunately, this relatively precise formulation of one of the conceptual changes involved in QM is not at present matched with an equally precise analysis of the remainder of QM (as related to its underlying mathematical structure). I will not be able to attain this goal here, but I hope to set out a general context in which the detailed work may be done in what I hope will be an informative way.

A striking feature of the usual discussions of the physical descriptive language and interpretation of QM is the relatively superficial level on which the discussion proceeds – one is taught that 'wave' and 'particle' descriptions vary along a continuum from appropriate to inappropriate, depending upon the experiment to be described. The foregoing discussion should have made it clear, by contrast, that the appropriate level of analysis is that concerning the fundamental features of the atomic and plenum schemas. One is led by the extant discussion, nevertheless, to wonder whether QM does not embody some marriage of the atomic and plenum points of view which accounts for this oscillation between descriptive types. Fortunately we may commence upon this investigation aided by one fundamental feature of QM which is relatively well understood, namely its mathematical structure (or at any rate the individual components of that structure). The obvious place to begin then exploring the possible dichotomous nature of QM is with its mathematical structure, to see how closely it resembles CPM and CFT. The appropriate formulation of QM to work with is the rigorous Hilbert space formulation due to von Neumann.[94] Since the structure is well known and the ground work for the analysis has been laid in the foregoing section, I shall be relatively brief in characterizing it.

The central object in this formulation is a complex, separable Hilbert space $\mathcal{H}$. The fundamental structure of QM is given by specifying a partial algebra of operators on $\mathcal{H}$ together with the dynamical group. In QM the dynamical group is a one-parameter group of unitary transformations on $\mathcal{H}$. According to a fundamental theorem of M. H. Stone every such group may be represented in the form

$$U_{t, t'} = e^{iH(t' - t)}$$

where H is an hermitian operator and in fact $H/\hbar$ is the Hamiltonian for QM.[95]

The operators represent the *observables* of QM. (Though strictly semantically improper, I shall hereafter mostly speak of observables and the operators which represent them as identical.) The basic observables are the linear, hypermaximal hermitian operators on $\mathcal{H}$. All other observables of QM can be expressed as functions of the basic observables. The partial algebra PA_0^{QM} of such operators has been carefully studied, it is known as a C^*-algebra.[96] The algebra may be characterized in a manner outlined in Section III. The key feature of such an algebra is that not all operators commute. For example, the basic position and momentum operators satisfy the canonical commutation relations

$$[\tilde{p}_i, \tilde{p}_j] = [\tilde{q}_i, \tilde{q}_j] = \tilde{0}, \quad [\tilde{p}_i, \tilde{q}_j] = i\hbar\delta_{ij}\tilde{I}$$

where $\tilde{0}$, $\tilde{I}$ are the zero and identity operators respectively, [ ] is the commutator bracket, $[\tilde{0}, \tilde{0}'] = \tilde{0}\tilde{0}' - \tilde{0}'\tilde{0}$, and is the analogue of the poisson bracket of CPM and CFT. It is a theorem that two observables may be jointly precisely specified (i.e. have precise values jointly attributed to them) or are compatible just when they commute. In either formulation we obtain the (quite general) result that *the idempotent observables of QM are just the projections (projection operators) of the algebra of observables and these stand in one-one correspondence with the closed linear subspaces of $\mathcal{H}$.* The idempotent elements of $PA_{\mathcal{O}}^{QM}$ form a partial *boolean* algebra $P\mathcal{B}_{I_0}^{QM}$. (The projections $P_{[\phi]}$ are idempotent, $P_{[\phi]}P_{[\phi]} = P_{[\phi]}$ and so take only the eigen values 0,1.)

Any two bounded, compatible operators will generate a complete or 'classical' subalgebra between them (because they, and functions of them, are all pair-wise compatible and hence behave 'classically', in particular products are unique because commutative). This notion can be extended to maximal sets of bounded operators, i.e. maximal sets such that any two operators in the set commute and no further operator in $PA_{\mathcal{O}}^{QM}$ commutes with all members of the set. In this case, a fortiori in the case of just two operators, there exists a member $\mathcal{O}$ of the set such that all operators in the set may be written as functions of $\mathcal{O}$; thus we obtain a maximal subalgebra $SA_{M\mathcal{O}}^{QM}$ of $PA_{\mathcal{O}}^{QM}$. These maximal subalgebras clearly do not exhaust $PA_{\mathcal{O}}^{QM}$ since not all observables commute; they are in fact just the subalgebras generated by the basic observables. The idempotent observables of these maximal subalgebras form maximal boolean subalgebras $SB_{MI_0}^{QM}$ of $P\mathcal{B}_{I_0}^{QM}$. (These notions continue to apply when the set is infinite.) The entire partial algebra $PA_{\mathcal{O}}^{QM}$ may be regarded as constructed from the $SA_{M\mathcal{O}}^{QM}$ by identifying certain of their elements with one another (i.e. by 'joining' the $SA_{M\mathcal{O}}^{QM}$ at certain points).[97]

A fundamental theorem for observables is the Spectral Theorem which states that for every observable there is a unique spectral measure, i.e. a map from the borel subsets of the real line into the projection operators such that

$$\mathcal{O} = \int_{-\infty}^{\infty} \lambda dP_{\mathcal{O}, \lambda} \quad \text{where } P_{\mathcal{O}, \lambda} = P_{\mathcal{O}}([-\infty, \lambda]).$$

Conversely, every such measure defines a corresponding observable. Thus the observables can be regarded as 'decomposable' into projections. To every borel subset E there corresponds a projection operator $P_{\mathcal{O}(E)}$ for the observable $\mathcal{O}$.[98] Each basic observable has a non-degenerate spectrum whose projections together span $\mathcal{H}$. Determining a particular value for a basic observable $\mathcal{O}$ determines, via the spectral theorem, a unique projection operator and a corresponding unique one-dimensional subspace of $\mathcal{H}$, hence determining values for each of the observables in the maximal subalgebra generated by $\mathcal{O}$.

Digression. Those interested in the quantum logic approach to quantum theory are interested in a lattice of propositions isomorphic to the lattice of subspaces of $\mathcal{H}$. This lattice ought not to be confused with the usual algebra of projection operators. Thus in the lattice, the supremum of two elements corresponds to the subspace spanned by the basis vectors of both of the elements, the infimum to the intersection of the subspaces and the ortho-complement to the orthogonal subspaces and so to each operation there corresponds again a projection operator. But these latter operators in general have no simple definition in terms of the operators corresponding to the individual lattice elements. For example, if P_1, P_2 are two *commuting* projections then $P_1 P_2$ is the projection operator for the intersection, and $P_1 + P_2 - P_1 P_2$ that for the span, of the subspaces corresponding to P_1, P_2, *but not otherwise.* Thus the algebra of projection operators is very different in structure from the lattice of subspaces of $\mathcal{H}$.[99] (The partial algebra of commuting projection operators is however related to the *PO* set of propositions usually considered instead of the full lattice – cf. n. 12.) *End of digression.*

In terms of the abstract structure given in Section III we can assess the role of $\mathcal{H}$ in QM by viewing QM in the usual Schrödinger representation where $\mathcal{H} = L^2(R^3)$, is the set of all functions ψ over the complex number field such that $\int_{-\infty}^{\infty} |\psi|^2 dx < \infty$, $dx = dx_1 dx_2 dx_3$. The observables are realized (represented) on this space as operators. (Most familiarly, position and momentum are represented by the operators $\tilde{x}$, $i\hbar\partial/\partial x$.) In this representation we have

$$\psi_t = U_{t,t'}\psi_0 = e^{iHt/\hbar}\psi_0$$

$$\text{Therefore } i\hbar\psi_t = i\hbar\frac{d\psi_t}{dt} = H\psi_t$$

which latter is the Schrödinger equation, solutions ψ_t of which belong to $\mathcal{H}$. The Hamiltonian $H/\hbar$ is thus an observable and its eigen values are the system energies. (Equivalently, we can let the $U_{t,t'}$ operate on the observables instead of on the ψ_t. In this case the observables retain their spectral values but transform unitarily under $U_{t,t'}$ as $\mathcal{O}_{t'} = U_{t,t'}\mathcal{O}_t U_{t,t'}^{-1}$. This is the 'Heisenberg picture'.) We see then that $\mathcal{H}$ plays the logical role of the (derived spacial) *function space* $\mathcal{S}_T$, it is the space of all functions to which the solution of the Schrödinger equation, describing the dynamical evolution of quantum systems, belongs. Denote this space by $\mathcal{S}_T^{QM}$. Finally we consider the set $\mathcal{S}_\mu^{QM}$ of probability measures on $\mathcal{H}$. These are given by the set of all positive definite, hermitian trace class operators of trace 1 on $\mathcal{H}$ and I shall refer to them as the statistical operators of the theory. This set of measures is both atomic and complete, i.e. every possible probability measure over $\mathcal{H}$ is included in this specification (Gleason[100]) and every measure is a convex combination of the extreme or pure measures. These latter are the projections on the closed one-dimensional linear subspaces of $\mathcal{H}$. The expectation value for the operator $\mathcal{O}$ given by the statistical operator W is

$$\langle \mathcal{O} \rangle = \mathrm{Tr}(W\mathcal{O})$$

where Tr is the trace and is defined by

$$\mathrm{Tr}(\mathcal{O}\mathcal{O}') = \sum_{n,m} (\phi_n, \mathcal{O}\phi_n)(\psi_m, \mathcal{O}'\psi_m)$$

where the ϕ_n, ψ_m are any two sets of normalized $((\phi_n, \phi_n) = (\psi_m, \psi_m) = 1$, $(\phi, \psi) = \int \phi^* \psi \, dx)$, pair-wise orthogonal $((\phi_n, \phi_k) = (\psi_n, \psi_k) = \delta_{nk})$ 1-dimensional elements of $\mathcal{H}$ that span $\mathcal{H}$ (i.e. such that every $\psi \in \mathcal{H}$ can be written as a linear combination of the ϕ_n or the ψ_m). If $\mathcal{O}$ is an observable and the ϕ_n are eigenvectors of $\mathcal{O}$, i.e. if

$$\mathcal{O}\phi_n = \alpha_n \phi_n$$

then the α_n give the permissible values of the observable $\mathcal{O}$. (In this special case $\mathcal{O}$ has the spectral decomposition $\mathcal{O} = \sum \alpha_n P_{\phi_n}$.) In the coordinate basis given by the ϕ_n the statistical operator W will have a general representation

$$W = \sum \omega_i P_{\phi_i}$$

where the ω_i give the weights of pure states P_{ϕ_i} in the mixture W. Then in this representation

$$\langle \mathscr{O} \rangle = \mathrm{Tr}\,(W\mathscr{O}) = \sum_n (\phi_n, W\mathscr{O}\phi_n) = \sum_n \omega_n \alpha_n$$

as required.[101] If $\omega_n = \delta_{ni}$, i.e. if $W = P_{\phi_i}$, then

$$\langle \mathscr{O} \rangle = (\phi_i, \mathscr{O}\phi_i).$$

In particular if $W = P_\phi$, for any ϕ, then

$$\langle \mathscr{O} \rangle = (\phi, \mathscr{O}\phi).$$

Now suppose that $\phi = \sum_n a_n \phi_n$ then

$$\langle \mathscr{O} \rangle = (\sum_n a_n \phi_n, \mathscr{O} \sum_i a_i \phi_i) = \sum_n |a_n|^2 \alpha_n.$$

Thus we interpret the $|a_n|^2$ as the probabilities, for element ϕ, that the observable $\mathscr{O}$ will display the value α_n. In this fashion the statistical interpretation of the theory is established.[102] (I mean, the *mathematical* statistical interpretation, for the *physical* significance of QM of course lies in the interpretations given to the elements ϕ of $\mathscr{H}$, to the notion of an 'observable' and to such phases as "display a value". It is just these latter which are unsettled as yet – getting the mathematics straight does nothing to settle the issue.)

Thus we have the following important correspondence: *The extremal pure statistical measures are exactly those idempotents of the algebra of observables standing in one-one correspondence with the closed one-dimensional linear subspaces of $\mathscr{H}$.* Moreover, the characteristic operators of the Hilbert space are just those operators mapping every vector in a closed linear subspace of $\mathscr{H}$ into itself and all other vectors into the null vector. But these operators are just the projection operators once again. Thus we have in QM that the idempotent observables and the characteristic operators coincide and both coincide with the pure statistical measures in the one-dimensional case. This set of relationships is an exact analogue of that found for CPM-CSM and for CFT. This is then sufficient to show that for QM $SPA_0^{QM} \approx PA_0^{QM}$, the statistical partial algebra and the partial algebra of observables are isomorphic.[103]

The mathematical characterization of the structure of QM is now essentially complete. The structure may be represented diagramatically as follows:

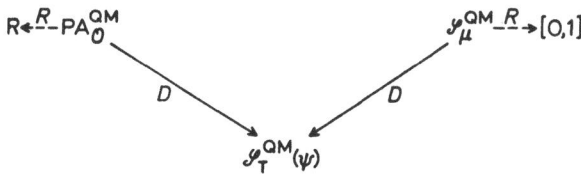

This structure is strikingly similar to that of CFT; it is quite unlike that of CPM. I shall comment now on a few of the pertinent comparisons between QM and CPM and CFT.

At the outset two obvious differences between QM and both CFT and CPM ought to be recorded. (1) The observables of QM are not all compatible, contra to CPM and CFT, so that one does not obtain a complete algebra. (Note, however, that the observables of all three theories can be construed as operators on their function spaces, though only for QM are these all *linear* operators. Moreover, in both CPM and CFT it is possible to choose other definitions of products of observables that would make the resulting algebras non-commutative, but in the classical context this would have to be interpreted by saying that the observable $\mathcal{O}_1 \cdot \mathcal{O}_2$ was not the same as the observable whose values were the point-wise numerical products of the values of $\mathcal{O}_1$ and $\mathcal{O}_2$, nor the same as $\mathcal{O}_2 \cdot \mathcal{O}_1$). (2) The elements of $\mathscr{S}^{QM}_\mu$ are *operators* on the elements of $\mathscr{S}^{QM}_T$ and not *set functions* (measures) on $\mathscr{S}^{CFT}_T$, respectively S^{6N}, as is the case with CFT, respectively CPM. These two features of the basic mathematical structure of QM constitute its most significant departures from classical thought. Nonetheless its links to classical physics are intimate.

Surely the most fundamental striking similarity between QM and CFT is the absence of a phase space to play the logical role of a state space in QM. Even more clearly so than in CFT such an entity is *logically* ruled out in the case of QM – this is the import of the theorems of Gleason and Kochen and Specker.[104] (On the other hand even phase spaces in CFT are artificial constructions and, to a degree, *logically* out of place. It would be of considerable interest to conduct a more thorough investigation into the possible similarities between the structure of QM and that of CFT in respect of the logical status of phase space between the two theories.)

Correspondingly, both CFFT and QM have linear vector spaces as their functions spaces, thus both share the Principal of Superposition and the correlative Principal of Resolution Ambiguity. Indeed, it is well known that QM exhibits the Principal of Superposition in a very strong way – it lies at the root of the beautiful symmetries which the theory displays[105] and, arising from that role, it is at the heart of the paradoxes of QM. (Thus, for example, it is just the ability to form such superpositions in Hilbert space that is at the root of the paradoxes of the two-slit experiment, the Einstein-Podolsky-Rosen argument and the peculiarities of the so-called identical particles of QM.[106]) The Principal of Superposition is grounded essentially on the *linear vector* space structure of its function space and it shares this fundamental characteristic with CFFT, and, basically, with CFT as a whole.

The situation with CPM is in marked contrast to this. The pivotal point of the structure of CPM is phase space in its logical role as state space but the essential characteristic of this space, a characteristic which it imparts to all other aspects of the structure, is its set-theoretic structure, in direct antithesis to any vector space structure. (This linear vector space structure is precisely the ground for the distinction between the logical structures arising in QM and those arising in CPM and CFT, cf. below.)

I reach here then my first tentative conclusion: *In its fundamental structure QM has the form of a field theory.* Thus in respect of its overall mathematical structure QM lies in the plenum tradition of physics. (It is needless to add that this runs counter to the traditional textbook emphasis on the akinness of QM to CPM, but reversing things in this way does help to explain a great deal, as we shall see.)

When we turn to the dynamics of QM we find basic, but different similarities to both CPM and CFT. These similarities lie on quite different structural levels. Fundamentally, all three share a Lagrangian dynamics; what this amounts to we have already seen (at least as far as it is clearly understood). Because of this common structure, however, it remains true in all three theories that invariances of the Lagrangians (symmetries of the systems) under linear transformations correspond to conserved observable quantities. Moreover, to the dynamical evolution of observables

$\mathcal{O}$ for both CPM and CFT

$$\frac{d\mathcal{O}}{dt} = \{\mathcal{O}, H\}$$

there corresponds in QM the dynamical equation

$$ih\frac{d\mathcal{O}}{dt} = [\mathcal{O}, H]$$

and to the dynamical evolution of probability measures in CPM

$$\frac{d\mu}{dt} = -\{\mu, H\}$$

(it is uncertain in my mind whether CFT also shares this equation of dynamical evolution and, if so, under which conditions), there corresponds

$$ih\frac{d\mu}{dt} = -[\mu, H].$$

On the other hand the construction of observables in CFT also permits the development of microscopic conservation equations such as the equations of internal continuity of charge current, energy-momentum density and so on which have no counterpart in CPM but do have counterparts in QM, for example, in the continuity equation expressing conservation of 'probability current density' (i.e. of the Schrödinger field intensity).[107] The dynamics of QM, like CFT, is described in the derived spacial function spaces of the theory, the Schrödinger equation may be obtained from a field-like Lagrangian (but not from a particle-like Lagrangian) and, as in CFT, is a partial differential equation whose solutions are functions of both space and time.

Unlike CFT and like CPM, however, the basic dynamical equation (the Schrödinger equation) is not covariant nor can it be put in covariant form because of the basic asymmetry between the roles of the space and time variables in it. (Time is not an observable of QM, spacial location is; the time variable appears as an independent parameter in QM while spacial location is represented by an hermitian operator.) Indeed, the *specific form* of the dynamical equation of QM displays a remarkable similarity to that of CPM. In the traditional presentation their correspondence is made clear in the comparison of the Hamilton-Jacobi characteristic equation and Schrödinger equation.[108] There is, however, a deeper way in which to state the matter. The quantum dynamics satisfies the following 'CPM Conservative Principal': QM preserves exactly the algebraic form of the Hamiltonian of CPM when it is expressed in the preferred coordinates of CPM, i.e. those generating the space-time symmetries (canonical momenta – translation symmetry, angular momenta – rotation symmetry, and positions – frame transformation equivalence) of both CPM and QM, when the mathematical representatives for observables are changed from real valued functions on phase space to hypermaximal hermitian operators on a Hilbert space in such a way that the Lie algebra satisfied by the preferred quantities is left invariant.[109] The dynamics of QM is then, in a certain sense, as close as possible to that of CPM. On the other hand it is, in this sense, far removed from CFT, since in CFT the Lagrangian structures are markedly different from those occuring in CPM.

On the basis of these considerations I reach my second tentative conclusion: *QM shares with both CPM and CFT the general Lagrangian-Hamiltonian dynamical structure and it shares with CFT the general formulation of the dynamical structure characteristic of a field theory, but, with respect to its specific dynamics QM retains as closely as possible the fundamental form of the dynamics of CPM.* (There remain here important questions for which I do not have an answer: Is there a sharp formulation of the question of whether the structural departures of QM from both CPM and CFT are precisely those required to accomodate the specific dynamical forms of CPM to the general field theoretic (plenum) context within which QM is formulated? What is the answer to this more precise question?)

Though these two tentative conclusions are a little crude, they certainly support the general conclusion to which the discussion leads: *QM, in respect of its mathematical character, has the form of a complex fusion (mixture, marriage) of CFT and CPM structures.*

Throughout this discussion I have deliberately refrained from introducing the notion of a QM *state*. Contrary to the common approach where talk of the *state vectors* of Hilbert space abound, I have deliberately refrained from introducing this idea. The last general conclusion, and the preceding considerations (especially those of ontology and conceptual structure) should have made the reason for this caution obvious.

Any move to introduce the notion of state to QM must proceed very cautiously unless all the very important distinctions are to be slurred over and all the interesting questions begged, for this concept is at the heart of any interpretation of QM. I shall postpone this consideration until the next section.

Of a piece with the view that the elements of the represented Hilbert space of QM are QM states is the view that this space is the analogue of the phase space of CPM. There are some analogies between the two, it is true, but these are born only by the narrow dynamical forms which CPM and QM share; by comparison with the deep going analogies with CFT they are relatively more shallow. Even more insidiously, there is the temptation to say that the two (Hilbert space and phase space) are analogues just because in both cases they are the domains of the elements of the sets of observables and probability measures. This comparison carries weight, however, if and only if the fundamental difference between CPM and CFT is ignored. In particular this conclusion can be reached only by ignoring the fundamental difference in logical roles played by the function space in both CFT and QM as the domain of the observables and probability measures as compared to that role played by the phase space in CPM. But I shall equally refrain from the temptation to speak of the elements of Hilbert space as states on the grounds that the function space of CFT is the state space of that theory.

For the sake of completing the formal description of QM I shall briefly discuss the propositional structures arising therein. Of course these are precisely the propositional structures considered by those interested in 'quantum logic' and its algebraic equivalents, so once again I can be brief.[110] This latter approach essentially deals with only two propositional forms – these are special cases of propositions of the form

> The value of observable $\mathcal{O}$ lies in E

which I denote by $P_{\mathcal{O}, E}^{QM}$, $E \subseteq R$ and is a borel set, and propositions of the form

> The probability is r that in an ensemble characterized by a statistical operator W the value of observable $\mathcal{O}$ lies in E

which I denote by $M_P^{QM}(W, \mathcal{O}, E, r)$.

Consider first the $P_{\mathcal{O}, E}^{QM}$ and to begin with restrict attention to just projections (i.e. to just the idempotents of PA_0^{QM}) and the special case $E = \{1\}$. These propositions, which have the symbolic form $P_{I_0, \{1\}}^{QM}$, are the analogues of the $P_{fc_s, \{1\}}^{CPM}$ of CPM and of the $P_{I_s, \{1\}}^{CFT}$ of CFT. They are often the only propositions referred to in the discussions in the literature, for example in the discussion by Kochen and Specker [74]. Now the projections are in 1-1 correspondence with the subspaces of $\mathscr{H}$ and take on only the values 0,1. Thus the propositions $P_{I_0, \{1\}}^{QM}$ stand in 1-1 correspondence with the subspaces of $\mathscr{H}$. Let us denote the subspace corresponding to the observable I_0 by $h(I_0)$ and rewrite the propositional symbol so as to indicate the subspace explicitly, thus $P_{h(I_0), \{1\}}^{QM}$. The correspondence between propositions and subspaces then forms the basis for the following definition of the logical connectives.

$$P_{h(I_0), \{1\}}^{QM} \wedge P_{h(I_0'), \{1\}}^{QM} =_{df} P_{h(I_0) \cap h(I_0'), \{1\}}^{QM}$$

$$P_{h(I_0), \{1\}}^{QM} \vee P_{h(I_0'), \{1\}}^{QM} =_{df} P_{h(I_0) \cup h(I_0'), \{1\}}^{QM}$$

$$\neg P_{h(I_0), \{1\}}^{QM} =_{df} P_{h(I_0)^\perp, \{1\}}^{QM}$$

where $h \cup h'$ is the span of the subspaces h, h'; $h \cap h'$ is the intersection of the subspaces h, h'; $h^\perp$ is the orthogonal subspace to h and each is again a subspace of $\mathscr{H}$. Under these definitions it can be shown that the $P_{h(I_0), \{1\}}^{QM}$ form themselves into an orthocomplemented, orthomodular lattice $L_{I_0}^{QM}$.

Now consider the $P_{0, E}^{QM}$ in general. By the spectral theorem we have that to every observable and borel subset of R there corresponds a set of projection operators on $\mathscr{H}$. Thus with $P_{0, E}^{QM}$ we may associate the subspace $h(0(E))$ of $\mathscr{H}$, where $P_{h(0(E))}$ is the projection (onto $h(0(E))$ associated with $\mathscr{H}$, E by the spectral theorem. Under this association and with the foregoing definitions of the connectives $\wedge$, $\vee$, $\neg$ the $P_{0, E}^{QM}$ achieve the same structure exactly as the $P_{h(I_0), \{1\}}^{QM}$, namely $L_{I_0}^{QM}$.[111]

The foregoing definitions define $\wedge$, $\vee$ for all propositions, even when $0, 0'$ do not commute. In the light of the alleged physical significance of compatibility (commutativity) as co-measurability it seems unreasonable to demand that such logical combinations of propositions be always defined and many writers have proposed to introduce only a weaker structure, namely that obtained when $\wedge$, $\vee$ are defined only for compatible $0, 0'$, e.g. Gudder and Greechie (cf. this volume), Kochen and Specker and others. In this way one obtains an orthocomplemented, orthomodular POset POS_0^{QM} rather than the lattice $L_{I_0}^{QM}$. This is surely the more correct approach (compatibility seems at least a *necessary* condition of co-measurability, if not a sufficient condition) and in what follows the stronger conditions (employed, e.g. by Jauch [68]) will be dropped.

Earlier on we saw that the full partial algebra $PA_{\mathcal{O}}^{QM}$ was constructed from a collection of subalgebras $SA_{\mathcal{O}}^{QM}$ generated by the basic observables $\mathcal{O}'$. In similar fashion the *PO*set $POS_{\mathcal{O}}^{QM}$ is constructed from a collection of maximal subalgebras $S\mathscr{B}_{M\mathcal{O}'}^{QM}$ these being exactly those boolean subalgebras corresponding to all propositions of the form $P_{\mathcal{O},E}^{QM}$ with a given maximal observable $\mathcal{O}'$ compatible with $\mathcal{O}$. (It is clear that we will obtain a *boolean* subalgebra since every proposition is compatible with every other, or, to put it another way, the associated subspaces either pair-wise coincide with, or are orthogonal to, each other and hence the foregoing subspace connectives assume a boolean algebraic structure.) From the earlier discussion of maximal subalgebras of $PA_{\mathcal{O}}^{QM}$ it is clear that a proposition $P_{h(I_{\mathcal{O}}),\{1\}}^{QM}$ determines the values of all observables in one such maximal subalgebra, since it determines a unique value for the basic observable whose spectral decomposition includes that particular 1-dimensional subspace. This idea may be represented explicitly in the present context for, reversing the above procedure, for every observable $\mathcal{O} \in S\mathscr{B}_{M\mathcal{O}'}^{QM}$ generated by the basic observable $\mathcal{O}'$ whose spectrum contains the 1-dimensional subspace ϕ let $E_{\mathcal{O}} \subseteq R$ be the unique borel set such that $h(\mathcal{O}(E)) = [\phi]$ (where $[\phi]$ is the 'ray' containing ϕ, i.e. the subspace of all constant multiples of ϕ). Then we have, for all $\mathcal{O} \in S\mathscr{B}_{M\mathcal{O}'}^{QM}$,

$$P_{P_{[\phi]},\{1\}}^{QM} \equiv P_{\mathcal{O}',E_{\mathcal{O}'}}^{QM} \text{ and } P_{\mathcal{O}'',E_{\mathcal{O}''}}^{QM}, \text{ and } \ldots$$

In short the $P_{P_{[\phi]},\{1\}}^{QM}$ ($= P_{I_{\mathcal{O}},\{1\}}^{QM} = P_{h(I_{\mathcal{O}}),\{1\}}^{QM}$) generate ultrafilters on, and are the atoms of, $S\mathscr{B}_{M\mathcal{O}'}^{QM}$ (recall again that the $S\mathscr{B}_{M\mathcal{O}}^{QM}$ do not exhaust $POS_{\mathcal{O}}^{QM}$).

Now let us turn to propositions of the form $M_P^{QM}(W,\mathcal{O},E,r)$. As previously, attention is restricted to the case $r = 1$ and then the $M_P^{QM}(W, \mathcal{O}, E, 1,)$ for fixed $W,\mathcal{O}$ and varying E can be made to formally correspond to a boolean algebra (with a suitable definition of a $\wedge$, $\vee$, cf. Section IV). The interest arises in QM, as elsewhere, in the attempt to combine such propositions for varying $\mathcal{O}$. In CPM and CFT this was accomplished relatively easily because the observables were themselves all compatible and indeed a unique value for each was determined by fixing a location in phase space for CPM and in function space for CFT. To obtain some structure in the present case, let us first specialize to $M_P^{QM}(W, P_{[h]}, \{1\}, 1)$, where the $P_{[h]}$ are projection operators on the subspace $[h]$. The QM algorithm for computing probabilities informs us that these latter propositions are true exactly if the statistical operator W can be expressed as a mixture of projections in the subspace $[h]$. Thus these special propositions correspond 1-1 to the propositions $P_{P_{[h]}, \{1\}}^{QM}$ and so form the same structure, namely $POS_\mathcal{O}^{QM}$. Now we can see how to generalize to the $M_P^{QM}(W, \mathcal{O}, E, 1)$. The spectral theorem and the QM algorithm for computing probabilities inform us that this proposition is true if and only if W can be expressed as a mixture of projection operators in the subspace $h(\mathcal{O}(E))$; thus the $h(\mathcal{O}(E))$ may also be associated 1-1 with these propositions and so the $M_P^{QM}(W, \mathcal{O}, E, 1)$ form a partial algebra $POS_{W, 1}^{QM} \approx POS_\mathcal{O}^{QM}$. (This last correspondence explains, I think, how it is that varying writers at varying times, and some writers in the same paper, can speak about the propositions $P_{\mathcal{O}, E}^{QM}$ and $M_P^{QM}(W, \mathcal{O}, E, 1)$ as if there were no distinction between them. And it might explain why no one has commented on the significance of the fact that in the classical context we can form these probabilistic propositions into a *formal* boolean algebra only by introducing logical connectives which behave exactly like those defined in terms of the relations among the subspaces of our linear vector space, whereas in QM where we have precisely to do with such subspace relations the possibility of constructing such a formal boolean algebra is largely irrelevant.) Of course we remain almost totally ignorant about the logical structure of propositions for which $r \neq 1$.

Just as for CFT, there are again no analogues to the P_S^{CPM}, $P_{S, t}^{CPM}$, $M_S^{CPM}(\mu, S, r)$ propositions of CPM. This leaves us with only the analgues of the T_S and $\mathcal{M}_S$ propositions of CPM and CFT to discuss. Then consider first the propositions T_S^{QM}

> The system selects an element ϕ of $\mathcal{H}$ contained in the subspace S.

It is easy and natural to turn the T_S^{QM} into the usual orthocomplemented *PO*set (or orthomodular lattice) by introducing the usual definitions of the connectives in terms of subspace operations. What then of the relation of the T_S^{QM} to the $P_{\mathcal{O}, E}^{QM}$? Suppose, for a given $\mathcal{O}$, $h(\mathcal{O}(E)) = S$, then T_S^{QM} is true exactly if $P_{\mathcal{O}, E}^{QM}$ is true. A slight change of formalism will simplify the general statement; let $\mathcal{O}(S)$ be the borel subset on the real line associated with the subspace S and observable $\mathcal{O}$ by the spectral theorem. Then we have

$$T_S^{OM} \equiv P_{\mathcal{O}, \mathcal{O}(S)}^{QM} \text{ and } P_{\mathcal{O}', \mathcal{O}'(S)}^{QM} \text{ and } \dots$$

just as we do in CPM and CFT (except that in this case the $\mathcal{O}(S)$ will not necessarily be defined for all cases so that we do not obtain the complete correspondence characteristic of CPM and CFT). The foregoing equivalence holds only for a particular time t, if we were to form the equivalent T_S^{QM} propositions for the full function space which included the time-dependent part of the functions then we would obtain a similar equivalence but with propositions of the form $P_{\mathcal{O}, \mathcal{O}(S), t}^{QM}$ and covering all times t.

We come now, finally, to propositions $\mathcal{M}_S^{QM}(W, S, r)$ of the form

> The probability is r that a member system of an ensemble characterized by a statistical operator W selects an element of $\mathcal{H}$ in the subspace S.

Because each W can be represented as a sum of projections, the coefficients of which give the statistical weightings of each projection in the sum, these coefficients give exactly the probability values r as S runs through the subspaces of $\mathcal{H}$. Thus in particular the $\mathcal{M}_S^{QM}(W, S, 1)$ are true just in case W is a mixture of projections in the subspace S and hence the $\mathcal{M}_S^{QM}(W, S, 1)$ once again form the POset $POS_{\mathcal{O}}^{QM}$ under the usual definitions of the logical operations. Slightly more generally, we have

$$\mathcal{M}_S^{QM}(\sum \alpha_i P_{[\phi_i]}, [\phi_i], \alpha_i), \text{ all } i.$$

Conversely, a consistent set of the $\mathcal{M}_S^{QM}(W, S, r)$ such that the subspaces span $\mathcal{H}$ is sufficient to determine W. Finally, we can obtain a relation between the $\mathcal{M}_S^{QM}(W, S, r)$ and the $M_P^{QM}(W, \mathcal{O}, E, r)$. For if $\mathcal{M}_S^{QM}(W, S, r)$ then $W = rP_{[S]} + W'$, where W' contains only projections $P_{[S']}$ such that $[S] \cap [S'] = \emptyset$, and in that case $M_P^{QM}(W, \mathcal{O}, \mathcal{O}(S), r)$ is true, and conversely. Moreover since

$$M_P^{QM}(W, \mathcal{O}, E, 1) \leftrightarrow P_{\mathcal{O}, E}^{QM}$$

we have

$$\mathcal{M}_S^{QM}(W, S, 1) \leftrightarrow P_{\mathcal{O}, \mathcal{O}(S)}^{QM}$$

Once again I shall refrain from further discussion of these propositional structures, though clearly the foregoing discussion barely scratches the surface. (There are, for example, subtle relationships between the $\mathcal{M}_S^{QM}(W, S, r)$ and the $P_{\mathcal{O}, E}^{QM}$ which may be explored, for example after the fashion of van Fraassen in this volume by introducing modal logical structures – though this exploration is itself in its infancy.)

Thus we have for QM:

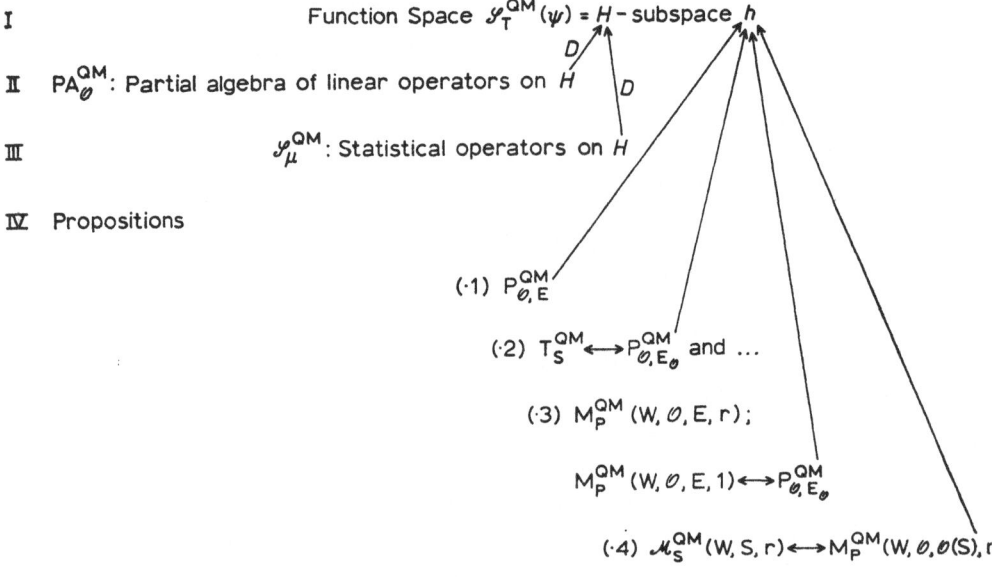

I   Function Space $\mathcal{S}_T^{QM}(\psi) = H$ - subspace h

II $PA_{\mathcal{O}}^{QM}$: Partial algebra of linear operators on H

III $\mathcal{S}_\mu^{QM}$: Statistical operators on H

IV   Propositions

         (·1) $P_{\mathcal{O}, E}^{QM}$

         (·2) $T_S^{QM} \longleftrightarrow P_{\mathcal{O}, E_{\mathcal{O}}}^{QM}$ and ...

         (·3) $M_P^{QM}(W, \mathcal{O}, E, r)$;

         $M_P^{QM}(W, \mathcal{O}, E, 1) \longleftrightarrow P_{\mathcal{O}, E_{\mathcal{O}}}^{QM}$

         (·4) $\mathcal{M}_S^{QM}(W, S, r) \longleftrightarrow M_P^{QM}(W, \mathcal{O}, \mathcal{O}(S), r$

         .... and $M_P^{QM}(W, \mathcal{O}', \mathcal{O}'(S), r)$ ..

Surveying the QM propositional structures we see that they also bear overall important similarities to those of CFT (for example, the absence of the $P_S^{CPM}, P_{S,t}^{CPM}, M_S^{CPM}(\mu, S, r)$ propositions) while displaying the profound dissimilarities to *both* CPM and CFT which arise form the departures from classical thought in QM and are reflected in the non-boolean propositional structures which recur in QM. (One might say, however, that since this logical structure is precisely that arising from the lattice of subspaces of an infinite dimensional vector space, QM was in this respect merely exploiting more fully the basic structure which it shared with CFT and which CFT, because of its classical (and/or non-linear?) character did not exploit.) Thus a review of the propositional structure of QM supports the general conclusion drawn earlier, namely, that QM has the form of an amalgam of the structures of CFT and CPM, but that of CFT dominates the structure.

XII. ON THE INTERPRETATION OF QM

Before any examination is undertaken of the possible interpretations of QM there are two kinds of considerations which have priority and of which I should like to remind the reader.

(1) The interpretational problem is *not* merely a question of how to correlate certain mathematical features of the formalism to observation (crassly so-called). As I pointed out earlier, scientists know how to do this already (though in something of an intuitive guess-heuristic way in the less familiar situations), the Born Rule is the working guide and one has only to figure out what to count to obtain appropriate statistics. The point of an interpretation is to tell us what is really going on when scientists are undertaking this activity. Ultimately, I have argued, this boils down to giving an account of the ontology of QM.[112]

(2) Let us put together the following propositions each of which has been discussed earlier:

P_1: The basic ontologies of CPM and CFT diverge radically from one another (Section VII).

P_2: The conceptual descriptive schemes truly appropriate to the description of the physical world in terms of the plenum and atomic ontologies have radically different concepts and structures (Section VIII).

P_3: The mathematical and propositional structures of CPM and CFT are markedly different (Sections IV, IX, X).

P_4: QM has the mathematical form of a fusion of the structures of CPM and CFT (Section XI).

P_4, in the light of P_1, P_2, P_3, suggests *great caution* in the discussion of any interpretation for QM.

In fact P_4 taken in this light leads us to expect that attempts to interpret QM, especially early attempts, will have the appearance of mixtures of components from the interpretational structures of both CPM and CFT. Such was indeed the case. The early discussions (I mean until at least

1965! – some physicists excepted) surround the problem of the "wave-particle duality". This duality is exactly one of field (plenum) and particle (atom) conceptual structures and the 'problem' was that QM seemed to demand features of both at once and yet resist an interpretation exclusively in terms of the one or the other. This feeling on the part of physicists struggling with the new formalism should no longer surprise us. What I hope the foregoing discussion has made clear is that *such an approach fails to appreciate the depth of the cleavage between the two schemes, fails to consider carefully enough the mathematical structures of particle and field physics, and fails to appreciate the intimate connections between the mathematical characteristics of the two structures and the corresponding conceptual structures; the approach is therefore doomed at the outset to be no more than a makeshift, more or less ad hoc, adjustment pressing various fragments of the particle and field conceptual schemes into service blindly as the interpretational occasion demands.*[113]

Recently, however, there have been more sophisticated studies undertaken on the interpretational problem. Von Neumann initiated the first rigorous program when he raised the question of supplementing QM with 'hidden variables', i.e. of reconstruing QM as a phase space theory. This line of argument was pursued by Gleason, Kochen and Specker and others, culminating in a clear demonstration that no such emendation is possible so long as the structure of QM is strictly preserved (cf. Bub in this volume and also further below). Again, from the present perspective this is not surprising (though it is not obvious and what Gleason and Kochen and Specker demonstrated were deep theorems). In the process there developed the propositional structure approach to QM which has to date proven one of the truly fruitful tools for the precise discussion of the structure of QM (cf. n. 97). Although quite sharp formulations of the structure of QM have been obtained which bear directly upon the question of its interpretation, for myself I still feel that we do not understand the theory yet, not its place in the historical scientific tradition, at all deeply. I shall try to make my reasons for thinking this clear in the remainder of the essay (if they are not clear enough already).

Once the peculiar structure of QM is clearly perceived, in particular once the distinctively non-boolean character of the propositional structure of QM is properly understood and its mathematical character as a fusion of the structures of CFT and CPM is properly grasped, it is clear

I think that there can be no interpretation of QM which (i) is a Realist interpretation, (ii) makes the QM description of the world complete, except that interpretation which asserts the objective non-boolean form of the world. All other interpretational attempts are essentially disguised proposals for a new physics displacing QM to the realm of the relatively superficial, and/or misleading and/or false. (I am tacitly assuming here that all such interpretations are to be given within the conceptual structures of the plenum and atomic schemas, for just these have dominated the entire scientific tradition to date. Of course it is possible to challenge this assumption and I shall discuss that challenge a little later on.)

We can understand this conclusion concerning QM from within the framework provided by the foregoing analysis. If QM is indeed a fusion of mathematical structures connected to incompatible ontologies and conceptual schemes then a *consistent* interpretation, restricted to the concepts of these conceptual schemes, *must* do one of the following:

(α) Refrain from treating QM Realisticly at all (no direct challenge to any preferred ontology arises),

(β) treat Realisticly only those portions of QM pertaining to the favoured ontology,

(γ) introduce special actions (e.g. uncontrollable disturbances-upon-measurement, creation-of-property-values-at-measurement, mental actions) external to QM to explain those features of QM lying outside the scope of the favoured ontology,

(δ) treat QM as a systematically misleading guide to ontology (to the favoured ontology), i.e. assert that what the QM instruments measure and to which the usual QM observables are related is itself related only indirectly to the real objects of the preferred ontology and their real properties,

(ϵ) introduce rules for the use of physical descriptions that are designed to logically segregate the concepts of the two ontologies,

(ι) claim that QM is strictly false and at best an approximation to a better theory.

These alternatives capture every attempt at 'interpreting' QM that I have been able to find in the literature. At least one of them is necessary under the foregoing assumptions because the hybrid character of the structure of QM assures us that no interpretation exclusively in terms of either the plenum or atomic schemas can succeed. A way must therefore be found to render innocuous the clash between part of the theory and the favorite view of the world. The six ways sketched out above are the six possible ways to render part of, or the whole of, a theory of the QM sort innocuous.[114]

It is clear that each of these alternatives, with the possible exception of (ε), explicitly or tacitly demands a new physics if a Realist account of the world is to be given. (α) demands a new physics simply because on this view there is at present no objective fundamental physics at all, (β) because the objective part of QM is incomplete, (γ) because we have as yet no physics of these special actions or relationships, (ε) because, if the rules are to be understood as referring to the physical structure, we have no physics of that structure, and the reason in case (ι) is obvious. (On the other hand alternative (ε) will escape if it is insisted that what is required is not new physics but new logic and/or new semantics.)

In particular, as Bub has emphasized,[115] each of these positions, with the possible exceptions of (α) and (ε), creates its own particular form of the 'measurement problem' in QM. (This problem initially arises, as the reader will recall, from the 'collapse of the wave packet' upon measurement, since prior to measurement all physical possibilities were open whereas after the measurement we know which of these was realized. The problem, however, goes deeper than this for it is not simply that a collapse occurs but the logical structure of the manner of calculating probabilities, especially conditional probabilities, that is ultimately at the root of the problem. See Bub's essay in this volume for a clear exposition and cf. also my treatment in [52, 57].) (α) may escape the difficulties simply by refusing to discuss the physical significance of measurement at all. In so far, it is a totally mysterious theory without real empirical foundation. To the extent it wishes to enter upon such a discussion of measurement it must give an account of how the representation of the *knowledge of possibilities* in QM is related to the *actualities* revealed in measurement – but this is just the measurement problem. (γ) attempts to retain a favoured ontology by introducing special physical actions. In this case the measurement problem appears as the problem of explaining, by introducing these special actions, the collapse of the ψ field to a localized particle-like result if the preferred ontology is of the plenum variety, or as the problem of the relations between measurement results in complementary situations (I use Bohr's terminology, especially pertinent here are those situations displaying 'self-interference' phenomena) if the preferred ontology is of the atomic variety. (β) is half-way between (α) and (γ) and so shares at least one form, and usually both forms, of the measurement problem. For (δ) the problem is even more sweeping: since QM offers an account of sorts of what is happening during a measurement, this account must be so understood in the light of the true ontology (i.e. the preferred ontology) that (i) we can say what is actually being determined in various measuring situations and (ii) the detail of such situations must explain why QM, based upon the usual measurement statistics, is a systematically misleading guide to the preferred ontology. (ε), if it is to be a significantly different position from the quantum logic approach, will also have its special form of the measurement problem, namely to show how, by using the new semantical rules it proposes, no descriptions of measurement situations based upon the QM account of them can give rise to contradictions, absurdities of mysteries. Finally (ι)'s problem is simply to give the true account of measurement and show how the QM account is relatively superficial (but, presumably, is retained as some sort of limiting case).

A deeper formal reason why there is a measurement problem is this: under the logical structure of QM one does not calculate probabilities according to the classical, boolean rules (cf. Bub) whereas in one way or another each of the preceding alternatives, (γ) excepted, tacitly or explicitly seeks to retain these latter rules. In general, attempts to return to an essentially classical ontology for QM may be perspicuously viewed as attempts to divorce the partial algebra of observables from the statistical partial algebra and to adopt a boolean structure for the former. This move raises the question of the relation between SPA_θ^{QM}, which cannot be imbedded into a boolean algebra as it stands, and the new algebra of observables, BA_θ^{QM} say. It is possible to find an appropriate imbedding of SPA_θ^{QM} into BA_θ^{QM} if the mapping effecting this is many-one from BA_θ^{QM} to SPA_θ^{QM}, in particular, if the degenerate observables of QM are associated with as many distinct elements of BA_θ^{QM} as there are maximal boolean subalgebras in which they occur (cf. Bub, van Fraassen in this volume).[116] This approach yields the most important cases of alternative (δ). (α) is a degenerate form of any proposal of this sort, while (β) attempts simply to retain only a boolean-imbeddable fragment of QM and (ι) to substitute a new theory with a classical logical structure. However, since the QM probability structure is not that of a classical theory, there will always arise the problem of giving an account of these probabilities (especially of conditional probabilities) and this becomes essentially the measurement problem in each case (cf. Bub). In this sense, of course, proposing a new *logical* form to the world – the QM logical structure – will solve this problem by *dissolving it*; for probability rules are founded on a logical structure, thus change that underlying structure and the calculations change, *logically* must change, accordingly.[117]

On the other hand I must insist that from my perspective the measurement problem is generated from the attempt at marriage of the plenum and atomic points of view. The Ψ functions generated by Schrödinger's equation have the structural role and logical form of fields. In the field schema there are no localized objects and hence no probabilities in that sense, there are only probability measures over global field states. On the other hand the fundamental probabilities of the atomic schema are just those of the localized events associated with the localized atomic objects. Put vaguely, the problem arises because in QM fields are made to do duty for the distribution of atomic events, but the method of combining field distributions is radically different from that of combining atomic events (for example, the former permit superposition, the latter does not).[118] Unfortunately, I have at the present time no sharper formulation of the situation. (I do hope that the reader will be goaded into searching for one and that by the end of the paper he will have a clearer idea of what to look for.)

In the face of the foregoing there are some, for example, Bub in this volume and Putnam [106], who boldly postulate that the logical form of the world is non-boolean, i.e. non-classical, and in fact just that logical form appearing in the propositional structure of QM. This claim is to be understood in a fashion analogous to the way in which we understand the claim that the geometrical form of the world in the theory of *general relativity* is non-Euclidean.[119] Moreover it is clear, at least in outline, how Bub's approach 'solves' the paradoxes of QM. First notice that a pure particle conceptual structure is retained – though the particle ontology must be somewhat modified, as we shall see. Now in the *classical* conceptual scheme associated with the *classical* particle ontology one is accustomed to reasoning in certain ways arising from the classical assumption that each situation is every way definite in the full logical complement of physical properties (cf. my comments in Section IV). Some of these ways of reasoning are incompatible with QM. In particular the rule for conditional probabilities which holds for CPM is incompatible with the method of calculating conditional probabilities in QM, i.e. using von Neumann's projection postulate. (Bub discusses this matter in detail in this volume, cf. also my discussion in [52] and [57].) What Bub's approach does is to declare these ways of reasoning *objectively* invalid. (That is, they do not conform to any of the valid argument forms compatible with the actual logical form of the world.) These discarded methods of reasoning are replaced by objectively valid forms (axiomatized by Kochen and Specker, see [72, 73], but for what now seems to be a more adequate approach see van Fraassen [34]). In this way any clash with QM is avoided and the necessity to treat any of its parts as other than objectively about the world is avoided.

The fact remains, however, that when we have $P_{\mathcal{O},\{r\}}^{QM}$ true, $\mathcal{O}$ a basic observable, $\{r\}$ a singleton set in R (that is, a precise value for $\mathcal{O}$), we have $P_{\mathcal{O}',E}^{QM}$ false for *every* $E \subset R$, for *every* $\mathcal{O}'$ which does not commute with $\mathcal{O}$. (For example, if the position observable has a precise value then every assertion of a finite value range for the momentum observable is false.) From this it is clear that the QM ontology cannot be for Bub a pure classical particle ontology. And he is in fact driven by the 'logic' of his own position to claim that the *facts* of QM are physical situations in which not all of the classical properties appear. This comes about as follows. In CPM and CFT propositions specifying states determined ultrafilters on the logics $\mathcal{B}_{\mathcal{O}}^{CPM}$, $\mathcal{B}_{\mathcal{O}}^{CFT}$. This was, indeed, precisely the role demanded of a state specification, namely to determine maximally informative (maximally precise) consistent descriptions of the world. But in QM we have only a *partial* boolean algebra $P\mathcal{B}_{\mathcal{O}}^{QM}$ so, if we insist on carrying this criterion of state description across to QM propositions, QM state descriptions will be precisely those propositions that generate ultrafilters on the maximal boolean subalgebras $S\mathcal{B}_{M\mathcal{O}}^{QM}$, i.e. exactly the $P_{\mathcal{O}',\{r\}}^{QM}$ or equivalently, the $P_{P_\phi,\{1\}}^{QM}$.[120] But these state descriptions, we have just seen, omit reference to any physical properties represented by non-commuting observables. QM states, under this approach, may entirely lack some of the basic properties of classical physics. Against the systematic structural criteria for state descriptions used to arrive at this position, the reader will naturally pit the argument of Section IV that what counts as an exhaustive description is determined by *logical* relations among the descriptive concepts and that on this latter criteria QM state descriptions are simply not complete. It is just at this point that the full force of the assertion that the logical form of the world is not classical (boolean) is more strongly felt – it must apply exactly to the appropriate logical relations among our old descriptive concepts. Now there is some grounding for this conclusion (outside of the desire to have a consistent interpretation of QM at any cost), namely in the reciprocal indeterminacies of description holding between canonically conjugate quantities which, unlike the cases of CPM and CFT, now cannot be made arbitrarily small because of the finite units (quanta) in which changes occur. (See the full argument in [61].)[121] Thus Bub's non-boolean world is one of partially denuded classical situations. One supposes, therefore, that its ontology is a partially denuded classical ontology. (Though which properties an 'object' lacks, if substantial identity has a clear sense in this context, change from situation to situation.) At all events this ontology is so structured as to make the QM descriptions exhaustive (surprise!).

I regard this approach to QM, as elegantly as it deals with the structure of that theory, as a failure – but let us be quite clear as to the reason. It is not a failure because this *kind* of approach is misguided or a priori wrong, nor is it a failure because it is a wrong approach to this particular theory, for this approach may be legitimately taken to every such theory, and neither is it a failure because it fails to deal with the mathematical structure of the theory, indeed it is the only approach among those commonly recognized that does throw a penetrating light on the structure of QM, other approaches tend to obscure its structure. It fails because the kind of ontology it is forced to adopt in order to achieve coherence among its conceptual-logical, mathematical and ontological structures is simply unbelievable (even unintelligible). In fact, one can scarcely say there is an ontology at all – for although, through parasitism upon the atomic ontology of CPM, one is inclined to suppose that one can conceive of particles that have position and do not have momentum one moment (and place), have momentum but do not have position the next (at no place), more careful thought leads to the conclusion that no notion of an individual is really constructible at all under these bizarre circumstances. The quantum logic approach 'succeeds' in bringing the conceptual and mathematical structures into coherence but succeeds with ontology as well only at the expense of an essentially unintelligible ontology. The blame for this unsatisfying situation is not in my view to be laid at the door of the approach however, but on the theory itself – it reflects a deep internal incoherence in QM, one cannot bring conceptual, mathematical and ontological structure simultaneously into mutual coherence.

At this point it is appropriate to mention briefly Bohr's approach. I have examined and developed Bohr's approach elsewhere (see n. 6), the point I wish to make here is just this: Bohr's approach, appropriately developed in a rigorous fashion, is essentially just that of the quantum logicians. Reason: I have argued that Bohr's approach is to be understood as a semantical thesis based upon a notion of presupposition; quantum logic can also be developed as a presuppositional logic and when cast in this form it comes very close to expressing Bohr's position. The prospects for bringing the two into coincidence look promising at the present time.

This completes my account of the extant views on the interpretation of QM. There are no others of which I am aware. But still I have not spoken of the notion of a QM *state*. Considering that states in CPM and CFT enter these theories in quite different ways, let us first inquire how talk of states arises in QM. The issue revolves around the quantization process, for otherwise the theory could only be understood as a version of CFT with CFT states (as indeed Schrödinger first tried to view it). Quantization introduces discrete spectra for the observables (which are *linear* on the function space, unlike the observables of CFT) and hence the possibility of introducing an atomism (i.e. particles). The field theoretic nature of QM makes an appearance in this new context in the fact that, not field intensities, but probability intensities, are defined over all space-time locations by each member of the function space.[122] These probabilities are taken to refer to events exemplifying a member of the discrete spectra of the observables.[123] Talk of states then arises directly out of the appearance of discrete values and simultaneously shares the mixed analogy with CPM states (borrowing the notion of determination of precise discrete values to observables) and with CFT states (borrowing the space-time wholeness of a single CFT state specified in function space). Thus the sources for the notion of state lead, as elsewhere in QM, at best to a mixed conception destined to remain unclear and unsatisfactory.

Historical Aside. I believe it is helpful to view the history of QM wearing the glasses provided by the point of view of this essay. Schrödinger's inclination, as I said, was to view QM as a version of field theory – and a natural inclination it is. But with the advent of Heisenberg's matrix mechanics, which abstracted and manipulated only the discrete eigenvalues as it were, there came powerful reason to view QM as a particle dynamics. (Or rather, to view these theories in this fashion, it was not for a little time that the mathematical connection between them was recognized.) Bohr tried to reconcile these two divergent approaches and to provide a coherent conceptual framework for the 'wave' and 'particle' (and 'wavicle') talk that began to proliferate. This move already obscured the crucial issue (as I see it now, with the wisdom of hindsight), namely that of coming properly to grips with the plenum/atom contrast in all its dimensions: conceptual, ontological, mathematical. A few short years later von Neumann published his elegant mathematical treatise and introduced the language of state vectors, wave packet collapses and measurement types. This succeeded in fixating attention on the structure of QM as if it were a coherent, complete thing in itself, in isolation from broader conceptual and ontological issues. The obscuring of the issues was now complete and the debates waged fruitlessly on (or almost so) among the alternatives sketched earlier in the essay. A crude version of Bohr and von Neumann is now the 'working physicists' diet. It is not that the original Schrödinger-Heisenberg debate was not to the point, nor that Bohr's effort was superficially misguided (in fact I still believe it to be a great and masterful accomplishment in semantical-theoretical analysis), nor that von Neumann did not provide a penetrating mathematical formulation of QM – but simply that these contributions turned out to have shifted attention away from the true historical foundations of the subject and hence from the deepest issue. End of digression.

And this historical judgement of this historical movement is borne out in practice. The confusion over the notion of state is all too obvious a consequence. There is no agreement on the idea of state by those who write using the notion of a QM state. In particular, a QM state for some writers clearly means a physical state of an individual system after the fashion of CPM states. For other writers it is either an ensemble state, or synonymous with the probability measure itself – in any event the logical role of the notion derives from the role of states in CFT. And for Bub the notion of a QM state can really mean nothing at all, or at best it is elliptical, referring to the QM lattice state of affairs described above.

Certainly the elements of $\mathscr{H}$ are not states in any *individual* sense. This possibility is clearly ruled out by the structure of QM, the elements of $\mathscr{H}$ could neither represent the states of individual systems as in CPM nor of individual global fields as in CFT. To call such elements statistical states is only to confuse the issue. This is so because of the ambiguity of resolution of statistical density operators which prevents an understanding of these statistics in terms of ensembles of well defined, physically independent individuals. (Coherent collectives are not statistical ensembles.)[124]

These strictures on the use of "states" apply exactly to the alternative standard interpretational attempts which I have described above. As soon as one says "this is the position observable", "that is the momentum observable", "those are interference terms", and so on one has begun incorporating willy nilly the conceptual structures of CPM and CFT into an amalgam 'interpretation'. Dynamical analogies in CPM (the broader analogy with CFT is usually ignored) then lead one to say that the elements of $\mathscr{H}$ are QM states and $\mathscr{H}$ is made to play a role analogous to phase space. I have already argued for the futility of this approach. That futility is readily apparent in the way in which one is quickly forced to talk of statistical states instead of individual states and then forced to assert that subdivision of ensembles is (mysteriously!) neither complete (no dispersion-freeness occurs) nor unique, on pain of contradiction.[124] (Indeed my own past work on QM is to be understood essentially as follows: it is an attempt to take the view that the elements of $\mathscr{H}$ are indeed in some sense state specifications to its logical conclusion, namely to frustration or mystery or rejection via (α) to (ι) above, or Bohr.) It is better not to use the word "state" than to clutch at its comfort and be cast into utter darkness.

Now I wish to go beyond the preceding discussion. For the most consistent interpretational position on QM, which retains faith in QM, is certainly that of the quantum logicians (e.g. Bub). And yet I find that position also unaware of the mathematical and conceptual tradition in which QM stands (namely, in its relations to the atomic and plenum schemas). Its implausible ontology of half-denuded situations bears eloquent witness to the attempt to restore one of the two ontology-conceptual schemas (namely the atomic schema) that the mathematical structure of QM partially reflects. (In addition, that approach has, by its very logical position, the methodological danger of stopping, on logical grounds, any further inquiry into the roots of QM – cf. my [52; 54], Section XII.)[125] I would argue instead for the view that *no consistent, plausible, Realist interpretation of QM is possible within our present conceptual resources.* This view is, I think, made quite plausible by my general conclusion concerning QM, P_4 above, when read in the light of its backing, i.e. in the light of P_1, P_2, P_3 and the discussion following them.

Certainly the particle scheme will not do – the best attempt is Bub's and it is unsatisfying. Certainly no naive field schema interpretation will do either, the failure of Schrödinger's matter wave interpretation demonstrates that. (Of course sophisticated field interpretations are not pursued because everyone is blinded by the alleged analogy of QM with CPM.) And the foregoing considerations give one good reason to believe that no interpretational approach confined to these alternatives can hope to succeed, or rather hope to succeed with a physics that is as straight forward, 'clean', clear and non-mysterious as those in the classical tradition. (Any one of (α) through (ι) might be trumped up so as to 'succeed'.) But confined we are at present to these conceptual-ontological alternatives. In particular, we must say, I believe, that the precise conceptual and physical significance of the quantization process is not yet fully, or even very clearly, understood. (Cf. also my comments in the next section.)

The view I am advancing suggests that QM demands either a new conceptual-ontological scheme (a revision of the two conceptual schemes more thoroughgoing even than their logic) or the abandonment of QM as a hopelessly bastard offspring of an attempted marriage of the two great classical theoretical structures, doomed forever to a jerrymandered interpretation in terms of one of them.

XIII. BEYOND QM

I have tried to place QM in historical context – mathematically, conceptually, ontologically – because I believe that this is the way to deeper insight. This exercise has not left much room for optimism about obtaining a clear understanding of QM in all *three* respects. One must therefore go beyond QM. At this point I should like to say several things all at once but am forced to say them 'linearly'.

First I shall say something about linearity. QM is the linear theory par excellence. Let me explain. Suppose we take a CPM Hamiltonian that leads to a non-linear equation for the q_i's, for example

$$H = \frac{p_p^2}{2M_p} + \frac{p_e^2}{2M_e} - \frac{e^2}{|q_p - q_e|}$$

which describes the proton-electron hydrogenic system. (It leads to a non-linear dynamics because of the $|q_p - q_e|^{-1}$ term.) In QM, however, this Hamiltonian becomes the Hamiltonian operator

$$-\frac{\hbar^2}{2M_p} \nabla_1^2 - \frac{\hbar^2}{2M_e} \nabla_2^2 - \frac{e^2}{|\mathbf{r}_p - \mathbf{r}_e|}$$

which acts upon the *linear* space $\mathcal{H}$ and still generates a *linear* dynamical evolution $\psi_t = e^{iHt/\hbar}\psi_0$. The Schrödinger equation is still linear in the time variable *no matter what the content of $\mathcal{H}$* (so long as it does not contain time explicitly).[126] Now this is a very special state of affairs quite remarkable from the point of view of the scientific tradition, with respect to either CFT or CPM, in which non-linearity abounds. In the light of the foregoing analysis it suggests a point of view on the 'crudity' of QM.

Earlier I pointed out that QM might be regarded as essentially a field theory with Lagrangian dynamics of a specificly particle theory form. We can understand how this is possible if we remember that QM is essentially a *linear* theory and that in CPM the dynamics of a class of *linear* theories could be represented in a *linear* Hilbert space also. This suggests that QM is essentially a crude theory, a first attempt at quantizing CFT by adjointing to the (generalized) structure of the *linear* part of CFT, i.e. CFFT, a first attempt at quantization, undertaken in CPM because of the greater familiarity of the ideas there. This adjunction worked, formally, essentially because of the linearity of the theories involved.

These considerations strongly suggest that the passage beyond QM should be to a *non-linear theory*. This suggestion is strongly reinforced by the following consideration: in a non-linear theory we can expect the Principle of Superposition to break down. But this very principle is what is at the root of the so called paradoxes of QM (we can now see that these 'paradoxes' have a deep origin within QM), so that we can hold out hope that in a non-linear theory the interpretational difficulties of QM will not recur.[127]

On the other hand the foregoing characterization of QM as basicly a field theory, together with acceptance of modern experimental findings as presenting *prima facie* evidence for the atomicity of the quantum domain, strongly suggests that we pass beyond QM to QFT. (The '*Q*' of *QFT* is to take care of the atomistic aspects, the '*F*' the field theoretic character of quantum theories.)

Moreover, there are two specific considerations that decisively favour the exploration of field theory rather than particle theory at this point. The *first* of these is the natural fashion in which the former may be given a relativistically covariant formulation, compared with the relative obscurity of the relativistic basis for particle theories. The *second* of these reasons arises, paradoxically enough, out of some of the evidence of fundamental 'particle' physics which suggests inter-particle conversion processes most easily comprehended within the field ontology. I am referring here to situations in which members of a set of particles, α, β, ... can be produced from other members of the same set under suitable conditions so that one has, for example, γ's being produced from α's and also α's produced from γ's.[128] In these situations it is impossible to regard an α as a complex of a γ and other particles and at the same time regard a γ as a complex of an α and other particles. On the other hand, no such problem arises in the plenum schema, where "particles" have no particular integrity being merely non-linear near-singularities of the field, for here it is not a question of impossible combinations of constituents but merely one of rearrangements of field intensity distributions.

These last remarks indicate also that a non-linear field theory could be adequate to the task of simulating the particle-like features of the world; particles appear as singularities, or near-singularities, of the field.[129] Moreover a non-linear theory would break part of the dynamical symmetry between QM and CPM, thus perhaps providing greater freedom for a coherent plenum-scheme interpretation.

But QFT is, like QM, a thoroughly linear theory. Just as in QM we may take a Lagrangian or Hamiltonian that leads to non-linear field equations in CFT and use it in QFT to generate a linear dynamical evolution in a linear function space. Briefly the situation is this. We commence from the Hilbert function space $\mathcal{H}_1$ generated by the appropriate linear equation of CFFT (Dirac or Klein-Gordon equation, depending upon spin, cf. below). Then we construct from this the Fock space $\mathcal{F}$, the direct sum of the spaces of all orders of symmetric tensor products of $\mathcal{H}_1$ (i.e. $\mathcal{F} \approx$

$\approx \mathcal{H}_1 \oplus (\mathcal{H}_1 \otimes_s \mathcal{H}_1) \oplus (\mathcal{H}_1 \otimes_s \mathcal{H}_1 \otimes_s \mathcal{H}_1) \oplus \ldots)$. The 'fields' are then regarded as operator valued distributions on the Fock space $\mathcal{F}$. The dynamical equation of evolution is then still expressible in the linear form $\psi_t = e^{iHt/\hbar}\psi_o$ on $\mathcal{F}$ and this remains true here, as in QM, even if H contains non-linear terms.[130] In this respect then QFT may perhaps also be regarded as a relatively primitive quantized theory.

Another reason to believe this lies, perhaps, in the well-known divergences which exist in QFT and which prevent a strictly consistent mathematical formulation of the theory (at least of its perturbation theory). Although these divergences can be removed by the process known as 'renormalization' this is at present certainly an ad hoc procedure – both mathematically and physically.[131] Despite intensive investigation it is a striking fact that very likely not one physically non-trivial QFT theory exists.[132] (The reader will not have missed the fact that these divergences arise essentially in the counterparts in QFT to those situations in classical physics where the attempt was made to join CPM and CFT together, namely in the interaction of field and particle, and which also was not amendable to consistent mathematical formulation.)

Moreover, QFT suffers from the peculiar disability that it seems extraordinarily difficult to obtain suitable expressions within the formalism for the basic observables: position and momentum. The reason for this hinges on the fact that in a field theory the position variable joins the time variable as an external, independent parameter and so cannot be directly represented by an operator within the theory. To my knowledge there is still no physically adequate fashion in which to introduce these observables (and no adequate argument, in a theory which is represented as treating of systems of particles, for the view that they ought not to be the fundamental observables)![133]

In this respect the precise mathematical investigation of an axiomatic foundation for QFT recently initiated by several schools of mathematicians and mathematical physicists represents the first serious attempt at full rigour in this area.[134] Such attempts are as yet a long way off the production of results comparable to the experiments. (Moreover they remain exclusively within the linear tradition of QM – for example, the observables, still linear on the function space, are generally taken to form a locally C^*-algebra. In this sense they may also ultimately have to be regarded as approximations.)

These unpromising remarks bring me to yet another. It seems to me that we have as yet no deep understanding of the significance of the procedure called 'quantization'. The one obvious way to state that significance, namely to say "Compare the structure and interpretation of quantized theory with classical theory, the significance lies just in the difference" is not open to us because we have no clear-cut understanding of quantized theories. Moreover I venture to suggest that the search for an answer, when it has been carried on at all, has largely been carried on in the wrong place. Those investigations, and this essay is intended as one of them, into the mathematical structure of QM (or QFT), are to the point, certainly. But if these investigations are isolated from an appreciation of the depth of the impact of mathematical changes on our conceptual descriptive forms and ontology then they remain blind. We speak glibly of quantization representing the change from continuity to discreteness without bothering to find out precisely *where* this change is made mathematically and what the conceptual ramifications are likely to be. In Sections VII, VIII I have tried to briefly point out the depth of those ramifications as I understand them now – and I have only begun to scratch the surface of a precise account.

In Sections VII, VIII I pointed out the sources of continuity and discontinuity in the atomic and plenum schemas. The 'natural' (coherent) condition of an atomic formulation is total discontinuity, continuity enters only through the incorporation of continuous space-time. By contrast, the 'natural' (coherent) plenum condition is total continuity, discontinuity enters actual field theories only derivatively at the level of object simulation through non-linear field equations and through field periodicities in certain circumstances. In the atomic schema there is no special connection between linearity and continuity/discontinuity, in CFT there is. Actually there are three ways to introduce discontinuity in a fundamental way to the plenum schema: (i) make explicit use of the 'normal modes' of a field produced by boundary conditions, (ii) introduce finite difference equations as dynamical equations, (iii) introduce a discrete space-time. In addition there is particle-like simulation through non-linear field dynamics. QM and QFT essentially exploit only the first of these. For example, the eigenvalues of the momentum operator in QM can be viewed as just the discrete frequencies of the normal vibrational modes under specific boundary conditions. (Indeed, Jost [70] remarks that the quantization of CFT in analogy with the procedure for CPM leads both to a restricted theory – Fermi-Dirac statistics do not occur – and to a theory whose relativistic basis is obscure, whereas if we follow the procedure of first fourier analyzing the field and then quantizing on the resulting components we remove both of these difficulties. Cf. the relevance of this to the remarks above concerning the status of position and momentum observables.) The other ways of introducing discontinuity, mentioned above, together with non-linearity, are ignored, Moreover, I do not believe we understand the relations among these possibilities very well. This is perhaps yet another reason to believe that theory is as yet at a primitive stage of development and the ramifications of quantization not yet clearly understood.

What are the sources of continuity and discontinuity in QM and QFT? There are, perhaps surprisingly, two strong sources of continuity in these theories. The first of these stems from the Principle of Superposition, for if ϕ_1, ϕ_2 are solutions of Schrödinger's equation so also is $\alpha\phi_1 + \beta\phi_2$, all $\alpha, \beta \in F$, thus allowing continuous superpositions of ϕ's (fields) to be formed. This continuity arises directly from the characteristics of the plenum schema in which all 'objects' are simply complexes of field distributions, freely superposable. It is embodied in CFFT exactly as the free superposition of field states. The way to remove this source of continuity is to pass to an appropriate non-linear theory, but all quantized theories are, we have seen, thoroughly linear.

The second source of continuity derives from the incorporation of continuous space-time into the theory. I pointed out that the plenum and atomic schemas make perhaps their closest approach to one another when space-time is quantized. Considering that we have found QM at least to be a hybrid of the mathematical structures expressing these two ontologies, there is promise in the idea that a coherently interpretable theory would develop a reformulation of QM within a quantized space-time. The notion of a fundamental length and a fundamental time have indeed been explored, especially in connection with the reaction times of fundamental processes and with preventing the divergences of QFT, but to my knowledge no formalism has yet been developed which is workable (probably largely because the formidable mathematical difficulties involved).[135] Nonetheless from a conceptual point of view the approach is very attractive and one can hope that it will be pursued in the future.

There are other features of physics – classical and quantal – that are not well understood. I am thinking particularly of the connection, if any, between the charge and mass parameters – which have naturally discrete values in the atomic schema – and the space-time parameters of CFT and QFT (and also, spacial variables excepted, of CPM and QM). The quantization process evidently introduces an intimate connection here. The situation in QFT is this: The 'canonical' quantization of CFFT, i.e. the process of quantization followed in analogy to the quantization of CPM with the field regarded as a linear CPM system of infinite degrees of freedom, leads only to the description of integral spin particles, while the anti-commutation, or Dirac commutation, is required to provide QFT with a description of half-integral spin particles; moreover this course is thrust upon us because the classical real or complex scalar field has positive definite energy density but indefinite 'charge density' and the classical Dirac field vice versa, the two differing quantization procedures are required to achieve a positive definiteness for both energy and charge densities. (Cf. comments 3 paragraphs back.) I do not believe that there has been any serious attempt yet to integrate this feature of quantization with the other aspects of it discussed above. On the basis of this brief review of the progress of our understanding of the significance of quantization, one can only conclude that we have as yet only a relative superficial grasp of the problem. The present essay was aimed at extending our grasp of how to *formulate* such problems in a more penetrating fashion, though it is only a preliminary study. I reserve for another occasion similar studies of QFT and Relativity Theory.

Moreover, in Section VIII I pointed out how deeply our descriptive schemes were attached to our mathematical theories and how deep a change in those concepts the new theories demand. Recall that the informal concept of signal is basic to Relativity Theory and cannot be precisely defined in QM, and the argument of Section VII concerning quantization and precise definability. But such investigations are largely ignored. Yet if we understood more clearly our classical physics – and I don't just mean a few equations, but its deep structure – we might perhaps understand quantized theories more clearly.

There is one group of workers who are sensitive to these questions of conceptual structure. Not surprisingly, I feel, they are also producing some of the most interesting new ideas. This is the group connected with Bohm and Penrose in London.[136] For some time now Bohm and others have been attempting to draw attention to the importance of informal conceptual structure in physics.[137] Although the ideas are as yet only vaguely expressed (what else could one expect? – especially when the scientific world offers no help, nor even encouragement) they reinforce the point of view being expressed here. Moreover, members of this group are exploring discrete, even combinatorial space-time as a possible tool in physics as well as generalized non-linear field theories sufficiently rich in formulation to perhaps include quantum phenomena as well.[138] Such research holds out great promise of new ideas and penetrating insights into existing theories. Also of great interest are such closely related developments as the non-linear spinor field theory by Sachs (as reported on in this volume). Sachs has shown that such a theory can predict a wide range of quantum phenomena, and even include QM as a kind of limiting case, yet it essentially belongs to CFT![139] Moreover, not all those who work within the logico-algebraic approach to QM wish to preserve this structure in its entirety, there is in fact a very strong tradition which seeks, by generalizing the structures uncovered, to formulate a more general (and hence hopefully more powerful) theory.[140]

Perhaps one of the alternatives outlined above that 'go beyond' QM, will prove to be the theory of the next generation. Doubtless thousands will then study its applications and connected mathematical theory. In the longer run, however, it may be of more importance to gain a more penetrating grasp of the nature of scientific theorizing, especially of those theories of mathematical physics already part of the scientific tradition.

XIV. CONCLUSION

I reserve this conclusion for a recitation of questions that have been raised in the body of the text. Many of them are somewhat technical in nature, indeed the answers to them may already be known (if so, I hope the knowers communicate them to me, the ignorant), but all bear upon the philosophical task of understanding clearly the foundations and significance of modern mathematical physics.

1. What is the most adequate descriptive conceptual structure for the plenum schema?

2. What is the exact relationship in classical science between the linearity/non-linearity dichotomy and the continuity/discontinuity dichotomy? How do these dichotomies and the relations between them apply to quantized theories?

3. Are there alternative nontrivial constructions of the algebra of observables in CFT which throw light on the nature of that theory? Are there constructions of a similar nature for the observables of both CPM and CFT in which the observables are construed as operators on the corresponding function spaces?

4. What is the nature and status of observables in QFT?

5. What are the exact relations among the algebras of observables of (i) CPM and CFT, (ii) QM and QFT, (iii) the theories of (i) and (ii)?

6. What is common to all theories sharing a Lagrangian-Hamiltonian dynamics?

7. Are there interesting QM-type theories with a less specifically CPM dynamics?

8. How is dynamical evolution to be precisely described in the non-linear cases of CPM and CFT, especially the interacting field cases of CFT?

9. What promise do theories hold whose dynamics are historically dependent (whether on the past or on the future)? What promise for theories that deal with multi-dimensional or branching times?

10. What is the precise nature of the relationship, if any, of the departure of QM from a boolean logical structure and the nature of QM as a fusion of the structures of CFT and CPM?

11. How are the sources of continuity and discontinuity related in physical theories? How do they tie into the conceptual-ontological schemas of western science?

12. Are there informative similarities between CFT and QM with respect to the logical status of a phase space within each?

13. What is the complete significance of the relation of canonical conjugateness in the light of the fact that the characteristic conceptual structure involved encompasses more relations than current physics explicitly recognizes?

14. What is the precise significance of the quantization procedure? (How are the charge and mass parameters connected to the space and time variables and what is the significance of quantization for that connection?)

15. What are the logical structures of the set of probabilistic propositions occurring in CPM, CFT, and QM? (What is the significance of the formal modeling of a boolean algebra obtainable for probability 1 propositions with QM-like connectives? Can van Fraassen's modal treatment of these latter propositions be extended to all probabilistic propositions?)

16. Can the probability measure of CFT be reconstrued so as to exploit the vector space structure of the function space?

17. What is the precise description of the general dynamical evolution of probability measures in CFT?

18. What are the conceptual and physical ramifications of quantizing space and time? Is a more coherent, mathematically unified theory constructable in this fashion?

Acknowledgement

I am indebted to Professors B. van Fraassen and S. Gudder for criticisms of an earlier draft which improved it at several points and to Professors J. Bub, D. Finkelstein and E. Prugeovecki for valuable discussions on several issues. My greatest practical indebtedness is of course to Mrs. A. Smith, my secretary, who typed and corrected the manuscript.

BIBLIOGRAPHICAL NOTE

The literature pertaining to modern theories of mathematical physics has suffered, so it seems to me (and to others to whom I have spoken) from the unfortunate bifurcation of mathematics and physics. When I began these investigations I had in the back of my mind (though the 'front' knew better!) a naive belief that there *must* exist a book containing all the answers to the mathematical questions which it seemed natural to pose (or containing at least all the known answers). No such book yet exists to my knowledge. What one finds instead are two distinct kinds of books. (1) Books written by, and for, physicists, containing excellent discussions of the key equations, techniques for obtaining solutions in physically important cases and so on. (2) Books by mathematicians investigating fields of mathematics, of varying degrees of relevance to physics. The former, by and large, do not contain rigorous mathematical formulations of physical theories and are largely uninterested in general structural questions. The latter contain a wealth of precise information, but in a form which makes it extremely difficult to extract general results of relevance to physical theories. If one wishes to utilize a study of the structure of mathematical theories to shed philosophical light on their interpretation one is forced to be one's own generalist and master and synthesize the separate works in physics and mathematics before commencing! Undoubtedly I have failed to do this completely. (The fields of physics and mathematics which it would be philosophically pertinent to master seem to stretch out before one in a never-ending vista, expanding, like the universe, much faster than one can hope to travel.) Nor can I resist adding that experience strongly suggests that this state of affairs is a creation of that anti-intellectual, anti-rational monster, the 'scientific establishment', where fashion and authority dictate and those who hope to earn a living and a little prestige stifle curiosity and criticism and obey. What follows is a brief bibliographical comment on reading which I found helpful and which may prove useful to others who, I hope, will follow on and develop in a thorough fashion what I have been able only to sketch.

One field in which excellent texts are not lacking is that of lattice theories and boolean algebras – see Abbott [1, 2], Bell and Slomson [8], Halmos [45] and Sikorsky [126] and many others – though of course the application of this approach to physical theory is still in its infancy. On this latter score see the article by Holland in [51] and the fundamental paper by Birkhoff and von Neumann [9]. A standard work in the field of great depth is the volumes by Varadarajen [133]. (For other work the bibliographies of the papers by Bub, van Fraassen, Gudder and Greechie in this volume and that of the paper by Hooker [52] may be consulted.)

The American Mathematical Society has sponsored some excellent sets of lecture notes which come closer than anything else to bridging the gap between mathematics and physics. I am thinking of the fundamentally important text by Mackey [83] and the sophisticated but extremely clear treatments of contemporary QFT in Jost [70] and Segal [122].

Two mathematically clear texts presenting a great deal of the fundamental mathematical material in a form more directly relevant to physics are those by Jauch [68] and Prugovecki [104] (though Bub and myself have been critical of Jauch's more 'philosophical' uses of the mathematics, see Bub in this volume and Hooker [53]). The classical Hilbert space exposition of QM, and still one of the best, is that by von Neumann [95]. For the foundations of classical mechanics Abraham and Marsden [3] is excellent with the later work of Marsden [86, 87] bridging the gap to QM dynamics. Rosen [111] is also an excellent introduction to the canonical formalism. CFT is virtually never treated as a subject in its own right as far as I know but always as a prelude to QFT (perhaps this is why we have only a sketchy account of it – an important loss to physics, I judge). The classical work on field theory is Wenzel [135]. Other useful though equally conventional treatments may be found in Bogoliubov and Shirkov [10] Landau and Lifschitz [78] and Rzewuski [114].

When it comes to modern algebraic developments of physics and the role of symmetry groups (both concerned with QFT almost entirely), besides the classical paper by Wigner [141] on the representations of the inhomogeneous Lorentz group (conveniently reprinted, along with a large number of more recent papers in Dyson [32]) and the books by Jost and Siegal already mentioned, most of the clearer accounts are again found in various lecture notes on physics. The reader might like to consult especially the excellent articles by Robinson [108] and Guenin [43, 44] on algebraic approaches to QFT as well as the analytic survey of axiomatic field theory by Tomozawa [131], and the standard text by Streater and Wightman [128].

November 1971

POSTSCRIPTED BIBLIOGRAPHY

The reader may wish to follow up the possibilities that the differential manifold approach holds out for precise philosophical analysis by adding to those references cited in the bibliography the following selection from among those I have discovered since the time of writing this essay.

[P1] Godbillon, C., *Géométrique Différentielle et Mécanique Analytique*, Hermann et Cie, Paris, 1969.
[P2] Hurt, N. E., 'Differential Geometry of Canonical Quantization', *Annales de l'Institut Henri Poincaré*. A, **14** (1971), 153–170.
[P3] Souriau, J. M., 'Quantification Géométrique', *Communications in Mathematical Physics*, III. **1** (1966), 374–98.
[P4] Souriau, J. M., *Structure des Systèmes Dynamiques*, Dunod, Paris, 1970,
[P5] Streater, R. F., 'Canonical Quantization', *Communications in Mathematical Physics*, **2** (1966), 354–74.
[P6] Taam, C. T. (ed.), *Lectures in Modern Analysis and Applications*, Vols. II and III, Springer-Verlag, New York, 1970.

Also I draw the reader's attention to a recent interesting article by Gudder on an elementary length in physics:

[P7]   Gudder, S., 'Elementary Length Topologies in Physics', *SIAM Journal of Applied Mathematics*, **16** (1968), 1011–9.

Finally I wish to correct my omission of the early work on a function space, though not one of the kind I was directly concerned with, in CPM:

[P8]   Koopman, *Proceedings of the National Academy of Sciences*, **17** (1931), 315–8.
This led on to the later work of von Neumann, and later many others, on the foundations of ergodic theory.

BIBLIOGRAPHY

[1]   Abbott, J. C., *Trends in Lattice Theory*, Van Nostrand, New York, 1970.
[2]   Abbott, J. C., *Sets, Lattices and Boolean Algebras*, Allyn and Bacon Inc., Boston, Massachusetts, 1969.
[3]   Abraham, R., and Marsden, J. E., *Foundations of Mechanics*, W. A. Benjamin, New York, 1967.
[4]   Aharanov, Y. and Bohm, D., 'Significance of Electromagnetic Potentials in the Quantum Theory', *Physical Review* **115** (1959), 485–91.
[5]   Aharanov, Y., Pendleton, H., and Petersen, A., 'Modular Variables in Quantum Theory', *International Journal of Theoretical Physics* **2** (1969), 213–30.
[6]   Ballantine, L. E., 'The Statistical Interpretation of Quantum Mechanics', *Reviews of Modern Physics* **42** (1970), 358–81.
[7]   Bastin, T. (ed.), *Quantum Theory and Beyond*, Cambridge University Press, Cambridge, 1971.
[8]   Bell, J. L. and Slomson, A. B., *Models and Ultraproducts*, North-Holland, Amsterdam, 1969.
[9]   Birkhoff, G. and von Neumann, J., 'The logic of Quantum Mechanics', *Annals of Mathematics* **37** (1936), 823–43.
[10]   Bogoliubov, N. N. and Shirkov, D. V., *Introduction to the Theory of Quantized Fields*, Interscience, New York, 1959.
[11]   Bohm, D., *Quantum Theory*, Prentice-Hall, Englewood Cliffs, New Jersey, 1951.
[12]   Bohm, D., 'A Suggested Interpretation of the Quantum Theory in Terms of "Hidden Variables"', I and II, *Physical Review* **85** (1952), 166–93.
[13]   Bohm D. and Bub, J., 'A proposed Solution of the Measurement Problem in Quantum Mechanics by a Hidden Variable Theory', *Review of Modern Physics* **38**(1966), 453–69.
[14]   Bohm, D., *Causality and Chance in Modern Physics*, Harper Torchbooks, New York, 1961.
[15]   Bohm, D. and Bub, J., 'A Refutation of the Proof by Jauch and Piron that Hidden Variables can be Excluded in Quantum Theory', *Reviews of Modern Physics* **38** (1966), 470–5.
[16]   Bohm, D., 'On the Role of Hidden Variables in the Fundamental Structure of Physics' in *Quantum Theory and Beyond* (ed. by T. Bastin), Cambridge University Press, Cambridge, 1971.
[17]   Bohm, D. and Schumacher, D. L., 'On the Failure of Communication between Bohr and Einstein', unpublished.
[18]   Bohm, D. and Vigier, J. P., 'Model of the Causal Interpretation of Quantum Theory in Terms of a Fluid with Irregular Fluctuations', *Physical Review* **96** (1954), 208–17.

[19] Bohm, D., 'Quantum Theory as an Indication of a New Order in Physics, Part A, The Development of New Orders as Shown Through the History of Physics', *Foundations of Physics* 1 (1971), 359-81.

[20] Broglie L. de, *Non-linear Wave Mechanics*, Elsevier, Amsterdam, 1960; and 'The Reinterpretation of Wave Mechanics', *Foundations of Physics* 1 (1970), 5-15.

[21] Bub, J., 'Hidden Variables and the Copenhagen Interpretation – A Reconciliation', *British Journal for the Philosophy of Science* 19 (1968), 185-210.

[22] Bub, J., 'What is a Hidden Variable Theory of Quantum Phenomena?', *International Journal of Theoretical Physics* 2 (1969), 101-24.

[23] Bub, J., 'Towards the Interpretation of Quantum Mechanics', to appear in the *British Journal for the Philosophy of Science*.

[24] Bub, J., 'Popper's Propensity Interpretation of Probability and Quantum Mechanics', to appear in *Minnesota Studies for the Philosophy of Science*.

[25] Cartan, E., *Sur la Structure des Groupes de Transformations Finis et Continus*, Thèse, Paris 1894.

[26] Catlin, D. E., 'Spectral Theory in Quantum Logics', *International Journal of Theoretical Physics* 1 (1968), 285-97.

[27] Chew, G. F., *The Analytic S-Matrix*, W. A. Benjamin, New York, 1966.

[28] Chew, G. F., 'The Bootstrap Idea and the Foundations of Quantum Theory', in *Quantum Theory and Beyond* (ed. by T. Bastin), Cambridge University Press, Cambridge, Massachusetts, 1971.

[29] Dirac, P. A. M., *Principles of Quantum Mechanics*, Clarendon Press, Oxford, 1958.

[30] Diximier, J., *Les algèbres d'opérateurs dans l'espace Hilbertien*, Gauthier-Villars, Paris, 1969.

[31] Diximier, J., *Les C*-algèbres et leur representations*, Gauthier-Villars, Paris, 1969,

[32] Dyson, F. J., *Symmetry Groups in Nuclear and Particle Physics*, W. A. Benjamin. New York, 1966.

[33] Feinberg, G., 'Physics and the Thales Problem', *The Journal of Philosophy* 63 (1966), 5-16.

[34] Fraassen, B. van, 'Incomplete Assertion and Belnap Connectives', presented at the Linguistics and Semantics Workshop, University of Western Ontario, Spring 1972. To be published in the proceedings.

[35] Frauenfelder, H. (ed.), *The Mossbauer Effect*, W. A. Benjamin, New York, 1961.

[36] Furry, W. H., 'Behaviour of de Broglie Waves and Wave Packets', in *Lectures in Theoretical Physics*, Vol. V. (ed. by W. E. Brittin), Interscience Publishers, New York, 1962.

[37] Giles, R., 'Foundations for Quantum Mechanics', *Journal of Mathematical Physics* 11 (1970), 2139-60.

[38] Gleason, A. M., 'Measures on the Closed Subspaces of a Hilbert Space', *Journal of Mathematics and Mechanics* 6 (1957), 885-93.

[39] Goldstein, H., *Classical Mechanics*, Addison-Wesley, Reading, Massachusetts, 1950.

[40] Graves, J. C., *The Conceptual Foundations of Contemporary Relativity Theory*, Massachusetts Institute of Technology Press, Cambridge, Massachusetts, 1971.

[41] Grünbaum, A., *Modern Science and Zeno's Paradoxes*, Wesleyan University Press, Middletown, Connecticut, 1967.

[42] Gudder, S., 'Partial Algebraic Structures', University of Denver, Department of Mathematics Publication MS-R 7102.

[43] Guernin, M., 'Axiomatic Foundations of Quantum Theories', *Journal of Mathematical Physics* **7** (1966), 271–82.

[44] Guernin, M., 'Algebraic Methods in Quantum Field Theory', in *Lectures in Theoretical Physics*, Vol. IXA, (ed. by W. E. Brittin, A. E. Barut, and M. Guernin), Gordon and Breach, New York, 1967.

[45] Halmos, P. R., *Lectures on Boolean Algebras*, Van Nostrand, Princeton, New Jersey, 1963.

[46] Hamermesh, M., *Group Theory and Its Application to Physical Problems*, Addison-Wesley, Reading, Massachusetts, 1962.

[47] Heisenberg, W., *Physics and Philosophy*, Harper Torchbooks, New York, 1966.

[48] Heisenberg, W., *Introduction to the Unified Field Theory of Elementary Particles*, Interscience Publishers, London, 1966.

[49] Herstein, I. N., *Tropics in Algebra*, Ginn and Co., Waltham, Massachusetts, 1964.

[50] Hesse, M. B., *Forces and Fields*, Littlefield, Adams and Co., New York, 1965.

[51] Holland, S. J., 'The Current Interest in Orthomodular Lattices', in *Trends in Lattice Theory* (ed. by Abbott, J. C.), Van Nostrand, New York, 1970.

[52] Hooker, C. A., 'The Nature of Quantum Mechanical Reality: Einstein Versus Bohr', in *The Pittsburgh Studies in the Philosophy of Science*, Vol. V, Pittsburgh: University of Pittsburgh Press, 1972.

[53] Hooker, C. A., 'Theories of Measurement in Quantum Mechanics: A Critique of a Recent Proposal', *The International Journal of Theoretical Physics*, **5** (1972), 231–250.

[54] Hooker, C. A., 'The Impact of Quantum Theory on the Conceptual Basis of the Classification of Knowledge', invited paper read at the Conference on Conceptual Bases for the Classification of Knowledge, Ottawa. To appear in the proceedings, 1972.

[55] Hooker, C. A., 'Energy and the Interpretation of Quantum Theory', *The Australasian Journal of Philosophy* **49** (1971), no. 3.

[56] Hooker, C. A., 'Sharp and the EPR Paradox', *Philosophy of Science* **38** (1971), 224–33.

[57] Hooker, C. A., 'The Statistical Interpretation of Quantum Mechanics', not yet published.

[58] Hooker, C. A., 'Presuppositional Logic, Quantum Logic and Niels Bohr', not yet published.

[59] Hooker, C. A., *Systematic Philosophy and the Secondary Qualities*. Ph.D. thesis, York University, Toronto, 1970.

[60] Hooker, C. A., 'The Relational Doctrines of Space and Time', *The British Journal for the Philosophy of Science* **22** (1971), 97–130.

[61] Hooker, C. A., 'Science and Metaphysics', To appear.

[62] Hooker, C. A., 'Defense of a Non-Conventional Status for Classical Mechanics', to appear in *The Boston Studies in the Philosophy of Science*.

[63] Hooker, C. A., 'Global Theories', to appear in *Symposium on Theories*.

[64] Hooker, C. A., 'Critical Notice: M. Radner and S. Winoker (eds.), *Analyses of Theories and Methods of Physics and Psychology*', *Canadian Journal of Philosophy* **1** (1972), 393–407.

[65] Hooker, C. A., 'Critical Notice: Against Method, P. K. Feyerabend', *The Canadian Journal of Philosophy* **1** (1972), 489–509.

[66] Hopf, E., 'Statistical Hydromechanics and Functional Calculus', *Journal of Rational Mechanics and Analysis* 1 (1952), 87–123.

[67] Jammer, M., *The Conceptual Development of Quantum Mechanics*, McGraw-Hill, New York, 1966.

[68] Jauch, J., *Foundations of Quantum Theory*, Addison-Wesley, New York, 1968.

[69] Jauch, J. M. and Piron, C., 'Quantal Propositional Systems', *Helvetia Physica Acta* 42 (1969), 842–7.

[70] Jost, R., *The General Theory of Quantized Fields*, American Mathematical Society, Providence, Rhode Island, 1965.

[71] Kingman, J. F. C. and Taylor, S. J., *Introduction to Measure and Probability* Cambridge University Press, Cambridge, Massachusetts, 1966.

[72] Kochen, S. and Specker, E., 'Logical Structures Arising in Quantum Theory', in *The Theory of Models* (ed. by J. Addison, L. Henkin, and A. Tarski), North-Holland Publishing Co., Amsterdam, 1965.

[73] Kochen, S. and Specker, E., 'The Calculus of Partial Propositional Functions', in *Logic, Methodology and the Philosophy of Science* (ed. by Y. Bar-Hillel), North-Holland, Amsterdam, 1965.

[74] Kochen, S. and Specker, E., 'The Problem of Hidden Variables in Quantum Mechanics', *Journal of Mathematics and Mechanics* 17 (1967), 59–87.

[75] Kronfli, N. S., 'Atomicity and Determinism in Boolean Systems', *International Journal of Theoretical Physics* 4 (1971), 141-3.

[76] Kronfli, N. S., *International Journal of Theoretical Physics* 3 (1970), 199.

[77] Krylor, N. M. and Bogoliubov, N. N., *Introduction to Non-Linear Mechanics*, Princeton University Press, Princeton, 1949.

[78] Landau, L. and Lifshitz, E., *The Classical Theory of Fields*, Addison-Wesley, Reading, Massachusetts, 1951.

[79] Leiter, D., 'Can Atomic Processes be Described by Non-Linear Wave Mechanics in Space-Time?', *International Journal of Theoretical Physics* 3 (1970), 205-31.

[80] Løs, J., 'Remarks on Foundations of Probability', *Proceedings of the International Congress of Mathematics* (1962) pp. 225-9.

[81] Ludwig, G., 'Versuch einer axiomatischen Grundlegung der Quantenmechanik und allgemeiner physikalischer Theorien', *Zeitschrift für Physik* 181 (1964), 233-60.

[82] Ludwig, G., 'Attempt of an Axiomatic Foundation of Quantum Mechanics and More General Theories II', *Communications in Mathematical Physics* 4 (1967), 331-48.

[83] Mackey, G. W., *The Mathematical Foundations of Quantum Mechanics*, W. A. Benjamin Inc., New York, 1963.

[84] Margenau, H., *The Nature of Physical Reality*, McGraw-Hill, New York, 1950.

[85] Margenau, H. and Park, J. L., 'Simultaneous Measureability in Quantum Theory', *International Journal of Theoretical Physics* 1 (1968), 211-84.

[86] Marsden, J. E., 'Generalized Hamiltonian Mechanics', *Archive for Rational Mechanics and Analysis* 28 (1968) 323-61.

[87] Marsden, J. E., 'Hamiltonian One Parameter Groups', *Archive for Rational Mechanics and Analysis* 28 (1968), 362-96.

[88] McKinley, W. A., 'The Search for a Fundamental Length in Microphysics', *American Journal of Physics* 28 (1960), 129.

[89] Melsen, A. G. van, *From Atomos to Atom*, Harper and Row, New York, 1960.

[90] Mercier, A., *Variational Principles of Physics*, Dover Publications, New York,

1963.

[91] Montague, R., 'Deterministic Theories', in *Decisions, Values and Groups* II (ed. by Washburne), Pergamon Press, New York, 1957.

[92] Muraskin, M., 'Particle-Like Objects in a Non-linear Field Theory', *International Journal of Theoretical Physics* 4 (1971), 49–53.

[93] Nagel, E., 'The Principles of the Theory of Probability', in *Foundations of the Unity of Science*, Vol. I (ed. by O. Neurath, C. Morris, and R. Carnap), University of Chicago Press, Chicago, 1955.

[94] Naimark, M. A., *Normed Rings*, P. Noordhoff N.V. Groningen, The Netherlands, 1959.

[95] Neumann, J. von, *Mathematische Grundlagen der Quanten Mechanik*, Verlag Julius Springer, Berlin, 1932.

[96] Newton, T. D. and Wigner, E. P., 'Localized States for Elementary Systems', *Reviews of Modern Physics* 21 (1949), 400–8.

[97] Park, J. L., 'Nature of Quantum States', *American Journal of Physics* 36 (1968), 211–26.

[98] Petersen, A., *Quantum Theory and the Philosophical Tradition*, Massachusetts Institute of Technology, Cambridge, 1968.

[99] Petersen, A. and Grgin, E., *Physical Review* 5 (1972), 300–4.

[100] Pool, J. C. T., 'Baer*-Semigroups and the Logic of Quantum Mechanics', *Communications in Mathematics and Physics*, 9 (1968), 118–41.

[101] Popper, K. R., *The Logic of Scientific Discovery*, Hutchinson, London, 1959.

[102] Popper, K. R., 'Quantum Mechanics without "The Observer"', in *Quantum Theory and Reality* (ed. by M. Bunge), Springer-Verlag, New York, 1967.

[103] Popper, K. R., 'The Propensity Interpretation of the Calculus of Probability, and the Quantum Theory', in *Observation and Interpretation in the Philosophy of Physics* (ed. by S. Korner), Dover Publications, New York, 1962.

[104] Prugovecki, E., *Quantum Mechanics in Hilbert Space*, Academic Press, New York 1971.

[105] Prugovecki, E., 'Measure-Theoretical Description of Klein-Gordon Multiparticle States', *Journal of Mathematical Physics* 10 (1959), 933–44.

[106] Putnam, H., 'Is Logic Empirical?', in *Proceedings of the Boston Colloquium in the Philosophy of Science*, 1966–1968 (ed. by R. S. Cohen), D. Reidel Publ. Co., Dordrecht, Holland, 1969.

[107] Quine, W. V. O., *Word and Object*, Massachusetts Institute of Technology Press, Cambridge, 1964.

[108] Robinson, A., 'Algebraic Aspects of Relativistic Quantum Field Theory', in *Axiomatic Field Theory* (ed. by M. Chretien and S. Deser), Gordon and Breach, New York, 1966.

[109] Rohrlich, F., *Classical Charged Particles*, Addison-Wesley, Reading, Massachusetts, 1965.

[110] Rosen, G., 'Dynamics of Probability Distributions Over Classical Fields', *International Journal of Theoretical Physics* 4 (1971), 189–95.

[111] Rosen, G., *Formulations of Classical and Quantum Dynamical Theory*, Academic Press, New York, 1969.

[112] Rosen, G., 'Hamilton-Jacobi Functional Theory', *International Journal of Theoretical Physics* 4 (1971), 281–5.

[113] Rosen, G., 'Conventional Formulations of Noether's Theorem in Classical Field Theory', *International Journal of Theoretical Physics* 4 (1971), 287–94.

[114] Rzewuski, J., *Field Theory*, 2 Vols. PWN-Polish Scientific Publishers, Warsaw, 1958.

[115] Sachs, M., 'A New Theory of Elementary Matter', *International Journal of Theoretical Physics* Parts I and II: **4** (1971), 433–51, 453–76; Parts III and IV: **5** (1972), 35–53, 161–197.

[116] Schiff, H. I., *Quantum Mechanics*, McGraw-Hill, New York, 1955.

[117] Schlegel, R., 'Space at Mid-Century', *American Journal of Physics* **20** (1952), 38.

[118] Schrödinger, E., 'Die gegenwartige Situation in der Quantenmechanik', *Naturwissenschaften* **23** (1935), I: 807–12; II: 823–8; III: 844–9.

[119] Schwartz, J. T., *Nonlinear Functional Analysis*, Courant Institute.

[120] Schwebel, 'Newtonian Gravitational Field Theory', *International Journal of Theoretical Physics* **4** (1971), 87–92.

[121] Scott, D. and Krauss, P., 'Assigning Probabilities to Logical Formulas', in *Aspects of Inductive Logic* (ed. by J. Hintikka and P. Suppes), North-Holland, Amsterdam, 1966.

[122] Segal, I. E., *Mathematical Problems of Relativistic Physics* (with an appendix on Group Representations in Hilbert Space by G. W. Mackey), American Mathematical Society, Providence, Rhode Island, 1963.

[123] Sellars, W., 'Time and the World Order', *Minnesota Studies in the Philosophy of Science*, Vol. 3, University of Minnesota Press, Minneapolis, 1962.

[124] Sen, D. K., *Fields and/or Particles*, Ryerson Press, Toronto, 1968.

[125] Schankara, T. S. and Srinivas, M. D., 'On Structure Preserving Quantizations', *International Journal of Theoretical Physics* **4** (1971), 395–401.

[126] Sikorski, R., *Boolean Algebras* (2nd ed.), Springer-Verlag, Berlin, 1964.

[127] Stone, M. H., *Linear Transformations in Hilbert Space*, American Mathematical Society, Providence, Rhode Island, 1932.

[128] Streater and Wightman, *PCT Spin and Statistics and All That*, Benjamin Inc., New York, 1964.

[129] Synge, J. L., *Relativity: The Special Theory*, North-Holland, Amsterdam, 1965.

[130] Tolman, R. C., *The Principles of Statistical Mechanics*, Oxford University Press, Oxford, 1938.

[131] Tomozawa, Y., 'Introduction to Axiomatic Field Theory', *Lectures in Theoretical Physics*, Vol. 7C (ed. by W. E. Brittin), (1965), pp. 438–88.

[132] Vainberg, M. M., *Variational Methods for the Study of Non-Linear Operators*, Holden-Day Inc., London, 1964.

[133] Varadarajan, V. S., *The Geometry of Quantum Mechanics*, Van Nostrand, Princeton, 1968.

[134] Visconti, A., *Quantum Field Theory*, 2 Vols., Pergamon Press, New York, 1969.

[135] Wentzel, G., *Quantum Theory of Fields*, Interscience, New York, 1959.

[136] Wheeler, J. A., *Geometrodynamics*, Academic Press, New York, 1962.

[137] Wheeler, J. A., 'The Universe in the Light of General Relativity', *Lectures in Theoretical Physics* Vol. V (ed. by W. E. Britten, B. W. Downs, and Y. J. Downs) Interscience, New York, 1963.

[138] Whyte, L. L., *Essay on Atomism*, Harper and Row, New York, 1963.

[139] Wightman, A. S., 'On the Localizability of Quantum Mechanical Systems', *Reviews of Modern Physics* **34** (1962), 845–72.

[140] Wigner, E., 'The Problem of Measurement', *American Journal of Physics* **31** (1963), 6–15.

[141] Wigner, E., 'On Unitary Representations of the Inhomogeneous Lorentz Group', *Annals of Mathematics* **40** (1939), 149–204.
[142] Wigner, E. P., 'Events, Laws of Nature and Invariance Principles', *Nobel Lecture*, December 12, 1963.
[143] Williams, L. P., *The Origins of Field Theory*, Random House, New York, 1966.

NOTES

[1] See Hooker [52].

[2] These evaluations are made in Sections V and XII. Further comment may be found in Hooker [54].

[3] The best presentation is still that of von Neumann [95], but see also Jauch [68], Prugovecki [104] for excellent treatments and a more recent and (hence) sophisticated mathematical perspective. Rosen [111] is also a helpful source on canonical formalism.

To the claim in the text two quite different qualifications need immediately to be added: (i) certain people, namely those following what can loosely be labelled a 'quantum logical' approach, may wish to claim that there is, or was, much about the formal structure of quantum mechanics that is, or was, not well understood, but rather awaited development of their approach; (ii) The basic theory is non-relativistic and one has great difficulty in developing a relativistic version of it, the reasons for this difficulty seeming to lie deep in the mathematical structure of the theory. Both of these points have merit and will be returned to later on.

[4] Of course there are some genuine difficulties of application here but these refer principally to the high-energy, sub-nuclear relativistic domain and suggest the limitation on the domain of theory alluded to in the previous paragraph. There are also some experimental situations which daunt applied quantum theorists because of their complexity or surprise them because of the new quantum effects they display (for example, the Mössbauer effect – see [35] – or, in reverse (it surprised the experimentalists, not the theoreticians]), the Bohm-Aharanov and linked effects, see [4,5]); but it is not on the presence of either of these that the philosophical puzzles posed by quantum theory focus.

[5] At least this is the conclusion to which my own extended examination and development of Bohr's position leads – cf. the chronological development in [52], Part II and bibliography therein, in [54] and [58]. Naturally I believe this to be the most insightful treatment of his position!

[6] A term drawn from Quine [107], Chapter 7.

[7] The basic representation space may be switched to momentum space (the basic independent variables become the momenta p_i, $i = 1, 2, 3$, and t) by a suitable fourier transformation.

[8] The transformations are postulated to form a one-parameter group for the following reason. Suppose $U_{t,t'}: \mathscr{S}_D \to \mathscr{S}_D$ gives us $\phi(t')$ $(\varepsilon \mathscr{S}_D)$ from $\phi(t)$, i.e. $U_{t,t'}(\phi(t)) = \phi(t')$. Then $U_{t'',t'}(U_{t,t'} \phi(t)) = \phi(t'')$, so that $U_{t,t'}$ followed by $U_{t'',t'}$ has the same effect as $U_{t'',t}$. We write $U_{t'',t'} \cdot U_{t'',t} = U_{t'',t}$. Moreover a unique inverse $U_{t,t'}^{-1} = U_{t',t}$ is assumed to exist giving 'reverse' dynamical evolution – i.e. we assume that the dynamical laws of physics are reversible in this special sense, and hence *dynamically deterministic*. This assumption is not trivial. Now the $U_{t,t'}$ form a one-parameter group. Notice that, just by the choice of the mathematical representation of dynamical evolution, we may also assume that the temporal dependence of these transformations involves only the time interval concerned – $U_{t,t'}$ depends only upon $t'-t$. This assumption, made in all fundamental theories of physics so far, is also non-trivial.

For a more precise treatment see Abraham and Marsden [3] and Marsden [86, 87].
[9] Thus in QM, for example, H must be a hypermaximal hermitian operator, but just this specification leaves the entire class of such functions from which to choose. Actually H is determined more strictly than this, namely as a dyadic tensor on Hilbert space, but this still leaves some freedom of choice.
[10] These are groups of linear transformations that leave invariant an antisymmetric bilinear form. They were first studied by Cartan [25]. For a more recent treatment see, eg., Hamermesh [46], and in particular Abraham and Marsden [3] and Marsden [86, 87].
[11] $P_{\emptyset,E} \leqslant P_{\emptyset,E}$; if $P_{\emptyset,E} \leqslant P_{\emptyset',E'}$ and $P_{\emptyset',E'} \leqslant P_{\emptyset,E}$, then $P_{\emptyset,E} = P_{\emptyset',E'}$ (we identify equivalent propositions); if $P_{\emptyset,E} \leqslant P_{\emptyset',E'}$ and $P_{\emptyset',E'} \leqslant P_{\emptyset'',E''}$ then $P_{\emptyset,E} \leqslant P_{\emptyset'',E''}$. More precisely $\leqslant$ is an ordering relation on the set of equivalence classes of propositions in $\mathscr{P}_0$ and corresponds to implication in the Lindenbaum-Tarski algebra of $P_{\emptyset}$. Notice that, unlike the cases which Gudder and Greechie consider elsewhere in this volume, the equivalence here is direct logical equivalence between propositions and not equivalence of values under all statistical measures; thus the difficulties they raise for the Jauch-Piron approach do not arise here.
[12] The connection between POsets and partial algebras is less obvious than perhaps many are inclined to suppose. Evidently there is no reason to suppose that, for example, the partial algebra for QM of Kochen and Specker [74] is the same as, or even closely related to, the orthocompleted POset familiar to another approach to QM (e.g. Gudder and Greechie in this volume). In particular, as van Fraassen has emphasized (private communication), there is no reason to suppose that one can extract a POset structure from Kochen and Specker's partial algebra at all. In fact the connecting link between the two structures has been established by Gudder [42], and it hinges on a property of associativity which, it happens, the partial boolean algebra (i.e. partial algebra of idempotents) for QM possesses but which is not directly reflected in Kochen and Specker's construction, and on a special feature to compatibility which, it happens, the compatibility relation in QM possesses.
[13] In QM these maximal observables are just the non-degenerate linear operators whose eigenvectors span the entire Hilbert space. In CPM fundamentally only *one* observable is maximal, the observable $M = \langle q, p \rangle$ consisting of the ordered pair of position and momentum. Of course the observables $M' = \langle f(q), g(p) \rangle$ could always replace M for this purpose as well. Moreover, under special circumstances other observables might replace q,p (e.g. whenever it happens that, in those circumstances, $\emptyset' = f(q)$ and/or $\emptyset'' = g(p)$ - for a free particle $E = p^2/2M = g(p)$).

In general the maximal observables may be defined as those observables sufficient to determine the values of all members of some maximal subalgebra.

The formulation in the text corrects an earlier, erroneous formulation and I have van Fraassen to thank for the correction. It is worth pointing this particular error out. I had inadvertently defined a maximal consistent set of observables as the set of all observables compatible with a given observable - thereby tacitly assuming that the compatibility relation was transitive. But there is nothing about the introduction of this notion in the formal algebraic context to suggest that it has this property and in fact in QM it is not transitive (a feature I had long ago drawn attention to in [52], n. 125). It is this fact which really prevents it from being physically interpreted as co-measurability, despite the acceptance of this terminology in the literature. It is also this feature which is distinctive of QM structures, for one of the ways to view van Fraassen's manoeuvre (in this volume) with degenerate observables so as to obtain the 'anti-Copenhagen interpretation' is precisely as the rendering of compatibility transitive. (Notice though that this is only a necessary, and *not* a sufficient, condition of obtaining a complete algebra).

[14] An *ultrafilter* $\mathscr{F}_u$ on a boolean algebra, or logic, $\mathscr{B}$ is a maximal filter, i.e. such that no proper filter on $\mathscr{B}$ contains $\mathscr{F}_u$, where a filter $\mathscr{F}$ on $\mathscr{B}$ is a non-empty subset of propositions of $\mathscr{B}$ such that (i) for all p, $q \in \mathscr{F}$, $p \cdot q \in \mathscr{F}$, (ii) for all $p \in \mathscr{F}$, $q \in \mathscr{B}$, if $p \supset q$ then $q \in \mathscr{F}$ also. A filter is proper if $\mathscr{F} \neq \mathscr{B}$. An ultrafilter can then be shown to be defined by (iii) for all $p \in \mathscr{B}$, $p \in \mathscr{F}_u$ or $\sim p \in \mathscr{F}_u$ but not both. See e.g., Bell and Slomson pp. 12–15. These definitions are naturally generalizable to lattices $\mathscr{L}$ (make the following replacements of lattice operations for logical operations in the above: $\mathscr{B} \to \mathscr{L}$, p, q, $\to x$, $y(\varepsilon \mathscr{L})$, $p \cdot q \to x \wedge y$, $p \supset q \to x \leqslant y$, $\sim p \to x$*) and hence will apply to QM as well. The distinction between the cases where the maximal consistent sets exhaust $PA_{\mathcal{O}}$, and where they do not, is pertinent to the distinction between CPM and QM.

[15] Here I ignore the presence of superselection rules.

[16] See Gleason [38]. An outline of the argument in the case of CPM-CSM is given in note 35.

[17] The formulation of $SPA_{\mathcal{O}}$ is due to Kochen and Specker [72]. With respect to $SP_{\mathcal{O}}{}^0$, cf. also the treatment of Gudder and Greechie in this volume.

The distinction between the two algebras is dealt with in more detail by Bub in this volume. The distinction between these two approaches was arrived at independently by Bub and myself, though I often follow his expression and development of it here since it is the more penetrating.

Kochen and Specker effectively choose the propositions $P_{I_{\mathcal{O}}, \{1\}}$ as corresponding to the idempotents of $PA_{\mathcal{O}}$. It seems to me that for consistency, however, the propositions

'The probability is 1 that any $\mathcal{O}' \in [\mathcal{O}]$ takes a value in E'

should rather be considered in connection with $SPA_{\mathcal{O}}$. Since it turns out that for CPM-CSM and QM $PA_{\mathcal{O}}$ and $SPA_{\mathcal{O}}$ are in each case isomprohic to one another, Kochen and Specker's choice leads to the fundamental logic of the theory, that corresponding to $PA_{\mathcal{O}}$. But even here the logical structure of the foregoing probabilistic propositions is not truly boolean (though it can be made to formally correspond to a boolean algebra by some 'trick' definitions). The traditional choice of propositional forms, typified by Kochen and Specker's choice, obscures the essential difference between $PA_{\mathcal{O}}$ and $SPA_{\mathcal{O}}$ and draws attention away from the logical structure of the probabilistic propositions and away from considering situations in which $PA_{\mathcal{O}}$ and $SPA_{\mathcal{O}}$ do not coincide. (Kochen and Specker do not themselves seem to appreciate this distinction between $PA_{\mathcal{O}}$ and $SPA_{\mathcal{O}}$ since in the discussion preceding their formula (3) they talk in terms of $PA_{\mathcal{O}}$, but their actual formula (3) is appropriate only for $SPA_{\mathcal{O}}$.

[18] This puts the matter very formally but, like most formalizations, it has the attractiveness of its precision. Incidentally the 'commonplace' mentioned in the text, like most of its kind, is surely only a half-truth – the tradition of mathematizing nature extends back to the Pythagorean school of ancient Greece, nor were the 'dark ages' totally dark in this respect. In other fields besides physics – astronomy it took some time to realize this ideal, e.g. chemistry, and in most others it is still not realized (e.g. biology, social sciences). It appears that scientists accept the assumption, moreover, that the quickest, surest way to achieve the ideal is to reduce each discipline to one where the ideal has been achieved.

[19] These remarks contain a great many qualifiers (some 'hidden') because of two unsolved sets of problems: (i) the exact accounting for biological and some chemical properties, (ii) the resolution of the nature and status of the secondary qualities. Qualifiers

such as "physical" and "relevant" are to be so read that the remarks in the text come out true vis-a-vis these problems. (I have made an effort to remove (ii) in my [59].)

[20] Because of the definiteness assumption the corresponding properties of all complex systems are thereby determined – 'is a definite complex of' is a transitive relation – and because the dynamics are essentially reducible to spatio-temporal terms and mass and charge are conserved, the spatio-temporal evolution of the system is thereby determined and thereby in turn the full temporal succession of exhaustive state descriptions generated for a given system.

A word of caution: it must not be assumed that a superficial classical conception of the properties discussed is always to be admitted, cf. the remarks in this section on the definition of canonically conjugate properties and the remarks on continuity in the atomic and plenum schemas, Section VII, and the remarks on dealing with rigid atoms, Section XII.

[21] A remark on the assumption of continuity is in order. Continuity for most physical magnitudes of theoretical interest was guaranteed by the classical assumption of the continuity of space and time. (Thus all geometric properties and all those kinematical and dynamical properties defined in spatio-temporal terms: velocity, acceleration etc.) This leaves only the so-called 'intrinsic c-numbers' of classical theory: mass and charge. These latter did not *actually* possess continuous values (at least if atomism be granted – cf. Section VII) but were acknowledged at least to be *in principle* continuous magnitudes, i.e. it was regarded as merely contingent that some values for masses and charges were not found – there were no theoretical limitations on continuity. Indeed, the assumption of continuity runs very deep in classical science. Thus all dynamical equations, being grounded on space and time, presuppose the use of continuous functions and all force functions introduced into CPM are in fact continuous, leading to such fundamental results as the equi-partition of energy, the denial of which by Plank was *perceived* as the first serious introduction of discontinuity into physics and also as the birth of quantum theory. But cf. Sections VII, XII.

[22] But there are important dynamical reasons why one might not want to countenance the non-punctiform ontology – cf. remarks on rigidity in Section XII.

[23] For penetrating discussions of the mathematical foundation of phase space see Mackey[83], Chapter 1, Abraham and Marsden [3], and Komar's paper in this volume.

[24] Alternatively, the transformations may by regarded as operations upon the coordinate axes themselves, transforming to new coordinatizations of the phase space in which the system locations do not alter. This alternative finds its counterpart in the 'Heisenberg picture' of QM whereas the traditional view corresponds to the 'Schrödinger picture'. For clear treatments of the canonical formalism see, for example, Goldstein [39] (who also discusses these alternatives for the special case of rotations – cf. Chapter 4), Mercier [90] and Rosen [111].

[25] Cf. also the discussion by Komar in this volume and the references of n. 10.

[26] See, e.g. Abraham and Marsden [3] and Marsden [86, 87].

[27] There are any number of excellent texts on boolean algebras, amongst them Bell and Slomson [8], Halmos [45] and Sikorski [126].

[28] Thus for every pair of functions $\mathcal{O}_1 = f(\langle q, p \rangle)$, $\mathcal{O}_2 = g(\langle q, p \rangle)$ we may define

$$\alpha \mathcal{O}_1 + \beta \mathcal{O}_2 = (\alpha f + \beta g)(\langle q, p \rangle)$$

$$\mathcal{O}_1 \mathcal{O}_2 = (fg)(\langle q, p \rangle)$$

with $\alpha, \beta \in \mathbf{R}$, hence the $\mathcal{O}_i$ form an associative commutative ring under '+' and a vector

space over R and hence an algebra in the usual sense (see, e.g., Herstein [49], p. 218).
[29] This is clear from the correspondence between the f_{C_S} and the subsets S of S^{6N}, the f_{C_S} provide a mapping of the S into $\{0,1\}$. More perspicuously, we have

$$f_{C_S} \cdot f_{C_{S'}} \leftrightarrow f_{C_{S \cap S'}}$$
$$f_{C_{S \cup S'}} \leftrightarrow f_{C_{S \Delta S'}} + f_{C_{S \cap S'}}$$

(hence)

$$f_{CS} \leftrightarrow 1 - f_{C_S}$$

where

$$\bar{S} = S^{6N} - S, \quad S \Delta S' = (S \cap \bar{S'}) \cup (S' \cap \bar{S}).$$

[30] On Zeno's paradoxes see, e.g. Grünbaum [41]. I am indebted at several points to one of my students, Mr. G. Foss, for an illuminating discussion of the matters discussed in this paragraph.
[31] As the reader will shortly see, most of the essay is preoccupied with the classical origins of features of QM. A precise version of the argument for the case of energy transitions in time is given in my [55]. This argument strengthens the conclusion that these are indeed canonically conjugate quantities and yet they strictly appear so in no theory of physics.
[32] One commences with a probability distribution over phase space $\mu_0(S)$ at $t = 0$. This evolves as follows:

$$\mu_t(S) = \mu_0(U_{t,0}^{-1}S)$$

where $U_{t,0}^{-1}S = \{\langle q, p \rangle : U_{t,0} \langle q,p \rangle \in S\}$. Thus the time dependence of μ_t is determined by the $U_{t,t'}$. In CSM the corresponding dynamical equation of evolution is Louiville's equation

$$\frac{d\mu_t}{dt} = - \{\mu_t, H\}$$

See, e.g. Tolman [130] for a detailed discussion.
[33] I eschew further discussion concerning the vexed question of the precise interpretation of the concept of probability; but see, e.g. Nagel [93], Popper [101, 103].
[34] See. e.g. Kingman and Taylor [71].
Simply, we have that (i) any measure on S^{6N} can be resolved into other such measures by the method of characteristic functions described below in the text if it is non-zero on at least 2 points of S^{6N}, (ii) clearly a measure concentrated at a single point is dispersion free and conversely, since

$$\langle \mathcal{O} \rangle = \int \mathcal{O} d\mu_{\{\langle q', p' \rangle\}} = \mathcal{O}(\langle q', p' \rangle)$$

$$\langle \mathcal{O}^2 \rangle = \int \mathcal{O}^2 d\mu_{\{\langle q', p' \rangle\}} = \mathcal{O}^2(\langle q', p' \rangle)$$

hence

$$(\langle \mathcal{O} \rangle)^2 - \langle \mathcal{O}^2 \rangle = 0, \text{ conversely if } (\langle \mathcal{O} \rangle)^2 - \langle \mathcal{O}^2 \rangle = 0$$

then

$$\left(\int \mathcal{O} d\mu \right)^2 = \int \mathcal{O}^2 d\mu$$

and since $(\Sigma a_i)^2 \geqslant \Sigma a_i^2$, all $a_i \geqslant 0$, equality is achieved only for a 1-term series, i.e. μ vanishes everywhere except at a point.

[35] It is difficult to find this argument set out explicitly (at least I found it so) and I am indebted to Professor E. Prugovecki for assistance at this point. In outline the argument runs as follows. With every measure μ on S^{6N} can be associated a linear functional ϕ_μ on some suitably chosen space of functions (say L^1_∞ – the space of all integrable, infinitely differentiable functions) as follows

$$\mu \leftrightarrow \phi_\mu(f) = \int f \mathrm{d}\mu', \ f \in L^1_\infty$$

The ϕ's can then be shown to form a convex, topological vector space $\mathscr{L}$. Then by applying the Krein-Millman theorem and supporting theorems (cf. e.g. Naimark [94], p. 62 and §§31, 32), one can show that *every* ϕ can be written

$$\phi = \int \Psi \mathrm{d}\nu$$

where the Ψ's are 'extreme points' of $\mathscr{L}$ (i.e. $\mathscr{L}$ has an appropriate basis) and ν is an appropriate measure on these extreme points. But $\Psi_{\mu'} \leftrightarrow \mu'$ an extreme measure in the set of μ's and so we may write every μ as

$$\mu = \int \mu' \mathrm{d}\nu'$$

where the μ' are the extreme measures $\mu_{\{\langle q, p \rangle\}}$.

[36] In *this* (Bub's) sense CPM-CSM is *complete*. The probability measures contain maximal expressive power. Indeed, one might even begin from CSM and construct CPM from within its resources! This order of construction is in fact more illuminating when comparing it with QM. Cf. Park [97].

In the terminology of Gudder and Greechie the situation is described by saying that CPM-CSM has a full set of states. In this case it does not matter whether one commences with the construction of $PA_\emptyset$ or of $SPA_\emptyset$ – the same structure, $\mathscr{B}^{CPM}S$, is reached. On this use of "*state*" see below in the text.

[37] The reader will find it illuminating to realize that the basis of much of the discussion of the logic of QM (e.g. that found elsewhere in this volume) is the taking of the elements of $\mathscr{S}_\mu^{QM}$ for that theory as (essentially) states – e.g. van Fraassen's modal semantics.

[38] The Lagrangians and Hamiltonians have a logical role in CPM which is quite different from that to be played by the members of $\mathscr{S}_T$ – the former are there to *generate* the dynamical evolution, the latter to describe the dynamical sequences of states. This is the fundamental reason for choosing the ϕ's, but it is reflected e.g. in (i) the fact that Lagrangians and Hamiltonians are not uniquely determined, (ii) regarded as functions of time they either do not in general change (the Hamiltonians) or change in ways irrelevant to this purpose (the Lagrangians).

[39] Considerable interest attaches to the study of developments of this type however, especially to the development of algebras with products other than the simple pointwise commutative produce defined here (e.g. a convolution), because of the work now being done on the relations between the algebraic structures of CPM and QM – see Shankara and Srinivas [125] and references cited therein and also the important work by Petersen and Grgin [99] – and because of the importance I attach to a comparison

of these structures with those arising in CFT (cf. below), a comparison whose point will, I hope, become more obvious as we proceed.

[40] Thus despite a certain superficial resemblance to the situation in QM the differences are fundamental. Indeed the mathematical operations of forming compound states are radically unlike at bottom, for the one (CPM) constructs point locations in a higher dimensional space whereas the other (QM) constructs vectors (rays) in a tensor product space of equipolent dimensionality, cf. also Section X.

The probability of the compound trajectory ϕ is given using the usual classical probability axiom, as

$$\text{Prob}(\phi) = \text{Prob}(\phi_1) \cdot \text{Prob}(\phi_2/\phi_1) = \text{Prob}(\phi_2) \cdot \text{Prob}(\phi_1/\phi_2)$$

because ϕ_1, ϕ_2 uniquely determine ϕ. In the interaction-free case we would have Prob $(\phi_1/\phi_2) = \text{Prob}(\phi_1)$, Prob $(\phi_2/\phi_1) = \text{Prob}(\phi_2)$. This would not be expected to hold in the interaction case.

Indeed with a *suitable* introduction of interactions one could reproduce any probabilities whatever via the consequent adjustment of conditional probabilities. This fact furnishes the basis of several attempts to interpret QM. But the key consideration in judging the acceptability of any such position is the case to be made for the real existence of such forces. For real forces have real sources and have a mode of action described by a definite law. No plausible account of QM has been given in these terms yet. And some accounts can be shown to be inadequate even on the basis of a vague outline – e.g. Heisenberg's disturbance of the system upon measurement (see Hooker [54]).

[41] See Montague [91] and van Fraassen [34], Part I.

[42] Specifying a functional form ϕ determines a family of functions in $\mathscr{S}_T{}^{\text{CPM}}$ in turn determining a family of trajectories in S^{6N}; specifying a particular ϕ (i.e. determining the 'initial condition') determines a particular trajectory. I pass over without comment that this latter determination lies outside the scope of the laws of CPM (cf. Petersen [98]).

CPM is deterministic precisely because it satisfies the following condition: If any two admissible trajectories in phase space share one point in common they are identical (share all points in common). Cf. Montague's formulation in [91]. Kronfli [75] has given an alternative characterization of classical determinism formulated for a theory with a boolean logical structure of finite degrees of freedom and shown determinism to be equivalent to the assertion that the logical structure is atomic (cf. below).

[43] For the definition of Ultrafilter see n. 12. An *atom* of a boolean logic or algebra $\mathscr{B}$ is a minimal, non-zero element of $\mathscr{B}$, i.e. an element p such that (i) $0 \supset p$ (0 is the zero of the algebra or the always false proposition), (ii) if $q \supset p$ then either $q \equiv 0$ or $q \equiv p$. It is then easy to show that every atom of $\mathscr{B}$ generates a unique ultrafilter on $\mathscr{B}$.

Note that not every boolean algebra need be atomic i.e. be such that for all $q \in \mathscr{B}$ there exists an atom p such that $p \supset q$, but every power set algebra is atomic, the atoms being the singleton sets. In our case then $\mathscr{B}^{\text{CPM}}{}_S$ is atomic and its atoms are the $\{\langle q, p \rangle\}$.

[44] This is obvious because of the correspondence $P^{\text{CPM}}{}_S \leftrightarrow S$.

[45] Cf. the treatments by Mackey [83], Kronfli [76] and Varadarajan [133].

[46] Thus even though we write $P^{\text{CPM}}{}_{\mathcal{O}, E} \leftrightarrow M^{\text{CPM}}{}_P (\mu, \mathcal{O}, E, 1)$, it cannot be concluded, eg. from

$$P^{\text{CPM}}_{\mathcal{O}, E_1 \cup E_2} \leftrightarrow M^{\text{CPM}}_P (\mu, \mathcal{O}, E_1 \cup E_2, 1)$$

$$P^{CPM}_{\mathcal{O}, E_1 \cup E_2} \leftrightarrow P^{CPM}_{\mathcal{O}, E_1} \text{ or } P^{CPM}_{\mathcal{O}, E_2}$$

$$P^{CPM}_{\mathcal{O}, E_1} \quad \leftrightarrow M^{CPM}_P(\mu, \mathcal{O}, E_1, 1)$$

$$P^{CPM}_{\mathcal{O}, E_2} \quad \leftrightarrow M^{CPM}_P(\mu, \mathcal{O}, E_2, 1)$$

$$M^{CPM}_P(\mu, \mathcal{O}, E_1 \cup E_2, 1) \leftrightarrow$$

$$\leftrightarrow M^{CP}_P M(\mu, \mathcal{O}, E_1, 1) \text{ or } M^{CPM}_P(\mu, \mathcal{O}, E_2, 1)$$

that

$$M^{CPM}_P(\mu, \mathcal{O}, E_1 \cup E_2, 1) \leftrightarrow$$

$$\leftrightarrow M^{CPM}_P(\mu, \mathcal{O}, E_1, 1) \text{ or } M^{CPM}_P(\mu, \mathcal{O}, E_2, 1)$$

and this is just the difference between 'or' and '*or*'. This is why the special sign '$\leftrightarrow$' has been introduced to herald this weaker correspondence. See van Fraassen's article in this volume, from which these points were originally drawn.

[47] These spring from the formal axioms of the probability calculus; e.g. from

$$\mu_{\mathcal{O}}(E_1 \cup E_2) = \mu_{\mathcal{O}}(E_1) + \mu_{\mathcal{O}}(E_2) - \mu_{\mathcal{O}}(E_1 \cap E_2)$$

we obtain

$$\mu(\mathcal{O}^{-1}(E_1 \cup E_2)) = \mu(\mathcal{O}^{-1}(E_1)) + \mu(\mathcal{O}^{-1}(E_2)) -$$

$$\mu(\mathcal{O}^{-1}(E_1 \cap E_2))$$

and hence

$$M^{CPM}_P(\mu, \mathcal{O}, E_1 \cup E_2, r) \quad \text{and} \quad M^{CPM}_P(\mu, \mathcal{O}, E_1, r_1) \quad \text{and}$$

$$M^{CPM}_P(\mu, \mathcal{O}, E_2, r_2) \to M^{CPM}_P(\mu, \mathcal{O}, E_1 \cap E_2, r_1 + r_2 - r)$$

or from set theory, e.g. from

$$\mu_{\mathcal{O}}(E) = r, \quad \mu_{\mathcal{O}'}(E') = r$$

then

$$\mu_{\mathcal{O}\mathcal{O}'}(E'') = r''$$

where $r'' \geqslant r, r'$ and $E'' = \{x : x = rr', r \in E \text{ and } r' \in E'\}$ and we assume $\mathcal{O}, \mathcal{O}'$ simultaneously definable. But these are extremely unhelpful fragments for founding a logical structure.

[48] The only *general* paper on this structure of which I know is that by Scott and Krauss [121], but their penetrating model – theoretic remarks are yet a good way off the development of an explicit algebraic structural characterization of the sort developed for the $P^{CPM}_{\mathcal{O}, E}$ (e.g. a *PO*set, lattice, algebra etc.)

[49] If clause (i) held but clause (ii) did not one would be tempted to say something like this: the theory applies to individual systems but each system is itself a complex of sub-systems, the measurable observables being really statistical averages over the properties of these sub-systems. This would be a plausible line to take since the entire structure could be re-built on the basis of these sub-system properties – i.e. a new phase space chosen based on them – in virtue of the fact that the present CPM superstructure has a boolean logical structure.

Alternatively, one may be inclined to say that the theory deals with individual systems each of whose properties is 'latent' or 'potential', or perhaps objectively spread over the range of values, but anyway such that interaction with a measuring instrument leads *the instrument* to read a particular value in the allowed range, either in virtue of a 'realization' of the latent or potential property, or because of some dynamics relating spread out properties to instrument reactions, such that the statistical dispersions are recovered from an ensemble of measurements. These alternatives, precisely because they are vague, continue to apply to quantum mechanics where neither clause (i) nor clause (ii) is satisfied.

The distinction between clauses (i) and (ii) represents a refinement of James L. Park's otherwise beautifully clear analysis of the nature of statistical theories, classical and quantum see [97].

[50] Cf. the discussion in n. 40.

[51] This is in direct contrast with QM where an initially pure ensemble of pairs of interacting systems evolves into a pure ensemble whose corresponding sub-ensembles are characterized by statistical *mixtures* (i.e. of weighted sums of pure ensembles). The situation in CPM-CSM has to do with the fact that each pair of trajectories in S^{6N} determines, under a given interaction law, a *unique* trajectory in S^{6N+6N} and *vice versa*. The latter condition fails to hold in QM. This difference between CPM-CSM and QM hinges on the absence of a superposition principle in $\mathscr{S}^{CPM}_T$.

[52] For example, Park (an ex-student and expositor of Margenau) says "Put simply, quantal laws govern the *statistics of measurement results* and that is all" [97], p. 219, his italics. Cf. also Margenau and Park [85] and most recently Ballantine [6] (and my comments on these in [52], notes 140, 78 and 24 and in [57]).

[53] See Schrödinger [118]. This article provides a beautifully clear discussion of the interpretive problems of QM in a philosophically as well as physically sensitive context – as of the 1930's. Notice that, in a rather extreme sense, this view thus incorporates the position α as well, since reality is 'created' whenever a measurement result appears. For an exposition of some of these early, heavily interwined, approaches, see Jammer [67].

[54] Cf. Hooker [55].

[55] I.e. those who believe that theories are there to tell us what the world contains and how it behaves, independently of our observation of it.

[56] Cf. my detailed exposition in [52], §§12, 13.

[57] This is the view taken by Putnam [106], and by Bub in this volume, for example. They base their case on a correspondence between logical form and geometrical form in theories.

[58] See, for example, Bohm [12, 13, 14, 15, 16, 18], Bub [21, 22], and in this volume, and other references cited in my [52].

[59] To the extend that one blindly pursues these alternatives then one might be guilty of Popper's "great quantum muddle" of confusing Ψ as (determining) a statistical state and hence applying it only to ensembles, with applying it to an individual – Popper [102]. Indeed, to believe Popper seems as if it would preclude speaking of any specific ontology, or even substratum, certainly any ontology of individuals, at the basis of QM at all. But to the extent that one is aware of the analogies of QM with CPM, and of the necessity to somehow achieve coherence between the three fundamental components of a theory, to that extent one might pursue this line of inquiry deliberately and legitimately without being muddled, i.e. one might deliberately set out to endow QM with an individual ontology. Cf. also remarks on divorcing SPA^{QM}_θ and PA^{QM}_θ in the text.

[60] For example, there is the work of Sachs discussed briefly in Section XIII below, see also his article in this volume, and Leiter [79], Heisenberg [48] and many others.

De Broglie has also pursued a non-linear field theory (the "Double Solution" theory) for a number of years and this has shown a recent resurgence – See [20].

[61] Incidentally, it would be of considerable interest, I think, to reconsider the ancient debates in the light of the divergences between these two great ontological schemes. Such philosophically sensitive examination might prove very illuminating of the 'logic' of those ancient debates.

[62] The interested reader will find much helpful historical material in Hesse [50], van Melsen [89], Whyte [138] and Williams [143].

[63] Cf. the analysis in Section IX below.

[64] The argument is given in detail in [61]. It rests essentially on the logical relations among canonically conjugate properties, though these are wider than present physics explicitly acknowledges, for example they contain also the energy-time relations. Cf. my comments in Section XIII and in [52], n. 31 and Appendix I.

[65] The question of identity under these circumstances becomes largely one of deciding whether to choose substantial or structure-preserving rules of identity. The latter, if chosen, are determined by the dynamics of the theory.

[66] Thus whether the individuals interpenetrate or not we count only sets of sets of fundamental individuals – cf. n. 65 on identity.

[67] Cf., for example, Sellars [123].

[68] See [17, 19, 14] and [7]. Cf. also my comments in Section XIII below.

[69] Such 'standard' presentations are not easy to find since prior to 1900 the field formalism was of interest only in the case of the electromagnetic field, between 1900 and about 1940 it was essentially of no interest at all (even those pursuing general relativity theory and the suggestion of a unified field approach seem not to have bothered developing the general theory of classical fields to any significant extent), but after 1940 with the development of QFT the general foundations of CFT, ironically, again became of some interest by way of heuristic. The clearest accounts of CFT are found then in the introductory chapters of books on QFT – see especially the classical work by Wentzel [135], the last chapter of Goldstein [39], and Rosen [111] as well as Landau and Lifshitz [78]. Cf. also Visconti [134] and Jost [70] for some penetrating observations on CFT.

[70] Actually, when one considers the greater penetration displayed by the Hamiltonian approach of Abraham and Marsden it begins to appear that a better foundation for the account of *dynamics* might be through this latter Hamiltonian theory. Cf. n. 8.

[71] One recognizes here the origin of the 'field individual' ontology discussed in [61] and briefly in [60].

[72] In a typical physics text – see, e.g., the account in Wentzel [135] – these two analogies are not distinguished, the index i is made to run over both fields and spatial volumes. In this way attention is diverted away from the significance of the logical and mathematical structures of physical theories.

[73] Functional differentiation is defined as follows

$$\frac{\delta \Psi(Q)}{\delta Q} \Big|_{Q'} = \frac{d}{d\varepsilon} [\Psi(Q' + \varepsilon Q)]_{\varepsilon = 0}$$

See Rosen [111] Appendix A.

[74] This procedure is dealt with, e.g., in Wentzel [135], Jost [70], Rosen [111], Rzewuski 114] and in many other places.

[75] Note that the x_k, like t, now play the role of *independent* parameters, just as was t in

CPM. Thus neither the x_k nor t will turn out, on the analogy with CPM, to be a static observable of the system! The gain in symmetry between the x_k and t, permitting the natural formulation of a relativistically invariant theory, is thus offset by the loss of the x_i (read q_i for CPM) as *dynamical variables*. (in QM, as in CPM, t is an independent parameter – it is not represented by an hermitian operator – while the x_i are dynamical variables.) This shift between CPM and CFT is at the heart of the contemporary difficulties in defining observables, especially the fundamental position observables, in QFT. Cf. Newton and Wigner [96], Wightman [139] and Prugovecki [104].

[76] This conclusion follows again simply from the observation that if $\phi_1(x_\mu)$, $\phi_2(x_\mu)$ are solutions of any of these equations then so also are $\alpha\phi_1 + \beta\phi_2$, $\alpha, \beta \in F$ (real or complex numbers, depending on the field).

[77] See, e.g., Prugovecki [104, 105]. To take an example, consider the solutions $\phi(x_i, t)$ to the Klein-Gordon free field equation

$$(\Box^2 - M^2)\phi = 0.$$

Taking the 4-space fourier transform, we find

$$(k_0^2 - \mathbf{k}^2 - m^2)\phi(k_0, \mathbf{k}) = 0$$

$k_0 \leftrightarrow ct$, $\mathbf{k} \leftrightarrow \mathbf{x}$, which has non-trivial solution only for $k_0^2 = \mathbf{k}^2 + m^2$. The 'positive energy' solutions $k_0 = +(\mathbf{k}^2 + m^2)^{1/2}$ can be written

$$\phi(x_i, t) = \frac{\sqrt{2}}{(2\pi)^{3/2}}\int dk_0 d\mathbf{k}\, e^{-ik_0x_0 + i\mathbf{k}\cdot\mathbf{x}}\, \delta(k_0^2 - (\mathbf{k}^2 + m^2)) \times$$
$$\times \theta_0(k_0)\, f(k_0, \mathbf{k})$$

$\theta_0(x)$ is the 1-0 step function at $x = 0$, and $x_0 = ct$, so

$$\phi(x_i, t) = \frac{1}{(2\pi)^{3/2}}\int \frac{d^3k}{[2(\mathbf{k}^2 + m^2)]^{1/2}}\, e^{i\mathbf{k}\cdot\mathbf{x} - i(\mathbf{k}^2+m^2)^{1/2}x_0} \times$$
$$\times \hat{f}(\mathbf{k}, (\mathbf{k}^2 + m^2)^{1/2})$$

and we require the fields to have 'regular' solutions at $x_i = \infty$, i.e. to be sufficiently fast decreasing at large x, that

$$\int \frac{d^3k}{(\mathbf{k}^2 + m^2)^{1/2}}|f(\mathbf{k}, (\mathbf{k}^2 + m^2)^{1/2})|^2 < \infty$$

Then the linear product is defined as

$$(\phi, \phi') = i\int \phi^*(x_{i,t})\, \frac{\overleftrightarrow{\partial}}{\partial x_0}\, \phi'(x_{i,t})\, d^3\mathbf{x}$$
$$= \int \frac{d^3k}{(\mathbf{k}^2 + m^2)^{1/2}}\hat{f}^*(\mathbf{k}, (\mathbf{k}^2 + m^2)^{1/2})\, \hat{f}'(\mathbf{k}, (\mathbf{k}^2 + m^2)^{1/2}).$$

The Hilbert space in question is then the completion of this linear function space of positive energy solutions with respect to the norm $\|\phi\| = (\phi, \phi)^{1/2}$.

[78] A few cases of non-linear field equations have been rigorously recently investigated – cf. e.g. Rosen [112], Muraskin [92], and Sachs [115] and Leiter [79] have also

studied such equations in a more general context as have, of course, those investigating the field equations of general relativity, see, e.g. Sen [124]. Also there are some general works on non-linear mathematical physics of relevance – see, e.g. [77, 119, 132]. Doubtless there is much more that has escaped my attention than I have recorded here. Even so, the results are extremely difficult to obtain, fragmentary and what results there are, are scattered through the journal literature. What it would be like to obtain a non-linear *space* can, however, be outlined, at least in abstract. Suppose that a metric could be introduced to a set of functions $\mathscr{S}'_T$ such that the metric relations among the set could be mapped homomorphically into metric relations among the points $\mathscr{P}_i$ of a non-euclidean geometry E under a mapping sending the elements $\phi_i \in \mathscr{S}'_T$ into the $\mathscr{P}_i$, then we could conclude that the $\mathscr{S}'_T$ formed a non-linear space homomorphic in structure to E. I am not familiar with cases where this correspondence has been shown, but I suppose that such cases exist.

[79] See my discussion in [62], §III.

[80] A clear account of the derivation of the conserved quantities may be found, for example, in Bogoliubov and Shirkov [10]. So that the reader will have examples before him, I give here the general expression for the energy-momentum tensor generated in this fashion

$$T^{kl} = g^{ll} \sum_j \frac{\partial L}{\partial(\partial \phi_i/\partial x^k)} \cdot \frac{\partial \phi_j}{\partial x^l} - L g^{kl}$$

expressed in the metric $g^{ij} = 0$, $i \neq j$, $g^{00} = -g^{11} = -g^{22} = -g^{33} = 1$,

$$x_k = x_1, x_2, x_3, k = 1, 2, 3, x_0 = ct.$$

For the case of the scalar Klein-Gordon field

$$(\square^2 - M^2)\phi = 0$$

we have

$$L = \tfrac{1}{2} \sum_k g^{kk} \left(\frac{\partial \phi}{\partial x}\right)^2 - \frac{M^2}{2} \phi^2$$

and

$$T^{kl} = g^{kk} g^{ll} \frac{\partial \phi}{\partial x^k} \cdot \frac{\partial \phi}{\partial x^l} - g^{kl} L$$

[81] This investigation has been carried out by Rosen [113].

[82] Rosen [111] offers some consideration of these matters. It is of some importance to consider the full algebras obtained, and in particular the detailed relationships among specific observables, since for those observables generated via Noether's theorem, i.e. those bilinear in the fields, CPM and CFT essentially coincide (the corresponding sympletic groups coincide). (I am indebted to Komar for emphasizing this.) For this purpose there may be more fruitful products which could be defined. For example, if we attempt to define a multiplication among the ϕ's as either successive application of the functions or as a convolution we can expect that it will not be commutative; $\mathscr{O}_2(\mathscr{O}_1(\phi)) \neq \mathscr{O}_1(\mathscr{O}_2(\phi))$, $\int \mathscr{O}_1{}^*(\phi)\mathscr{O}_2(\phi)\mathrm{d}x\mathrm{d}t \neq \int \mathscr{O}_2{}^*(\phi)\mathscr{O}_1(\phi)\mathrm{d}x\mathrm{d}t$. No matter in which of these ways we introduce the product, however, we still achieve the structure of an algebra, $A^{CFT}{}_\mathscr{O}$, since $\mathscr{O}_1(\phi) + \mathscr{O}_2(\phi) = (\mathscr{O}_1 + \mathscr{O}_2)(\phi)$ has all of the properties of ordinary addition, $\alpha\mathscr{O}_1(\phi)$, $\alpha \in F$, defined as ordinary multiplication satisfies all the axioms of a

vector space over F and both of the products introduced above yield the distributive laws $[(\mathcal{O}_1 \cdot (\mathcal{O}_2 + \mathcal{O}_3)](\phi) = [(\mathcal{O}_1 \cdot \mathcal{O}_2) + (\mathcal{O}_1 \cdot \mathcal{O}_3)](\phi)$, $[(\mathcal{O}_2 + \mathcal{O}_3) \cdot \mathcal{O}_1](\phi) = [(\mathcal{O}_2 \cdot \mathcal{O}_1) + (\mathcal{O}_3 \cdot \mathcal{O}_1)](\phi)$. Unlike CPM, however, $A^{CFT}_{\mathcal{O}}$ will not in this case be a commutative algebra. Since we are here dealing with the non-probabilistic classical case, the non-commutability would be interpreted as follows: the observables $\mathcal{O}_1 \cdot \mathcal{O}_2$ and $\mathcal{O}_2 \cdot \mathcal{O}_1$ are in general distinct observables, i.e. distinct from each other and from $\mathcal{O}_1$ and $\mathcal{O}_2$.

I have not had the opportunity to study these algebras. However comparable work done on CPM has proven extremely fruitful – see the references of n. 39 – and certainly the present field can prove no less important. In fact its central importance will become clear (if it is not clear already) as the essay proceeds.

[83] Equations of this sort have been studied by Rosen [112] and he shows that their asymptotic solutions are given by the asymptotic solutions to the corresponding Hamilton's characteristic equation.

[84] See, for example, Bogoliubov and Shirkov [10], §8. The simple interaction terms representing world scalars are, for example,

$$\phi^n \phi^k, \; \psi^* \psi \phi_1, \; \sum_n \psi^* \psi \phi_1, \; \sum_n \psi^* \gamma^n \psi \; \frac{\partial \phi_1}{\partial x^n} \; \text{etc}.$$

where the ϕ are scalar fields and ψ a spinor field and the first two interactions are scalar-scalar type while the second is vector-vector type. Each interaction, save the first for $n = k = 1$, leads to non-linear coupled equations.

[85] Unfortunately I am not able to direct the reader to a source where such tensor product sets or spaces are constructed explicitly for some classical coupled field equations. The notion of a tensor product space *is*, however, well known to mathematicians and widely used in QFT. Indeed in QFT the reader will at least find the tensor product space of the function spaces of each of the fundamental *free* field equations (Klein-Gordon, Dirac) constructed, for that is exactly the construction Fock space demands. Cunningly (or dangerously, depending upon viewpoint, cf. Sections XII, XIII) QFT does not consider the construction of non-linear function sets in the case of interacting fields, only the dynamical evolution in the linear space is affected (see Section XII).

For a mathematical treatment of tensor product spaces there are any number of books, but since we are interested primarily in their relevance to physics, the reader is directed to Diximier [30], Jauch [68], Prugovecki [104].

[86] The dynamical analogy with CPM may suggest we attempt to define probability measures in some way over sets of values for fields at particular spatio-temporal locations, or perhaps even over sets of spatio-temporal locations as well. Reflection on the globalness of the field states – their dependence upon one another from location to location – shows that these attempts lead either to nonsense or, where suitable constraints are placed, to essentially the proposal in the text.

[87] The following approach may, however, be a possibility. Define instead of $\mathscr{S}^{CFT}_{ST}$, the corresponding set $\mathscr{S}^{CFT}_{ST}(\phi, \pi)$ of ordered pairs of elements (ϕ, π) consisting of a field and its conjugate momentum at various, fixed times. Now in (weak) analogy with CPM we define probability measures over $\mathscr{S}^{CFT}_{ST}(\phi, \pi)$ instead of over $\mathscr{S}^{CFT}_{ST}$. In this way it could perhaps be shown that the probability measures obey

$$\frac{dP_t}{dt} = - \{P_t, H\}_{CFT}$$

analogously with the situation in CPM. At the present time it is not obvious to me, in view of the serious differences between the characteristics of $\mathscr{S}^{\mathrm{CFT}}{}_{ST}(\phi, \pi)$ and phase space S^{6N} of CPM how such a construction would proceed, but it does seem an obvious general line to try.

[88] The non-linear gravitation field equations have been thoroughly studied, see for example Sen [124], Graves [40] and Wheeler [136, 137]. For an interesting recent classical development see Schwebel [120].

[89] The approach of Marsden [86, 87], however, looks as if it will afford precisely this deeper insight.

[90] See [83], Chapter 1.

[91] See Marsden [86, 87].

These references came to my attention thanks to Dr. G. McLelland, but only very late in the process of correcting the first proofs. I have not had sufficient opportunity to study them in depth, but it appears likely that some rewarding generalizations concerning logical spaces for physical theories and the relations between quantum and classical theories will emerge from this approach.

On the other hand they may create the impression that the only distinction between field and particle theories is the finite dimensionality or otherwise of the underlying manifold. This judgement may have some merit to it if dynamical form alone is considered (how much merit remains to be seen) but this essay amounts to an extended argument that if the full content of theories is considered, including their logical structure and ontology, then the differences go far deeper than this.

[92] See, for example, the critique by Boscovitch and others in the accounts found in the references of n. 62.

[93] On this section see the remarks in Sen [124] and the recent thorough treatment of the classical electron by Rohrlich [109]. Cf. Synge [129] for example, on rigidity and relativity theory.

[94] See von Neumann [95]. Both Jauch [68] and Progovecki [104], for example, also offer clear and detailed presentations of this formulation of QM.

[95] See Stone [127].

[96] On these structures see for example, Diximier [30, 31], Guernin [43, 44], Naimark [94], Progovecki [104], Robinson [108] and many others.

[97] On these matters see especially von Neumann [95], Birkhoff and von Neumann [9], Mackey [83], Varadarajan [133] and the essay of Gudder and Greechie in this volume.

[98] For if the family $P_\phi((-\infty, \lambda])$ is known we may find the

$$P_{\mathcal{O}(E)} = \int_E \mathrm{d}P_\phi([-\infty, \lambda])$$

See, for example, Jauch [68], p. 54.

[99] Though I "knew better" an earlier draft tacitly contained just such a confusion and I am grateful to Professors van Fraassen and Gudder for pointing it out to me.

[100] See Gleason [38].

[101] We have

$$\sum_n (\phi_n, WO\phi_n) = \sum_n (\phi_n, W\alpha_n\phi_n) = \sum_n (\phi_n, \omega_n\alpha_n\phi_n)$$
$$= \sum_n \omega_n\alpha_n (\phi_n, \phi_n) = \sum \omega_n\alpha_n$$

[102] This gives the Born Rule for interpreting the theory, cf. any standard QM text.

[103] Once again, in this sense (Bub's sense) QM is complete. It has a 'full set of states' I cannot stress too heavily the importance of this structural coincidence for the viability of what many writers write – cf. n. 17 and also Bub's comments in this volume.

[104] See Gleason [38] and Kochen and Sprecker [74]. In this context it is assumed of course that the structure PA^{QM}_0 would be faithfully preserved in the reconstructed theory – or rather, Kochen and Specker assume this only for $PB^{QM}_{I_0}$, but this, we have seen, is essentially equivalent to assuming it for the entire $PB^{QM}_{I_0}$. But once one abandons the preservation of even this fundamental structure (as, for example, Popper evidently intends us to – cf. Bub [24]) then the proofs fail to apply and the question remains open. In this case, however, any serious attempt to retain QM as *the* physics is abandoned – and there seem better ways to abandon QM, e.g. in the ways discussed in Section XIII.

[105] See, for example, Wigner [142].

[106] Cf. the expositions and discussion in Hooker [52].

[107] See, for example, Schiff [116], §7.

[108] See the discussion in Goldstein [39], Chapter 9 for example, also that by de Broglie [20].

[109] This penetrating formulation will be found in Komar's paper in this volume. Notice that there can be no requirement that the form of the Hamiltonian be preserved for all coordinate representations. The reason lies in the problems discussed by Cohen in this volume.

[110] See the references of n. 104, 97.

[111] This is essentially because we have

$$P^{QM}_{\phi, E} \leftrightarrow P^{QM}_{P[\phi_1], \{1\}} \vee P^{QM}_{P[\phi_2], \{1\}} \vee \cdots$$

Where the $[\phi_i]$ are 1-dimensional subspaces spanning $h(\phi(E))$.

[112] Thus, for example, until we have an account of the theoretical view of the world at this level we cannot even make sense of the measurement results since in this case it is this very same theory that is the basic theory of the instruments themselves. Extant epistemology of science has largely overlooked such facts – cf. my [63]. On the simplistic approach to science generated by the Positivism in the background here cf. also [64, 65].

[113] With this remark I set aside most of the discussions in the common physics texts which to-day largely set the mental horizon for our student's minds. No wonder we produce many good computor technicians and few imaginative physicists!

[114] With (α) we associate the early debates over whether QM was purely a representation of our state of knowledge, with (β) those views, expressed by Heisenberg and numerous other physicists, that there is a 'subjective part' and 'objective part' to QM, e.g. that the 'reduction of the wave packet' is an epistemological affair, with (γ) the Heisenberg Doctrine of the uncontrollable disturbances of the system on measurement – an objectively inadequate doctrine, see my [55] – early Bohm [11], and later Heisenberg [47], doctrine of the creation of the measurement result at measurement and Wigner's view concerning the physical influence of mental acts of cognition (cf. his paper in this volume, as well as [140], with (δ) the (apparent) late Popper view that we must abandon the structure of the quantum partial algebra of observables as representing the real order of nature – see [102], with (ε) Bohr's doctrine, but see [52] and my comments below and with (ι) all the various alternative theories that reduce to QM in some suitable limit, cf. my comments in Section XIII below.

[115] See his paper in this volume, and also [23].

[116] Van Fraassen recently pointed out to me that one may also circumvent Kochen and Sprecker's argument that no mapping of this kind permits a satisfactory imbedding of $SPA^{QM}_\emptyset$ into $BA^{QM}_\emptyset$ (see [74], §5) if one distinguishes between a physical notion of simultaneous measurability which is transitive and the mathematical notion of compatibility (criterion: commutativity of operators) which is not. (Already earlier in [52] I had drawn attention to the peculiarity of the notion of compatibility as a reading of simultaneous measurability in the context of Bohr's warning not to uncritically distort physical concepts to suit mathematical constructions.) One simply requires that sums and products of observables be defined only whenever they are physically simultaneously measurable (e.g. are regarded as both functions of the same maximal observable) and not whenever they are compatible.

[117] Already two years ago when I first wrote [52], I was attempting to formulate this point about probability theories that Bub has now expressed very clearly. But then I saw a connection which I think Bub still does not see clearly enough, namely *the intimate relationship that exists between the formal structure of the classical probability calculus and the ontology of the atomic domain to which it is intended to apply*. Classical probability measures are measures on boolean logics and this, I stressed already in [52], is no accident, it is because probabilities were intended to be founded on *counting individuals* which requires exactly a set-theoretic structure. Thus when we arbitrarily change that basic logical foundation, as Bub wishes to do, we must expect severe dislocations in the corresponding ontology. This, I argue below, is exactly what happens. Bub seeks to blunt this criticism by speaking only of 'facts' not of ontology at all, but I think this is not a convincing position in the long run.

[118] Cf. the discussion of Sections VII, X above. The way in which this role of fields arises can be clearly seen in the limiting – geometrical optics approach to the Schrödinger equation (cf. the references of n. 108). Here the *set* of trajectories in phase space corresponding to Hamilton's characteristic equation represents the *possible* particle motions and it is this set, described by the characteristic function W, that becomes the Ψ function of Schrödinger's equation.

[119] And in fact we may extend the analogy by introducing in both cases a 'form-free' formulation of theory. In the geometrical case we formulate our fundamental physical laws in covariant fashion, in this form they are consistent with any continuous Riemannian geometry and specific geometrical forms may be 'built in' to them by choosing particular representations of the laws (i.e. particular differential forms and so on). Though *logical* form-free formulations of theory have not been explicitly studied as yet, we already have available, courtesy of the algebraists, one partial example of such a formulation – the C^*-algebra; for the algebras of bounded observables of CPM, CFT and QM all satisfy the axioms of a C^*-algebra, formally it all depends upon which mathematical representation one chooses as to which algebra results, someone of Bub's persuasion would say "As to which logical form results". Cf. in this respect the interesting paper by Giles [37].

[120] Jauch and Piron evidently also arrive at effectively this conception, though they do not mention the ultrafilter construction – see [69].

[121] This argument, as I pointed out earlier, derives from the mutual indeterminacies of precise definition holding between canonically conjugate quantities – cf. Section IV.

A precise form of the argument for the case of energy transitions in time is given in [55]. Cf. also n. 31.

[122] Cf. the remarks in n. 118 and text.

[123] Obviously I simplify for the sake of dramatic presentation – some observables have partially continuous spectra – but the point remains.

[124] Cf. my remarks in [52], §6 and those of van Fraassen in this volume. Cf. also [53, 56]. This point remains true even for Bub's reconstrual of conditional probabilities since the theorem on which that reconstrual is based depends essentially on the ability to form in QM coherent superpositions of 'independent' particle-like states.

[125] Of course it is proper to enter here the usual disclaimer on behalf of many examining what might be termed the 'quantum logic' approach who do not have any particular philosophic axe to grind but wish only to pursue the structure as pure algebraists, logicians etc. and perhaps even generalize them as amusing studies or for others to attempt to interpret as possible superior versions of QM – such activities are valuable in themselves and I do not wish to saddle every such practitioner with the Bub-Putnam philosophy.

[126] Of course, the time-*independent* part of Ψ will be given by a nonlinear function of x, but the time-dependent part remains linear. Reflection shows this to be a necessary condition, under the Born Rule, for conserving total probability (normalization) invariant through time when the observables are linear operators.

[127] In [52] I have drawn attention to the importance of the sense of non-locality that arises in QM because of the unrestricted application of the superposition principle and its central role in the Einstein-Podolsky-Rosen paradox. This problem would immediately be resolved with the breakdown of the principle (though, as I show, consistent preservation of experimental findings will not be easy).

[128] These are the experiments that lead to the "bootstrap" idea – cf. Chew [27, 28]. Cf. also Feinberg [33].

[129] This approach lay behind Einstein's unified field research programme – see Sen [124] and Heisenberg [49].

[130] In fact the dynamics works exactly as in QM. I am indebted to Prugovecki for emphasizing to me the systematic importance of this point.

[131] Cf. Bogoliubov and Shirkov [10], for an account of this procedure. Especially see Dirac's comments in [29] at the end of the book. Cf. also Jost [70].

[132] See Guernin [44], Jost [70], Segal [122] and Tomozawa [131].

[133] See Newton and Wigner [96], and Prugovecki [104] for discussions of the problem.

[134] See the references to the work of Araki, Haag, Wightman and others in the references of n. 132.

[135] On fundamental lengths and times see, for example [88, 117].

[136] An excellent coverage of their ideas is given in the volume edited by Bastin – see the papers by Bohm, Penrose, Hiley, Atkin and Bastin in [7].

[137] See Bohm [17, 19] and discussion in the Bastin volume [7].

[138] Peat's paper in this volume is part of this exploration – cf. also the closely related work of Penrose (as reported in Peat and in Bastin [7]) and the others mentioned in n. 136.

[139] Cf. also the remarks on the significance of spinors made by Peat in this volume and also the related classical gravitational field of Schwebel [120].

[140] Such is the approach taken in the important developments reported in the papers by Gudder and Greechie (see their paper in this volume), by Catlin [26], by Ludwig [81, 82] and by many others.

ARTHUR KOMAR

THE GENERAL RELATIVISTIC QUANTIZATION
PROGRAM

I. INTRODUCTION

The field equations of the classical theory of gravitation, the Einstein Theory of General Relativity, admit solutions which can be interpreted as gravitational radiation. Experiments currently in progress seem to indicate the existence of such gravitational radiation. It would be remarkable if the energy transported by the gravitational waves were not emitted and absorbed in quanta in conformity with the well known Einstein relation, $E = h\nu$. Yet, to date, we do not have a satisfactory theory for the quantized transfer of gravitational energy whose classical limit corresponds to the generally accepted Einstein Theory of Gravitation.

The intention of this paper is to present both the progress and the obstacles encountered in the course of the search for a Quantum Theory of Gravitation. However, in order to expose the essence of the difficulties which have thus far plagued this quantization program it is necessary first to probe deeply into the foundations of classical mechanics and thereby reveal the essential features which are necessarily preserved in the transition to the corresponding quantum theory. We shall find as a consequence of this investigation that there are fundamental impediments to the development of an unambiguous quantum theory of gravitation.

II. PHYSICAL THEORIES

Briefly stated, the role of a physical theory is to account for natural phenomena and to aid in predicting their evolution. In order to accomplish this, every physical theory must have two distinct aspects. It must on the one hand be a consistent mathematical theory capable of rigorous formulation, whereas on the other hand there must be a procedure for relating at least some of the mathematical symbols which occur in the formalism to quantities which one can measure in the laboratory; for, we must on the one hand have the possibility of determining whether a given assertion is in fact a consequence of the theory, and on the other hand be capable of

Hooker (ed.), *Contemporary Research in the Foundations and Philosophy of Quantum Theory*, 305–327. *All rights reserved* verifying the validity of the assertion within the realm of natural phenomena.

Although the classical physicist thought that the 'laws of nature' which emerged from his endeavors in some sense mirrored the true structure of nature, from our modern perspective we can see that such a viewpoint is naive and vain. Any formal mathematical theory of sufficient complexity to account for a reasonable portion of natural phenomena will necessarily admit many possible solutions, at most one of which would correspond to our own universe. In fact, should we attempt to formalize the relationship between the abstract symbols occurring in the theory and the corresponding measurements performed in the laboratory we would quickly find that the resulting structure would have infinitely many non-intended interpretations for which there would be no reason even to expect any correspondence with nature. On the other hand, there being but one objective reality, it is rather evident that infinitely many inequivalent physical theories could be devised which coincide in their predictions on this one model of interest. In some sense the structure of objective reality can be viewed, not as given by (i.e. isomorphic to) some one physical theory, but rather as that unique model which simultaneously realizes infinitely many inequivalent valid theories.

In any theory there must occur the assertion that some set of measurable parameters are related by specific functional forms; the theory, to be verified, must allow for the parameters to be varied in order to ascertain that they do in fact relate in the predicted manner. And yet the universe in its entirety is given to us but once, with no variable parameters available. Thus, from a yet deeper point of view, considering the universe in its entirety, the very concept of a verifiable theory to account for this phenomenon does not appear to be tenable. Except for occasional somewhat manic flights of optimism, the physicist is generally quite conscious of the fact that the range of phenomena under investigation, however large, is necessarily delimited by virtue of this necessity to sample a range of parameter values. There appears to be a basic complementarity between the scope of a physical theory and the accuracy to which it can be verified.

III. CLASSICAL SYSTEMS

In this section we shall narrow our sights to the modest proportions of a laboratory, for even here there is a considerable degree of subtlety in relating the abstract mathematical symbols of a theory to measureable quantities. The curious feature about known theories of classical systems is that although from the physical point of view there exist many inequivalent physical systems, say corresponding to the motions of particles under differing force laws, the theories of such systems are locally homomorphic, and in fact, are isomorphic for systems having the same number of degrees of freedom. The morphisms are readily exhibited by substitutions of variables. Thus, even on the more simple level we find infinitely many physical systems describable by the same theory, and at the same time infinitely many theories adequate for the description of one physical system. The problem with which we are naturally confronted is that of theoretically categorizing a particular physical system. It is instructive to see how this is in fact done.

Classical mechanical systems are most readily described by giving their dynamics in phase space. The particular power of this canonical formalism is that morphisms between physically inequivalent systems can be readily established which preserved the form of the dynamical laws, the Hamilton equations of motion. These mappings, the canonical transformations, form a function group which, as we shall see, plays a critical role in determining the relationship between the abstract mathematical symbols of the theory and quantities which are measured in the laboratory. Consider, for simplicity, a mechanical system consisting of a single particle. The customary way of characterizing such a system is to introduce the phase space variables $\mathbf{x}$ and $\mathbf{p}$, which are to denote respectively the position and momentum of the particle, and then to state that the energy (Hamiltonian) of the system is a particular function of these dynamical variables. Thus for example, if we are given that $E = \mathbf{p}^2/2m$, we say that our system is a free particle of mass m, while if we are given $E = \mathbf{p}^2/2m + \frac{1}{2}k\mathbf{x}^2$ we say that our system is a simple harmonic oscillator of mass m and spring constant k. Although at this stage of the discussion we seem to have succeeded within mathematical formalism in readily distinguishing between two such radically different dynamical systems, we must point out that our success is heavily contingent upon unstated, but implied, semantical interpretations of the symbols $\mathbf{x}$ and $\mathbf{p}$. For, as is well known, there exist mathematical transformations of the original variables $\mathbf{x}$ and $\mathbf{p}$, into new variables, $\mathbf{X}$ and $\mathbf{P}$, the canonical transformations, which have the property that the laws of motion obeyed by **X** and **P** (Hamilton's equations) are precisely the same as those obeyed by **x** and **p**, the only change being that in terms of the new variables the Hamiltonian alters its functional form. In particular, canonical transformations exist such that the Hamiltonian of the oscillator may be taken to have the form $E = \mathbf{P}^2/2m$, or that of the free particle may be taken to have the form $E = \mathbf{P}^2/2m + \frac{1}{2}k\mathbf{X}^2$. This being the case, it is evident that we have not yet succeeded in characterizing and distinguishing within our mathematical theory such grossly disparate physical systems as the free particle and the oscillator.

In order to distinguish the various physical systems we are compelled at some stage of the discussion to point to some set of **x** and **p**, and say that by these symbols we mean the 'real' position and the 'real' momentum respectively. In other words, we must in principle exhibit mechanical devices in the laboratory and arbitrarily assert that by the symbol **x** we shall mean the numerical output of a particular such mechanical device, dubbed a 'position measuring device', and by the symbol '**p**' we shall mean the numerical output of a different mechanical device, dubbed the 'momentum measuring device', etc. It is only after such a procedure has, at least in principle, been carried out that the abstract symbols **x**, **p**, and E attain an unequivocal semantic content and may then be employed to characterize a dynamical system. We discover in this fashion that the very concepts such as 'position', 'momentum', 'energy', attain meaning only to the extent that the corresponding measuring devices can in principle be constructed and operated. (Should the material objects of our physical universe have the property that the proper functioning of some of these measuring devices inhibit or interfere with the operation of others of these measuring devices, it follows that the corresponding physical concepts cannot have simultaneous meaning. There is nothing in classical physics which leads us to suspect that in nature this is in fact the case, however, the fundamental limitation is ultimately placed by nature itself rather than by the structure of a particular theory.)

In view of the evident fact that there are in general many different measuring devices and procedures which can be employed to measure, or equivalently, to define a given physical concept, it is necessary to determine the essential features which these varying devices must have in common in order to assert that they specify the same concept. Let us, for the sake of illustration, examine the concept of angular momentum, for it is a sufficiently complicated concept to be quite unintuitive. Basically, the reason why a concept as obscure as angular momentum was introduced into physics was that, for a wide variety of physical systems, it was a constant of the motion; that is, it provided a first integral to the equations of motion of those physical systems for which it was conserved. It remains to characterize the class of physical systems for which the angular momentum is conserved. This was accomplished by a theorem of Noether [1], which asserted that whenever a physical system, whose equations of motion are derivable from an action principle, exhibits a symmetry, there is a conserved quantity associated with that symmetry, and conversely each conserved quantity has an associated symmetry.

In somewhat greater detail, the dynamical variables of a classical theory play a dual role. On the one hand they are used attributively, that is, they lable attributes of the physical system, such as position, momentum, energy, etc. On the other hand they are used as operators, that is, they may be used as generators to realize the infinitesimal canonical transformations of the phase space. The attributive usage is the most immediate and natural, and consequently it was employed long before the operator aspect of the dynamical variables was realized. However, it is the operator aspect of a dynamical variable which enables us to relate it to a measuring device in a laboratory. This is a consequence of the Noether theorem, which asserts that the necessary and sufficient condition for a dynamical variable to be a conserved quantity is that it generate a canonical transformation which leaves invariant the functional form of the Hamiltonian and/or Lagrangian of the system. For the case in question, the associated transformation generated by the abstract symbol which realizes the angular momentum attributatively within classical mechanics is the rotation of the spatial coordinate axes. Since for an isolated system we do not expect the physical laws which it satisfies to depend on the particular orientation of the coordinate axes which we use to describe the system, to the extent which this expectation is fulfilled, the Hamiltonian for the isolated system will exhibit rotational symmetry and will have the angular momentum as the associated conserved quantity. This observation provides the essential clue to the measurement of angular momentum. For example if the system to be measured is coupled to a measuring device in such a way that the combined system is rotationally invariant, but such that each part individually is not, it will then follow that the total angular momentum of the combined system will be conserved, whereas the angular momentum of each part will not. We can then speak of a transfer of angular momentum between the original system under study and the measuring device. Subsequently, by further examination of the structure and functioning of the particular measuring device employed, we can then conclude information about the angular momentum of the system. In effect, angular momentum can only be measured subject to the understanding that it is precisely that quantity which is conserved for systems which are symmetric under rotations, or equivalently, that it is precisely that quantity which is realized in any theory by the generator of the invariant mapping which is induced by a rotation of the spatial coordinates.

What we have said for angular momentum applies to the other basic dynamical concepts as well. The semantic content of all these concepts are contained in their relationship to those measurements which depend for their interpretation on the existence of corresponding quantities which are conserved for the combined system consisting of the object being measured and the measuring apparatus. To define fully a physical concept, it is necessary, and frequently sufficient, to give the associated symmetry. Thus, associated with the linear momentum is the translation symmetry, or equivalently the insensitivity of the physical laws of an isolated system to the choice of origin of the coordinate system used to describe these laws; the energy is associated with the time-translation symmetry, or equivalently, the insensitivity of the physical laws to the choice of an origin in time. The position is somewhat more difficult to relate to a symmetry. We can, if we wish, regard it as an underlying primitive element of the arena of physics, to be measured by means of a standard measuring device, the microscope. However there is a somewhat more complicated symmetry to which it can be related, namely the insensitivity of the physical laws to the state of uniform, rectilinear motion of the origin of the coordinate system used to describe the physics. The associated conserved quantity is the center of mass of the isolated system. However, this latter characterization of the coordinates is somewhat circular, since the concepts 'uniform' and 'rectilinear' can only be well defined once we have established our coordinate system.

It is evident that isolated systems can only be expected to have the above symmetries provided the space-time itself has these symmetries. That is, the basic dynamical quantities depend critically for their existence on the particular symmetries of the space-time. Should the space-time itself lack a preferred status for certain of these symmetries, as is the case in the general theory of relativity, the corresponding physical concepts will also cease to have preferred status.

Some confusion is occasionally engendered by the failure to realize that the above discussion on the measureability of functions of dynamical variables applies only to direct measurements of these variables, as opposed to derived measurements. An example of a derived measurement would be the 'measurement' of energy E, by using a device which actually measures the momentum p, and employing the theoretical relation $E = p^2/2m$, under the implied assumption that the system under consideration is that of a free particle. In fact if it were not possible to give an independent, direct measurement of the energy, the relation $E = p^2/2m$, would be tautologically true as a definition and not subject to verification. Another common example of a derived measurement which is a frequent source of confusion, is the determination of the momentum of a particle by means of measuring its de Broglie wave length, λ, and then employing the relation $p = h/\lambda$. Here, again, a presumption of the theoretical dynamical nature of the system must be made in order to relate the two independent concepts, p and λ, by means of a non-tautological relationship subject to verification. In this latter example the implied assumption is that the physical system is not subject to velocity-dependent forces mediated by the existence of vector potentials.

IV. QUANTIZATION

The critical step in the transition from classical mechanics to quantum mechanics was the realization by Niels Bohr that the basic dynamical concepts with which one must deal in the new theory must be measured by the same 'classical' mechanical devices which one employed for the earlier theory. Although at first sight this might have seemed an odd requirement, in the light of the discussion given above it is clear that it could not be otherwise. For, the unambiguous employment of expressions such as 'energy', 'momentum', etc. demands that their semantic content be undisturbed and, in particular, insensitive to the vagaries of the particular theory which exploits these concepts, this semantic content being totally contained in the output of the mechanical devices which measure the quantities in question. In fact, to refer to a mechanical device which for example measures position, as 'classical', is a misnomer, for a device is an object of nature while expressions such as 'classical' or 'quantum' are meaningful only in a theoretical context.

The power of this qualitative statement of the 'correspondence' between certain dynamical concepts of the classical theory and those of the sought for quantum theory becomes conspicuous when we extract its full quantitative consequences. We have already indicated the association of symmetries with the basic dynamical concepts. This association imposes an algebra upon the symbols which are to represent these concepts in any theory. For example, the rotation group is an abstract mathematical object whose structure bears no immediate relationship to the underlying physical laws of the universe. The structure of the rotation group (its Lie Algebra) imposes very definite algebraic relations on the components of the angular momentum if the angular momentum is to be that quantity which is conserved modulo the rotational symmetry of the theory under consideration. These algebraic relations (commutation relations) must be precisely satisfied by the abstract mathematical symbols which are to represent the concept of angular momentum in any physical theory, otherwise the symbols will not have the epistemological content of 'angular momentum' – that quantity which is conserved for rotationally symmetric systems.

The particular choice of algebra to be satisfied by the basic dynamical variables depends critically on the supposed symmetries of space and time. Thus, in non-relativistic theory, space and time are presumed to have the symmetry of the Galilean group, while in special relativity space-time is presumed to have the symmetry of the inhomogeneous Lorentz, or Poincaré, group. In either case, the group is ten-parametric, and to the ten independent elements of its Lie algebra we must associate the ten physical dynamical concepts: 3 components of 'position', 3 components of 'momentum', 3 components of 'angular momentum', and the 1 component of 'energy'. In every theory the algebra to be satisfied by the symbols for these 10 concepts are precisely the Lie algebra of the relevant symmetry group. To have it otherwise would simply be an improper semantic usage of the terms 'position', 'momentum', 'angular momentum' and 'energy'.

Having established the quantitative identification of the basic dynamical concepts and the basis for forming the correspondence between the symbols for these concepts in varying physical theories, it remains to elucidate the role of the dynamical laws before we can complete the analysis of the transition from classical to quantum theory. Our most primitive intuition of dynamics presumes the existence of a constraining relationship between motion in time and motion in space. In a formal mathematical theory this is accomplished by equating the generator of the infinitesimal time translations (the energy) to some specified function of the generators of the infinitesimal space translations (the momentum). This function is in general dependent on the position as well. Such a relationship, of the form $E = h(\mathbf{p}, \mathbf{x})$, is known as the Hamiltonian constraint. As we have observed earlier, the particular functional form of the Hamiltonian constraint formally characterizes the physical system under consideration, provided we have already determined the relationship between the primitive dynamical variables E, $\mathbf{p}$, $\mathbf{x}$ and the measuring system in our laboratory.

Having specified the classical dynamical system by exhibiting the Hamiltonian constraint in terms of the generators of the prefered symmetries of the space-time, the transition to the quantum theory of this same mechanical system is accomplished by requiring that the functional form of the Hamiltonian constraint is preserved as nearly as possible, when the symbols for the generators of the infinitesimal symmetries of the space-time are realized by (algebra preserving) Hermitian operators on a Hilbert space. It is essential to recognize that the critical feature of the preservation of the functional form of the Hamiltonian should only be accomplished in the limited class of variables determined by the generators of the space-time symmetries. This is due to the fact that the transformation law for the Hamiltonian is radically different in classical and quantum theory. In classical mechanics the Hamiltonian behaves (under time-independent transformations) as a scalar field in the phase space, while in quantum mechanics it behaves as a Hermitian tensor in the Hilbert space. Thus, although classical canonical transformations have isomorphic images in the Hilbert space [2], the functional form of the Hamiltonian constraint cannot be preserved under such mappings [3].

We have already noted that the symbols for the dynamical variables of classical theories have dual aspects, attributive and operator. In quantum theory, the distinction between the operator and the attribute aspect of the dynamical variables is more transparant, for, the attributes of quantum systems cannot be realized by the same symbols which realize the operators. Rather, the attributes, that is, the numerical consequences resulting from the performance of measurements, are now realized by the eigenvalues of the associated operators.

At this point let us summarize the above discussion of the program of canonical quantization in precise quantitative form. The space-time which is the arena of the physics has a preferred isometry group, the Galilean group for non-relativistic physics or the Poincaré group for special relativistic physics. The given isometry group singles out a preferred set of vector fields on the space-time which are solutions of Killing's equation

$$(4.1) \qquad \xi_{\mu; \, \nu} + \xi_{\nu; \, \mu} = 0$$

(Greek indices run from 0 to 3, Latin from 1 to 3, the semi-colon denotes covariant differentiation relative to the space-time metric.)

By selecting the four commuting Killing directions we obtain a preferred set of space-time coordinates determined up to the homogeneous transformations available in the isometry group. The three spatial coordinates so obtained induce a preferred coordination of the classical configuration space, and the preferred time-like coordinate yields a corresponding coordination of the tangent bundle over the configuration space. Finally, if we are given a dynamics in the form of a Lagrangian, $L(q_i, \dot{q}_i)$, we can construct the momenta p_i canonically conjugate to the chosen set of q_i via

$$(4.2) \qquad p_i = \frac{\partial L}{\partial \dot{q}_i}$$

thereby obtaining a preferred coordination of the cotangent bundle, or classical phase space. The preferred sub-group of the canonical group whose Lie algebra is preserved under quantization is the sub-group of linear transformations of the phase space coordinates constructed above, that is, a sub-group isomorphic to the symplectic group. The quantization is completed by requiring that this symplectic group be represented by unitary mappings of a Hilbert space in such a fashion that the specific algebraic form of the mapping between the tangent bundle and the cotangent bundle (essentially the dynamics), when expressed as a function of the preferred phase space coordinates, be preserved. The essential point of this discussion is that, for purposes of quantization, a vital role is played by the assumed isometries of the space-time.

V. GENERAL RELATIVITY

The basic field variables of the general theory of relativity are a second order symmetric tensor field of Minkowskian signature over a four-dimensional manifold $g_{\mu\nu}$ (x^α). The ten independent components of $g_{\mu\nu}$ which realizes the free gravitational field satisfies ten field equations (deriveable from a variational principle),

$$(5.1) \qquad G_{\mu\nu} \equiv R_{\mu\nu} - \tfrac{1}{2}g_{\mu\nu}R = 0,$$

where $R_{\mu\nu}$ is a second-order, non-linear differential form constructed exclusively from the $g_{\mu\nu}$, known as the Ricci tensor [4], and R is the Ricci scalar obtained by tracing $R_{\mu\nu}$ with $g_{\mu\nu}$. Since Equations (5.1) provide ten second order hyperbolic differential equations for the ten field quantities $g_{\mu\nu}$, it would appear that the solutions of (5.1) are uniquely given by assigning as Cauchy data the functional form of $g_{\mu\nu}$ and its first time derivative on an initial surface. However the Equations (5.1) are tensor equations under the group of four-dimensional curvilinear coordinate transformations, the Einstein group. It is therefore possible, by means of such a transformation, to alter the form of a solution of Equation (5.1) off the initial surface without altering the Cauchy data. It follows that these field equations must have a rather intricate structure. In fact, four of the equations,

$$(5.2) \qquad G_\mu^0 = 0$$

are constraint equations upon the Cauchy data, preventing their arbitrary assignment, while the remaining six equations are not sufficient to provide a propagation of the field off the initial surface. The degree of arbitrariness in propagations corresponds precisely to the available coordinate freedom off the initial system.

It would appear that the existence of constraints, Equations (5.2), and more particularly the lack of a unique propagation of the field variables off an initial surface, would prevent one from finding a canonical, Hamiltonian realization of Einstein's gravitation theory. However, Dirac [5] succeeded in circumventing many of the consequent impediments and established a canonical formalism for the General Theory of Relativity, at the expense of having a Hamiltonian containing four arbitrary functions of the space-time variables. The canonical variables of the Dirac formalism are the spatial components of the metric tensor, g_{ij}, and their canonical conjugate momenta, p^{ij}, obtained in the usual manner from the appropriate Langrangian. These canonical variables must satisfy the four constraint equations

(5.3) $\mathcal{H}_s \equiv -2p^m_{s;\,m} = 0$

(5.4) $\mathcal{H}_L \equiv |g_{ab}|^{-1/2} [p^{mn}p_{mn} - \tfrac{1}{2}p^2] + |g_{ab}|^{+1/2} R = 0$

(where all the operations of differential geometry, such as raising and lowering of indices, covariant differention, formation of curvature scalers, employ exclusively the spatial metric, g_{ij}, in their formation). These four Equations (5.3) and (5.4), are simply the original four constraint Equations (5.2) realized in the notation of phase space variables. An arbitrary infinitesimal coordinate transformation

(5.5) $x^{\mu'} = x^\mu + \xi^\mu$

described by the descriptor

(5.6) $\xi^\mu = \delta^\mu_s \xi^s + l^\mu \xi^L$

(where l^μ is the unit normal to the space-like surface on which the canonical variables are defined) is generated by the Hamiltonian

(5.7) $H(\xi) = -\displaystyle\int (\xi^s \mathcal{H}_s + \xi^L \mathcal{H}_L)\, \mathrm{d}^3 x.$

Turning to the question of constructing a quantum theory of gravitation, the preferred space-time group is no longer the Galilean or Poincaré group, but rather the Einstein group. It would therefore appear that the Einstein group should now play the preferred role in the construction of the desired quantum theory. The Einstein group is generated by the constraints of the theory via Equation (5.7) (although it should be noted that the algebra of the Einstein group cannot be realized by the commutator algebra of $H(\xi)$ [6]). The dynamical variables which are to realize observable quantities of gravitation theory must be consistent with these constraints; that is, the infinitesimal canonical transformations generated by observables must leave invariant Equations (5.3) and (5.4). As a consequence, the observables of classical gravitation theory must commute with $H(\xi)$ and therefore be invariants of the Einstein group. In effect, the Einstein group must be eliminated from the theory in order to obtain invariant dynamical observables suitable for quantization [7].

VI. ASYMPTOTIC QUANTIZATION OF GRAVITATION

Having thereby lost the preferred status of the space-time group, one could investigate the possibility of preserving some arbitrarily chosen sub-algebra of the classical observables as a set of linear mappings of a vector space. However, such a procedure is hardly unique, and it is not clear what relation it bears to quantization as we previously understood the procedure.

Rather than abandon all hope of a unique quantization of gravitation we can reimpose a preferred space-time group by the following considerations. In view of the fact that general relativity is a non-linear theory, we cannot be too casual about the statement of the boundary conditions which we require of the class of admissable solutions. Since we are properly concerned with the quantization of the radiation modes of the field only, we should limit our consideration to the class of solutions which are asymptotically flat at infinity so that an unambiguous identification of gravitational radiation would be possible. If we impose such boundary conditions we recover a preferred space-time subgroup for our restricted class of solutions. This group, the Bondi-Metzner-Sachs group, which is somewhat larger than the Poincaré group, can readily be realized by the algebra of the Cauchy data for the Einstein field at retarded (or advanced) infinity [8]. It therefore seems to be ideally suited for the quantization of classical gravitational radiation.

The Einstein field may be viewed as a self-coupled, spin 2, zero rest mass field. In order to illustrate some of the novel features which occur when quantizing a theory employing a null Cauchy surface at infinity rather than the usual space-like Cauchy surface at finite times, let us suppress the added complication of the spin 2 nature of the gravitational field and consider for the moment two Lorentz covariant spin 0, mass 0 classical fields, ϕ, assumed real, and ψ, assumed complex, interacting via the Lagrangian

$$(6.1) \qquad L = \tfrac{1}{2}\phi^{,\mu}\phi_{,\mu} + \tfrac{1}{2}\psi^{*,\mu}\psi_{,\mu} + e\phi\psi^*\psi.$$

If we employ polar coordinates and define the retarded time

(6.2) $u = t - r$

and the advanced time

(6.3) $v = t + r$,

we can consider the behavior of the fields ϕ and ψ in the limit

(6.4) $u \to -\infty, v < \infty$.

(In order to satisfy both of these conditions it is evident that we must have $r \to \infty, t \to -\infty$. We obtain in this fashion a null surface at infinity, denoted by $\mathscr{I}^-$, which is coordinatized by the variables v, θ, ϕ).

In order to separate the radiation modes of the fields ϕ and ψ, one customarily considers those solutions which fall off as r^{-1} at infinity. Thus in the limit of Equation (6.4) we expect

(6.5) $\lim_{\mathscr{I}^-} \phi = \lim_{\mathscr{I}^-} \psi = 0$

(6.6) $\lim_{\mathscr{I}^-} r\phi = \rho(v, \theta, \phi)$

$\lim_{\mathscr{I}^-} r\psi = \sigma(v, \theta, \phi)$,

where $\rho(v, \theta, \phi)$ and $\sigma(v, \theta, \phi)$ may be specified arbitrarily.

We are now in a position to summarize the fundamental ideas in the asymptotic quantization program as follows:

(a) For the free fields ($e=0$ in Equation (6.1)) there is a one-to-one correspondence between regular solutions of the mass-zero Klein-Gordon equation, ϕ and ψ, and their null data on $\mathscr{I}^-$, ρ and σ, respectively [8].

(b) The standard free field commutation relations for ϕ and ψ induce commutation relations for the corresponding null data, ρ and σ, namely,

$[\rho(v, \theta, \phi), \rho(v', \theta', \phi')] = -2iS(v - v')\delta(\Omega - \Omega')$

(6.7) $[\sigma(v, \theta, \phi), \sigma(v', \theta', \phi')] = 0$

$[\sigma(v, \theta, \phi), \sigma^*(v', \theta', \phi')] = -2iS(v - v')\delta(\Omega - \Omega')$,

where $S(v-v')$ is the unit anti-symmetric step function,

$$\delta(\Omega - \Omega') = \frac{1}{\sin\theta}\delta(\theta - \theta')\delta(\phi - \phi')$$

is the invariant spherical δ-function, and * denotes complex conjugation.

(c) On account of the one-to-one correspondence between the infinite null Cauchy data and the free fields, if we impose the null data commutation relations, Equations (6.7), we recover the standard commutation relations for the fields. (It is in this sense that we can speak of 'quantization at infinity' in contrast to the more customary procedure of quantization of the Cauchy data on a space-like surface at a finite time).

(d) The full algebra of the Bondi-Metzner-Sachs group (and, *a fortiori*, the Poincaré group) can be realized by employing the algebra of Equation (6.7). Thus the preferred asymptotic symmetry group can be preserved under quantization.

(e) The surface $\mathscr{I}^-$ is an orbit of the full BMS group. Thus the procedure for integration off the initial surface $\mathscr{I}^-$ cannot be accomplished by employing any element of the preferred space-time group.

(f) Returning to the coupled theory ($e \neq 0$), we observe that as we approach the surface $\mathscr{I}^-$ the coupling terms of Equation (6.1) vanish faster (by one power of r^{-1}) than do the free parts. Thus, in the limit of $\mathscr{I}^-$ the interacting fields decouple naturally and we have no need to introduce an adiabatic decoupling term for the purpose of obtaining asymptotic free fields.

(g) The coupled fields are determined by the same set of Cauchy data at $\mathscr{I}^-$ (defined by precisely the same limiting procedure – Equations (6.6)) as the free fields. The integration off the Cauchy surface, however, proceeds via the coupled set of field equations obtained from the Lagrangian of Equation (6.1). The coupled fields are therefore very different functionals of the null-data, ρ and σ, than were the free fields.

(h) The customary quantization procedure imposes the free-field commutation relations upon the Cauchy data for the coupled fields on a space-like surface at a finite time. This then determines the commutation relations for the fields everywhere, since the fields are unique functionals of their Cauchy data. We know that coupled quantum field theories obtained in this fashion are inconsistent and require renormalization. Our proposed quantization scheme preserves instead the free-field commutation relations on $\mathscr{I}^-$, Equation (6.7), where the fields in question are more nearly free! By quantizing on $\mathscr{I}^-$ we are employing at all times the 'dressed' fields and should not expect our results to require renormalization. As before, the commutation relations for the fields at finite points are obtained by integration of the correct field equations off $\mathscr{I}^-$.

(i) Although the proposed alteration in the quantization procedure seems modest, the new commutation relations we obtain on space-like surfaces at finite times are radically different from the traditional ones, for although their support is the same as before, the coefficients are operators [9]. (The actual integration to finite points, required to verify this result, has only been accomplished by means of a perturbation expansion in the coupling constant, e.)

(j) Finally, the 'in'-Fock-space is easily constructed using the operators ρ and σ on $\mathscr{I}^-$, and the corresponding 'out'-Fock-space can be constructed by operators at $\mathscr{I}^+$, obtained by taking the limit $v \to \infty$, $u < \infty$. In this form the theory is particularly suited for the handling of scattering problems.

The procedure outlined above can be applied directly, with little alteration, to the quantization of the gravitational field. We obtain in this fashion a formalism particularly suited for the description of gravitons as the excitations of the independent transverse modes of the radiation (i.e. r^{-1}) portion of the gravitational field, as well as a very natural procedure for the description of graviton-scattering phenomena. Apart from computational difficulties in performing the required integrations from $\mathscr{I}^-$, the outstanding problem is the theoretical one of providing a consistent factor ordering for the (non-linear) functional dependence of the fields on their respective null Cauchy data. (This problem becomes particularly conspicuous as one treats the higher orders of the perturbation expansion.)

VII. CANONICAL QUANTIZATION OF GRAVITATION

If one wishes to obtain a local quantization of gravitation theory, that is, if one wishes to avoid essential reference to asymptotic boundary conditions, one must abandon tying the quantization procedure so intimately to the space-time symmetries. There occurs an inevitable lack of uniqueness of the resulting quantum theory in view of our earlier observation of the lack of commutativity between the performance of canonical transformations (always available in the classical phase space) and the process of canonical quantization. Nevertheless, assuming that we have settled upon some preferred set of canonical variables, say the Dirac variables of Section V, let us now indicate several different approaches to canonical quantization currently available in the literature.

As we have observed, within the classical theory 'observables' are defined as functionals of the dynamical variables which commute with the constraints, Equations (5.3) and (5.4). Since these constraints generate the coordinate transformations within each space-time manifold, the observables so defined are constant over an entire spacetime. In principle they can be grouped into two canonically conjugate pairs $\alpha_A(\mathbf{x})$, $\beta^A(\mathbf{x})$ ($A = 1, 2$), which can be freely assigned, and such that to each assignment there corresponds a particular solution of the Einstein field equations [10].

Case (a): In what we may call the phase space method of quantization one seeks to replace the classical canonical variables g_{mn}, p^{mn} by Hermitian operators satisfying the standard canonical commutation relations in such a way that, with suitable choice of factor ordering, the constraints (5.3) and (5.4) can be realized in a consistent fashion as operator relations on a linear vector space [11]. Among other things, consistency entails that the commutator algebra of the constraints give back a linear combination of the constraints themselves, with the added proviso that the coefficients can be brought exclusively to the left side of the expressions. Assuming that such a factor ordering could be found (a decidedly non-trivial task), the vector space of physical states would be constrained by the operator relations

$$(7.1) \qquad H_S \psi = 0$$

$$(7.2) \qquad H_L \psi = 0$$

(In the configuration space representation these operators are formed by replacing p^{mn} in Equations (5.3) and (5.4) by $(1/i)(\delta/\delta g_{mn})$.) It is in this sense that Equations (7.1) and (7.2) serve as the analogue of the Schrödinger equation of ordinary quantum mechanics.

A difficulty immediately arises should we attempt to interpret the operators g_{mn}, p^{mn}, or even the states ψ which are solutions of the constraint Equations (7.1), (7.2). For, although with a given state we could presumably define mean values $\langle g_{mn} \rangle$, $\langle p^{mn} \rangle$ of the corresponding operators, in view of the non-linearity of the constraints, these mean values would not satisfy the classical constraint Equations (5.3) and (5.4). If wave packets of such states are to be related to some mean classical solution of the Einstein field equations (as we must expect from the correspondence principle), it follows that $\langle g_{mn} \rangle$, $\langle p^{mn} \rangle$ cannot be interpreted as a mean value for the spatial metric and second fundamental form, respectively, of such a classical solution, for the constraint equations are four of the ten field equations of the theory.

The observables of the quantum theory are to be realized by self-adjoint operators on the constrained vector space. Thus if A is an observable and ψ a solution of Equations (7.1) and (7.2) we must require that $A \cdot \psi$ is also a solution of Equations (7.1) and (7.2). It readily follows that

(7.3) $[A, H_S] = 0$

(7.4) $[A, H_L] = 0$.

We see that the quantum observables must commute with the generators of the space-time coordinate transformations and therefore correspond to classical observables as previously defined. If we can select a complete non-redundant set of quantum observables corresponding in the mean to the classical observable Cauchy data, α_A, β^A we can recover in an evident way a means by which a wave packet can be understood to correspond in the mean to a classical state. In view the structure of Equations (7.3), (7.4), it is a non-trivial problem to exhibit functionals of the phase space variables which are solutions, and thereby observables.

Case (b): The method of superspace [12] is based on the observation that in view of the fact that H_S generates the spatial translations, Equation (7.1) implies that the state vector ψ is invariant under spatial translations. It is then a function exclusively of the invariant spatial geometry, and cannot depend on the fortuitous choice of coordinates which could yield a specific metric tensor $g_{mn}(x)$. Should one succeed in topologizing the space of three-dimensional Riemannian geometries and denote its elements by ^3G, one could then introduce the canonically conjugate momentum $\pi \equiv (1/i)(\delta/\delta^3 G)$. Assuming this accomplished we have in effect solved Equation (7.1), and shall denote the solution by $\psi\,(^3G)$. The remaining Equation (7.2), is now our sole Schrödinger equation, with a Hamiltonian qualitatively of the form (suitably symmetrized for Hermiticity).

(7.5) $H_L = A\,(^3G)\,\pi^2 + V\,(^3G)$

Although it would initially appear that if we could achieve the above choice of variables and the required form of the Hamiltonian, we would have a perfect correspondence between quantum wave packets and classical solutions of general relativity, further reflection shows that this is not the case. The basic difficulty is that, as in case (a), we are unable to conclude from Equations (7.2) and (7.5) that the mean values of the basic quantum dynamical variables $^3\bar{G}$, $\bar{\pi}$, satisfy the classical constraint Equation (5.4).

If we confine our attention solely to the space of physical states, that is, to the linear vector space of states $\psi(^3G)$ which satisfy the constraint Equation (7.2), then, as in case (a), the operators on this vector space, i.e., the observables, must satisfy Equation (7.4). Since it was assumed *ab initio* that Equation (7.1) and (7.3) were satisfied, we return to the same class of observables as in the method of phase space and the entire discussion given above becomes immediately applicable here. We may note in passing that the operators ^3G and π in the method of superspace are not observables for they throw a permissable state vector off the constraint hypersurface.

Case (c): In the method of observables [10] one proposes to employ directly in the procedure of quantization the complete set of canonically conjugate pairs of functionals of the canonical variables, $\alpha_A(\mathbf{x})$, $\beta^A(\mathbf{x})$ whose Poisson brackets with the classical constraints, Equation (5.3) and (5.4), vanish. The classical canonical commutation relations between the α's and the β's would be taken over directly as quantum commutators and the corresponding operators would immediately be defined on the constraint hypersurface in the linear vector space. Having solved for, and thereby eliminated all the constraints, the problem of consistent factor ordering need not arise. The discussion of the correspondence of wave packet solutions of the quantum system with mean classical solutions would again proceed exactly as in case (a). With this approach however, we abandon all attempts at following the evolution of the system in time, and resign ourselves *ab initio* to a frozen formalism.

These various canonical routes toward the quantization of the general theory of relativity become equivalent if we insist that the wave packet quantum states have an interpretation as mean classical solutions of the Einstein field equations. No matter which route one chooses, one is forced ultimately to consider as the observables the space-time constants of the motion. This invariably leads to a frozen formalism in which nothing appears to evolve in time. The three different canonical approaches discussed seem to trade off one set of difficulties for another. In the phase space approach the operators for the basic dynamical variable are readily exhibited but there exists a non-trivial factor ordering problem for the constraints. In the method of observables, the factor-ordering problem is obviated, but the operators corresponding to the observables are not readily exhibited. Even their classical expressions are as yet unknown. The method of superspace seems to form a compromise between the above two extremes, eliminating many, but perhaps not all, factor ordering problems, but not fully facilitating the exhibition of the basic dynamical variables. In the final analysis the choice of approach would appear to be a matter of efficiency, convenience and intuition, rather than a question of principle.

VIII. HAMILTON-JACOBI THEORY

It is of interest to examine the Hamilton-Jacobi realization of Einstein's gravitation theory with an eye toward determining those features of the quantum state vector which presumably will reduce to the Hamilton-Jacobi functional in the limit of the WKB approximation. A great virtue of Hamilton-Jacobi theory is that its considerations are entirely classical, and therefore require fewer speculative considerations.

The General-Relativistic Hamilton-Jacobi functional, $S(g_{mn})$, may be defined as a functional whose domain is the set of tensor functions g_{mn} (x^s) which are of signature 3 and characterize three-dimensional Riemannian manifolds which are either closed, asymptotically flat, or satisfy some other well-stated boundary condition of interest, whose range is the real numbers, and which satisfies the four functional equations

$$(8.1) \qquad \mathscr{H}_s\left(g_{mn}, \frac{\delta S}{\delta g_{pq}}\right) = 0$$

$$\mathscr{H}_L\left(g_{mn}, \frac{\delta S}{\delta g_{pq}}\right) = 0$$

obtained by substituting

$$(8.2) \qquad p^{mn} = \delta S/\delta g_{mn}$$

into Equations (5.3) and (5.4).

The following properties of Hamilton-Jacobi functionals have been proven (the interested reader is referred to the original papers).

(1) $S(g_{mn})$ is not an explicit function of the space-time coordinates, but depends exclusively on the configuration variables, g_{mn} [13].

(2) Under the action of the space-time group generated by the constraints, the functional form of $S(g_{mn})$ is invariant, although its value will in general alter under the action of the time-like translation [13].

(3) As a corollary of (2) it follows that within a family of classical trajectories determined by a given $S(g_{mn})$ (via employing g_{ab}, $p^{mn} = (\delta S/\delta g_{mn})$ (g_{ab}) as Cauchy data) occur all canonical pairs g_{ab}, p^{cd}, which can be obtained from a given one by means of a space-time transformation [13].

(4) The complete family of classical trajectories determined by a given $S(g_{mn})$ is fully characterized by $2 \times \infty^3$ commuting constants of the motion $\alpha_A(x^s)$, which do not commute with $g_{mn}(x^s)$ [14]. To elaborate, the family of Ricci flat four dimensional Riemannian manifolds determined by a given $S(g_{mn})$ have in common a complete, commuting set of observables, and conversely, given the set of all Ricci-flat 4-dimensional Riemannian manifolds having in common such a complete commuting set of observables $\alpha_A(x^s)$, there exists a unique Hamilton-Jacobi functional, which we shall denote by $S(g_{mn}, \alpha_A)$, which yields back precisely this family of manifolds.

(5) The invariance group of the Hamilton-Jacobi theory is isomorphic to the proper canonical group of general relativity (i.e. the canonical group modulo the space-time subgroup generated by the constraints); or, equivalently, the functionals, $S(g_{mn}, \alpha_A)$, provide a space for the realization of the proper canonical group as a transformation group [10]. It is at this point that the isomorphism between the theory given by the four functional Equation (8.1), and the theory given by the ten differential Equations (5.1) is established; that is, the Hamilton-Jacobi theory given by Equation (8.1) is an alternate realization of the Einstein Theory of gravitation.

(6) The observables $\beta^A(x^s)$ which are canonically conjugate to $\alpha_A(x^s)$ satisfy the familiar relation $\beta^A(x^s) = \delta S/\delta \alpha_A(x^s)$. [10].

(7) The $4 \times \infty^3$ observables $\alpha_A(x^s)$, $\beta^A(x^s)$ forms a complete independent (but not commuting) set of invariants which determine a unique Ricci-flat four dimensional Riemannian manifold [10].

To the extent that the functional $S(g_{mn}, \alpha_A)$ does in fact correspond to the WKB limit of the Schrödinger state vector of the quantized theory we can conclude that to a Schrödinger state will correspond not a single four-dimensional geometry, but rather an infinitely large family of four-geometries which have in common a complete commuting set of observables, but disagree on the canonically conjugate set of observables. We could then expect that wave packets could be constructed peaked about the expectation values $\langle \alpha_A(x^s) \rangle$, of the corresponding quantum observables. Since the assignment of the complete set of classical observables, $\langle \alpha_A(x^s) \rangle$, $\langle \beta^A(x^s) \rangle$, uniquely determines a classical four-geometry, we can in this way obtain a quantum state which can be interpreted as being peaked about a mean Ricci-flat four-geometry.

A most startling qualitative conclusion follows from the above properties of solutions of the Hamilton-Jacobi equations of General Relativity. Should one obtain a quantum theory of gravitation, the most startling feature it would have is the disolution of the space-time manifold. Thus, in the description of an experimental situation the theory would establish a contingency relationship between the out-come of an experiment and the geometry of the space-time in which the experiment was performed! This would be a most encouraging result for there is reason to believe that it might provide a resolution to the age-old paradox of 'the reduction of the wave-packet' [15].

In brief, half of the classically required canonical variables must be employed to specify the frame of reference, leaving the remaining half available for unequivocal observation. If the observer chooses to alter his procedure for the specification of the frame of reference of his apparatus he is at liberty to do so and obtain thereby a consistent, but complementary description of the empirical situation. The penalty, however is that he must conclude that the space-time structure was also altered by the experimental rearrangement! This conclusion follows from point (7) above. For, whereas a knowledge of both $\alpha_A(x^s)$ and its canonical conjugate, $\beta^A(x^s)$ is required for a complete determination of a classical space-time, to the extent that a quantum state for the gravitational field approximates a classical Hamilton-Jacobi functional, specification of $\alpha_A(x^s)$ requires total uncertainty of $\beta^A(x^s)$. However such considerations are far too speculative and premature to warrant further pursuit at this time.

Belfer Graduate School of Science,
Yeshiva University

BIBLIOGRAPHY

[1] Noether, E., *Göttingen Nachr.* 235 (1918).

[2] von Neumann, J., *Mathematical Foundations of Quantum Mechanics*, Princeton University Press, Princeton, 1955.

[3] Komar, A., *Studies in the Foundations, Methodology and Philosophy of Science* 4 (1971), 111.

[4] Eisenhart, L. P., *Riemannian Geometry*, Princeton University Press, Princeton, 1926.

[5] Dirac, P. A. M., *Physical Review* 114 (1959), 924.

[6] Bergmann, P. G. and Komar, A., *International Journal of Theoretical Physics* 5 (1972), 15.

[7] Bergmann, P. G. and Schiller, R., *Physical Review* 89 (1953), 4.

[8] Sachs, R., *Physical Review* 128 (1962), 2851.

[9] Klarfeld, J. and Komar, A., *Physical Review D.* 4 (1971), 978.

[10] Komar, A., *Physical Review* 170 (1968), 1195.

[11] Schwinger, J., *Physical Review* 132 (1963), 1317.

[12] Wheeler, J. A., *Relativity* (ed. by M. Carmeli, S. Fickler, and L. Witten), Plenum Press, New York, 1970, p. 19.

[13] Bergmann, P. G., *Physical Review* 114 (1966), 1078.

[14] Komar, A., *Physical Review* 153 (1967), 1385.

[15] Komar, A., *International Journal of Theoretical Physics* 2 (1969), 157.

F. DAVID PEAT

QUANTUM PHYSICS AND GENERAL RELATIVITY;
THE SEARCH FOR A DEEPER THEORY

I. INTRODUCTION

Upon reading a collection of research papers on the Foundations of Quantum Theory, it becomes apparent that opinions differ as to the relative importance of the problems which are outstanding and upon the particular methods which may be employed in their solution. Indeed in certain cases the existence of problems proposed by some, is denied by others. This divergence of opinion reflects the attitudes which are held on the nature of scientific theory and investigation. It seems apposite in an article of this nature for the writer to outline the views which lead him to adopt a given line of scientific enquiry, particularly when this line represents a divergence from those generally adopted.

While attention has been drawn to the diversity of opinion which exists within the interpretation and investigation of the foundations of physics, it will be argued that in general there exist features in common which are manifestations of a particular tendency in modern thought. The implications of this tendency will be examined critically and an alternative approach indicated which, it is suggested, is more in keeping with the results of modern physics and with research in related fields. Within the context of the quantum theory it is proposed that research into a wider theory be extended and, in the final section, some results which may be of use in such a program are discussed.

II. CLASSICAL VIEW OF NATURE

The successes of Classical Mechanics stemmed from a systematic methodology in which hypothesis and theory evolved hand in hand with investigation and experiment. Controlled experimental conditions led to repeatable phenomena and the possibility of the prediction of events under altered conditions.

The world view necessary for such an advance in knowledge assumed

Hooker (ed.), Contemporary Research in the Foundations and Philosophy of Quantum Theory, 328–345. All rights reserved

the existence of an objective nature which was supposed to operate in a fashion comprehensible to man's intellect. Systems could be analysed in terms of more fundamental entities, the classical bodies and their interactions. Such bodies were objective in the sense that their properties could be defined in isolation from any particular system, the effects of the experimental apparatus being used to measure such properties could be arranged to be negligible and one could confidently expect to obtain identical results if the properties were to be measured in some future time under some alternative experimental conditions. The classical bodies and their interactions therefore were held to be responsible for the behaviour of the myriad systems of nature. If the laws for the behaviour of bodies under the influence of interactions were elucidated, generally in a fairly simple mathematical form, then the properties of complex systems of bodies could, in principle, be numerically predicted.

The ability to provide a quantitative description of most of the physical world by means of simple laws, which were frequently independent of the range and magnitude of the phenomena involved, naturally endowed classical physics with considerable confidence.

Alongside this scientific view of nature must be placed the view of several schools of philosophy whose origins could be said to be somewhat influenced by the early successes of the natural sciences. In the view of these philosophers the Problem of Knowledge dealt with the problem of obtaining information concerning the external world. Such information being presented in the form of sense data, reflection upon the sense data enabled one to arrive at certain propositions about the world which would be admitted to be true. The process of reflection and perception was taken to be a passive occupation and gave rise to problems as to the duality of mind and matter. Within this century science entered new and unsuspected areas and at the same time the philosophical theories of knowledge and of the methodology of science came into criticism. In the philosophical field it was felt that the theories of perception were based upon an oversimplified, and in some cases erroneous, account of the process of perception. The act of perception, it was asserted, could not be described as totally passive, indeed it was proposed that consciousness had an intentional activity being a phenomena within nature and not standing outside.[1] It was felt that the world view of the previous centuries had lead to a fragmented or bifurcated view of the world[2] and should be replaced by an approach in which a holistic view of nature, and man's position in nature, had their part.

Also it may be noted that a fragmentary view of natural phenomena is inappropriate to a treatment of the newer areas of physics. That is, the world view which proved fortuitious in the development of a highly successful account of the 'classical world of physics' loses its *raison d'être* when one enters the world of twentieth century physics and, indeed, some of the other fields into which scientific method has recently made inroads. It becomes necessary, in the case of those who accept a newer viewpoint, to exercise care in designing their scientific research so that the techniques, language and methods employed are in harmony with their philosophy. A program in which the attitude and method are inconsistent will meet with confusion when it attempts to tackle the outstanding problems of modern physics.

III. AN HOLISTIC APPROACH IN MODERN PHYSICS

Quantum theory may be said to point towards the holistic view of nature. Bohr provided what is probably the most consistent interpretation of the quantum mechanics in a series of essays[3]. He showed that the indivisibility of the quantum of action and the Uncertainty Principal of Heisenberg are inconsistent with a 'fragmentary' approach to physics. No longer could a phenomenon be discussed as existing apart from the particular experimental arrangement under which the phenomenon is manifest. Apparatus may no longer be considered as representing a residual interaction, which may be reduced ideally to zero magnitude, acting upon an objective system. Rather the process of observation is no longer passive but an act of interference and paradoxically the observer is no longer objective.

Quantum mechanics, in the interpretation promulgated by Bohr, dictate that nature should be considered as a whole; the classical picture of isolatable bodies casually influenced by various interactions with other bodies may be abstracted from the wider picture only in an approximation which depends upon the size of the bodies. The consistency of the Bohr interpretation was obtained at the expense of delimiting certain areas of enquiry or of demanding no deeper analysis of one of several complementary descriptions. This state of affairs has proved unacceptable to many physicists who have turned to an interpretation based upon the formalism of von Neumann. In such an alternative, the wave function is endowed with a certain objectivity and the evolution of a particular physical phenomena may be discussed in terms of what to Bohr was only a convenient mathematical device. The von Neumann formalism however introduces its own problems of interpretation, for difficulties concerned with the Theory of Measurement and the Collapse of the Wave Function are created.

The other theory of the twentieth century which revolutionised our world-view of nature is of course the Theory of Relativity, which likewise points to a holistic approach. The motion of free bodies is supposed to be determined by the geometry of the space-time in which these bodies are defined. In turn this geometry is determined by the presence of the material bodies, this reciprocal arrangement being described by means of nonlinear field equations. From the nature of the field equations it is clear that all bodies affect each other in a fashion which is not adequately described in terms of the simple pair interactions of classical physics. Further, one discovers that the concept of the classical rigid body must be abandoned and that many of the well used invariants of classical mechanics lose their generality and transparency of definition. One must be content in place of the classical body with a bunch of world lines which preserve, more or less, their congruence. In his unsuccessful search for a Unified Field theory, Einstein was guided by a holistic attitude for he wished to describe the universe as a field governed by nonlinear equations which determined the existence and behaviour of the singularities of the field, from which singularities would be abstracted the appearance and properties of the classical bodies. Einstein's vision has not completely faded, for example it is in this spirit that much of the imaginative work of Wheeler is carried out[4].

IV. PROBLEMS OF MODERN PHYSICS

There are several problems outstanding in modern physics which may serve as pointers to some deeper and more embracing theory of phenomena. In the light of the above discussion it is proposed that an attempt to resolve these problems within the framework of a 'fragmentary' approach may not lead to a satisfying and entirely consistent solution. For example to account for the existence of the great diversity of elementary particles gives rise to considerable difficulties for the elementary particle theorist. Possibly an attempt to interpret the results of high energy scattering experiments in a language other than that of 'elementary entities' or particles would shed some light on this problem.

Another outstanding failure of modern theoretical physics is represented by the continued existence of Quantum Theory and General Relativity as separate theories. Attempts to rectify this state of affairs are usually denoted as Quantised General Relativity and this choice of name betrays a particular preconception of some, but by no means all, investigators. It is supposed that the order of magnitude of the gravitational coupling constant indicates that effects of gravitation will be negligible within the domain of quantum physics. A successful combination of General Relativity and Quantum physics is expected, therefore, to cause little modification to the quantum mechanical formalism; rather it will treat gravitation as a quantised field within the framework of Quantum Field Theory.

Some doubt is cast upon this order of magnitude estimate of the effect of gravitation in quantum physics by work which is reviewed in Section VIII. In addition it is not unreasonable to expect the ideas of General Relativity to penetrate Quantum Mechanics in a fashion which causes mutual modification, since the Quantum Mechanical formalism is grounded in the properties of Minkowski space. For example the light cone structure of such a space is invoked when constructing commutation relations in Quantum Field Theory and equivalently, through the theorem dealing with Dispersion Relations, the Unitarity of the S-matrix. In addition to obtaining a modified formalism for Quantum Mechanics it would be hoped that some theory for the field equations of General Relativity would be forthcoming.

The practitioners of Axiomatic Field Theory and axiomatised versions of Quantum Mechanics perform an extremely useful task in that they expose and seek to correct any inconsistencies within the formalism of quantum physics. Similarly those working towards quantizing General Relativity clarify and explore some of the troublesome and difficult areas of relativity theory. It has not been the purpose of this article to suggest that all trends in modern theoretical physics are the result of a 'fragmented' view of nature and that they must necessarily lead to a confused state of affairs. Rather it is emphasized that any uncompensated motivation for scientific research will give rise to serious interpretational problems and inconsistencies if it is not in harmony with the phenomena under study. It is suggested that a holistic approach involves various approaches which may illuminate certain of the difficulties of modern physics.

V. TOWARDS THE HOLISTIC APPROACH

It is unlikely that the secrets of nature will be revealed by the facile employment of a new technique or world view. Rather it is only after deep and persistent investigation within this framework that significant progress will be made towards a clarification of the problems of modern physics[5]. It therefore seems rather futile at this juncture to point towards what one would feel should be the features of such approaches. However it does not seem unreasonable that one should begin by attempting to work within a wide field; for example in an area embracing General Relativity and Quantum Theory, attempting to understand the problems which are common, or peculiar, to each. The underlying mathematical and formal similarities of both theories may serve to provide a clue to the structure of some deeper theory. In the context of an holistic method one would not so much be concerned with the mechanics of elementary entities or particles and their interaction, rather one would search for the invariants, orders and symmetries of such a theory and the laws governing their relationships. At some suitable degree of abstraction the properties of the classical bodies and their laws of motion would be obtained. At other approximations the results of quantum theory would be expected together with some explanation of charge, spin and the other parameters of elementary particle physics, and a description of their appearance as quasi invariants according to various symmetry classifications. In addition the relationship between these invariants should give rise to a picture in which the properties of space and time emerge in a natural fashion. The search for a deeper theory which is holistic in nature is fraught with pitfalls since our ways of thinking, the concepts and language we employ, are steeped in the ideas and methods of the last few centuries. (Even within a theory such as Statistical Mechanics difficulties occur as a result of the usage of such words as disorder, probability, random phase approximation, etc. Feeling such difficulties to be overwhelming Bohr suggested that the interpretation of quantum theory should be couched in everyday language, any attempt to pass beyond such a language would not be worthwhile.)

The remainder of this article will restrict itself to the area covered by General Relativity, Quantum Mechanics and Particle Physics. By examining some problems and some recent work in these fields it is hoped that an indication of the underlying structure of a more embracing theory may be given.

VI. QUANTUM THEORY AND GENERAL RELATIVITY

It is informative to reflect upon those problems of Relativity Theory which occur also in Quantum Physics since a synthesis of the similarities between theories may be just as valuable as the attempted resolution of their differences[6].

Many of the equations of physics are invariant under the conformal group of transformations but it is those equations which are not invariant which may prove to be the most interesting in the present context. The conformal group contains, in addition to the ten parameter Poincaré group, a transformation representing the dilatation, or scaling, of space-time and four operations which have the appearance of acceleration transformations. Electromagnetism and weak gravitational fields are invariant under all these operations but the addition of non-linear terms into the field or the appearance of massive sources breaks the conformal invariance. It may be easily appreciated that a material body does not preserve its properties under the scaling of its dimensions unless its ultimate constituents and the forces holding it together all scale in the correct fashion. In the quantum domain the presence of elementary particles having a discrete mass spectrum prevents the scaling of S-matrix amplitudes. In General Relativity the occurence of non-linearities in the field equations also breaks conformal invariance. The free massless spinor fields of Quantum Field Theory are invariant under the conformal group but this property is destroyed if a wide class of interactions are introduced. Hence conformal invariance appears to be a symmetry property of space-time which is disrupted upon the appearance of mass, interactions or non-linearities and a deeper understanding of conformal invariance and its breaking may throw further light therefore upon the relationship between these factors.

In addition to mass; charge, spin, etc. are parameters which characterise an elementary particle and are not yet fully understood. It is of interest that the spin and charge which determine the behaviour of an elementary particle also determine the world line of a body in General Relativity[7]. Also, as it will be mentioned, the spinor is as important in General Relativity as it is in Quantum Theory.

In the following sections of this article two ideas will be pursued in greater details. One of these is connected with the problems of symmetry breaking, the properties of space-time and the form of the Lagrangians used in physical theory. The other approach is more formal and deals with those geometrical properties of General Relativity which may be relevant to Quantum Theory. While these ideas have not been much extended outside their original domains and are not free from difficulties it is hoped that their outline in a volume of this nature will cause a wider audience to direct their attention to the underlying problems.

VII. SYMMETRY BREAKING[8]

The breaking of any symmetry is of course significant in physics, since it may generally be taken as an indication that an additional interaction, which is not invariant under all the operations of the symmetry group, has made its appearance. When this interaction acts to remove degeneracies within the original system new physical properties may be manifest. For example, the spectra of the transition metal ions are changed when these ions are found in various compounds as a result of the deviation from spherical symmetry of the field due to the ligands. The removal of degeneracies amongst the 'd' electrons of the metal ion results in rather small energy changes yet is responsible for the rich chemistry of the transition metal compounds.

The breaking of symmetry may be looked upon in an alternative fashion however, namely that it is possible for the Lagrangian to be invariant under the symmetry group while a non-symmetric ground state manifests itself. A well discussed situation of this nature is that of the violation of chiral ($SU(2) \times SU(2)$) symmetry. According to the universal V-A theory of the weak interaction a weak, non-strange hadronic current is invariant under the chiral $SU(2) \times SU(2)$ group and it is tempting to extend such invariance to the electromagnetic and strong interactions as some universal symmetry of interactions. While such interactions are invariant under the isospin group SU(2), unfortunately the particle multiplets are not invariant under the full chiral group. As an alternative to the introduction of symmetry violating terms into the interaction one may attempt a *nonlinear* representation of the chiral group. Such a realisation is carried out at the expense of introducing a number of 'preferred' fields; while they make their appearance in a rather ad hoc fashion it is hoped that they may either be identified with some additionally known particles or shown to be 'unobservables'. If a particle multiplet is invariant under the operations of the subgroup g to a required group G then the number of preferred fields which must be introduced is equal to the difference in the number of generators of the groups G and g. With the aid of these new fields a Lagrangian may be written down according to a simple prescription, a feature of which is the replacement of field derivatives. The form of this new Lagrangian is non-polynomial. In addition, mass terms for the preferred fields do not occur and the massless particles corresponding to these fields are generally referred to as Goldstone bosons. In the case of a non-linear realisation of the chiral group, three massless particles are introduced: the 'soft pions'.

In addition to carrying out these realisations for the internal symmetry groups of the elementary particles it is possible to investigate other broken symmetries of physics. For example, since the massive particles may be taken to form a linear realisation of the Poincaré group, it is possible to examine a non-linear realisation of these particles under the 15 parameter conformal group upon the introduction of $15-10=5$ preferred fields. These fields may be taken to correspond to a vector and a scalar massless particle. It is amusing but not necessary, to identify the vector field with the photon and the scalar field with a 'scalar graviton'.

Of even more interest is a discussion of the sixteen parameter group of general coordinate transformations, for this is the group under which, according to the theory of General Relativity, all equations of physics must transform. Since the laws of physics are today formulated in an invariant fashion under the Lorentz group one must introduce ten preferred fields in order that a non-linear realisation of the group GL(4, R) may be achieved. According to this formalism one therefore regards general covariance as a symmetry which is spontaneously broken with the appearance of the graviton as a Goldstone boson. The non-linear realisation may be carried out in the presence of other fields, for example the photon field when a 'gravity modified' electrodynamics results from the formalism. While these treatments of the symmetries of physics in the language of spontaneously broken symmetry and non-linear realisations are not particularly revealing at present they do indicate that, at least on a formal level, an interconnection exists between the properties of the elementary particles, their interactions and the space-time of which they are a part.

VIII. ORDER OF MAGNITUDE OF THE GRAVITATIONAL EFFECT WITHIN THE QUANTUM DOMAIN[9]

The above discussion on the construction of non-polynomial Lagrangians provides a convenient device for the consideration of the order of magnitude of effects which may result from the inclusion of gravitational effects within the quantum domain. For the quantised Lagrangian has the important property which in effect removes the logarithmic divergencies from the perturbation expansion of the self-mass and self-charge of the electron. That is the effect of using the non-polynomial Lagrangian derived in this way is to introduce a cut-off in energy which corresponds to a length equal to the Schwartzchild radius of the electron.

As has been mentioned earlier the magnitude of the gravitational coupling constant has frequently been taken as an indication of the relative unimportance of General Relativity within quantum physics. While the treatment discussed above does not pretend to represent a quantisation of General Relativity, nor should it be supposed that a successful combination of General Relativity and Quantum Physics should necessarily proceed along these lines, it indicates that the effect of introducing certain elements from General Relativity in this fashion are certainly not negligible. Indeed if one is prepared to make a further speculative proposal then the magnitude of the effect may increase. The phenomena referred to is 'Strong Gravitation' which may be introduced by analogy to 'Strong Electromagnetism'.

The photon is the carrier of the electromagnetic interaction between leptons and which, it has been suggested, does not couple directly to the hadronic particles. Instead a neutral massive particle, the ρ° meson, having identical quantum numbers as the photon, is responsible for the strong electromagnetic current, the photon interacting indirectly via $\gamma - \rho^\circ$

mixing. In a similar fashion it is proposed that while the graviton couples directly to leptons it may couple indirectly to hadronic matter via g–$f°$ mixing, where the $f°$ meson is a neutral, spin 2, massive meson. Therefore in the neighbourhood of hadronic matter a strong short ranged gravitational interaction operates which approaches the usual 'Einstein' gravitational interaction at larger distances. If non-polynomial Lagrangians are constructed corresponding to this universal strong gravitational coupling then it is found that an ultra-violet cut off occurs at much lower energies than that found using the infinite ranged gravitational interaction.

IX. SPINORS AND TWISTORS[10]

Spinors and twistors along with world tensors are the natural mathematical entities employed in a formal treatment of the properties of space-time; when it is the case that properties of these entities also reflect some of the formalism of quantum mechanics then one may be led to consider this as an example of some deeper underlying structure.

As a result of the Dirac equation, the importance of the spinor formalism is well established within quantum theory and needs no discussion here. However, the spinor algebra also occurs in a natural fashion within General Relativity and simplification of many problems is achieved by its use.

A two valued spinor is denoted by ξ^A, where the index A takes on the values 1, 2. As the world tensors are defined on a four dimensional real vector space so the space appropriate to the two valued spinor is of two complex dimensions. The spinor undergoes a spin transformation

$$\xi^A \rightarrow L_B^A \xi^B,$$

where L_B^A is complex and unimodular, which is associated with a restricted Lorentz transformation in Minkowski space. (Of course the use of spinors is not restricted to flat spaces).

A spin transformation is similarly associated with the complex conjugate and indicated by

$$\overline{\xi^A} = \overline{L_B^A}\,\overline{\xi^B} \text{ or for convenience } \xi^{A'} = L_{B'}^{A'}\xi^{B'}.$$

By means of the connecting quantity $\sigma_a^{AA'}$, and its inverse $\sigma_{AA'}^a$, which transforms like a spinor in the indices A, A' and a space-time vector in the index a, the spinor corresponding to any tensor may be defined

$$X_c^{ab} \leftrightarrow X_{CC'}^{AA'BB'} = X_c^{ab} \sigma_a^{AA'} \sigma^{A'} \sigma_b^{BB'} \sigma_{CC'}^c .$$

If ξ^a is null then its corresponding spinor $\xi^{AA'}$ has the important property that it may be written in the form,

$$\xi^{AA'} = \pm \, \pi^A \bar{\pi}^{A'},$$

the positive sign being used if ξ^A is future pointing and the negative sign if ξ^A is past pointing.

This two valuedness of the spinor is of importance in that a manifold constructed from spinors will be orientable, that is there will be a distinction between past and future light cones which is not the case for all manifolds constructed from vectors. It is tempting to propose that spinors are in some way more fundamental than tensors and that all physically realisable manifolds must be constructable from spinors.

The spinor formalism is of great use in a discussion of null vector fields, for example in gravitational optics or the classification of gravitational radiation.

The concern in this section is not so much with spinors but with their generalisation, the twistors. While spinors form a two valued representation of the homogeneous Lorentz group the twistors form a four valued representation of the 15 parameter conformal group. The twistor forms a geometrical entity which is a line rather than a point, that is, it is connected directly with the global properties of space-time rather than with the local properties.

It is convenient to introduce the twistor in the following fashion: Define a point particle in Minkowski space M by its linear momentum P_a and its angular momentum M^{ab} relative to an origin O. Supposing the total momentum to be null and future pointing, then the normal convention may be adopted that the spin vector P_d is parallel to P_a, that is

$$P^a M^{bc} e_{abcd} = 2sP_d,$$

where e_{abcd} is the alternating tensor and s is the intrinsic spin. Changing to spinor notation,

$$P_{AA'} = \bar{\pi}_A \pi_{A'}$$

and by virtue of the proportionality of the spin vector to the linear momentum vector

$$M^{AA'BB'} = i\omega^{(A}\bar{\pi}^{B)}\varepsilon^{A'B'} - \bar{\omega}^{(A'}\pi^{B')}\varepsilon^{AB},$$

where ε^{AB} is the skew symmetric Levi-Civita symbol. Thus the pair of spinors ω^A, π^A define the angular and linear momentum of the massless particle. Relative to O this ordered pair defines a twistor Z^α,

$$Z^\alpha \leftrightarrow (\pi_A, \omega^{A'})$$

and a complex conjugate twistor,

$$Z_\alpha \leftrightarrow (\bar{\pi}_A, \bar{\omega}^{A'}).$$

The product $Z^\alpha Z_\alpha = 2s$ gives the total intrinsic spin. In the case $s = 0$ the twistor is referred to as null and has an immediate geometrical interpretation as defining a null straight line in M. In the case $s \lessgtr 0$ the geometrical interpretation is not as simple, it is possible to view the non-null twistor as representing in a sense a 'complexified' null line. It is more useful however to represent this twistor by a congruence of null lines in M. (A congruence is a system of curves such that there is only one member, or a discrete number of members, of the system through a general point in the space.) When the twistor product is positive then the null lines twist around each other in a righthanded sense, in the case of a negative product the twisting is in a left hand sense.

The global geometry of M may therefore be discussed using the twistors, that is in terms of null lines and congruences of null lines. A disadvantage of the formalism arises as a result of its intimate connection with the conformal group; that is, it is not possible to represent the world lines of massive particles in any simple geometrical fashion. Whether this is an overwhelming objection or not depends to some extent upon one's attitude to the possible physical reasons behind the breaking of the symmetry of the conformal group.

Of course the twistor formalism is of use when discussing fields corresponding to massless particles. The general analytic solution ϕ_r of the spin s, zero-rest mass, free-field equation may be written in terms of a contour integral of a function $f(Z^\alpha)$ which is analytic and homogeneous of degree $-2s-2$ in Z; that is

$$\phi_r(X^\alpha Y^\alpha) = \frac{1}{2\pi i} \int \lambda^r f(\lambda X^\alpha + Y^\alpha) \, d\lambda \qquad (r = 0, 1, \ldots 2s).$$

The general form of the field equations is

$$\frac{\partial \phi_r}{\partial X^\alpha} = \frac{\partial \phi_{r+1}}{\partial Y^\alpha} \quad r = 0, 1 \dots 2s - 1, s \neq 0$$

and for $s = 0$

$$\frac{\partial^2 \phi_0}{\partial X^\alpha \partial Y^\beta} = \frac{\partial^2 \phi_0}{\partial Y^\alpha \partial X^\beta}.$$

The discussion of the null field solutions reduces to a discussion of the complex analytic structure of $f(Z^\alpha)$. Similarly the problem of finding all null solutions to a zero rest-mass free field equation reduces to a problem of finding all the twistor congruences which are shear free. Discussions of this nature are more properly carried out in projective space C which provides an alternative picture to that given by M space. To form C it is necessary to 'compactify' or complete M. (This is achieved by adding a light cone at infinity and the point at infinity of the light cone. It will be recalled that in projective geometry in order that theorems concerning lines (or intersections of lines) may be translated into theorems concerning points (or lines joining points) it is necessary for certain points at infinity to be added to the space.) The null lines and 'complexified' null lines in this 'compactified' M form an ∞^6 system and by giving them complex projective coordinates we may construct a complex three dimensional projective space C. Within C may be distinguished N, a ∞^5 system of null lines, that is N is a real five dimensional submanifold of C. N has the important property that a point in M is represented, in the C picture, as a complex projective line lying entirely on N. Similarly a null twistor Z^α is represented by a point lying in N. The right handed complexified null lines are points in C above this plane and the left hand complexified null lines are points below N.

The space C has a rich complex analytic structure and the properties of the zero rest-mass fields may now be discussed in terms of this structure. For example the problem of the discovery of all null solutions to a zero mass free field equation which has been translated into finding all the shear free twistor congruences now reduces to a consideration of the properties of the intersection of N with a complex analytic surface in C. Similarly the splitting up of field amplitudes into positive and negative frequency parts has a particularly simple geometric interpretation in terms of whether the singularities of $f(Z^{\alpha})$ occur above or below N in C.

The twistors therefore appear as a natural device for the description of Minkowski space-time and for the discussion of the solution of zero mass fields. It is, however, when we come to a consideration of more general space-times than M that the twistor structure takes on features which hint at a connection with quantum mechanics and suggest underlying structure common to both theories.

Consider the effect of allowing some curvature into the M picture, that is allow a gravitational wave to enter a region of flat space. Since we are dealing with a representation of the conformal group we must confine ourselves to conformal curvature, that is the weak field limit. The effect upon a shear free congruence of null lines will be to induce some shear, hence on re-entering a region of flat space-time the congruence will no longer be shear free. However it will be recalled that this shear free nature of the null geodesics in M is related to the complex analytic structure of C. Therefore the effect of a gravitational wave in M is to destroy the complex global analyticity of C and leave it defined only locally. However it is precisely this global analyticity which provides the power of the twistor method in discussing the properties of null fields, and at first sight the usefulness of the formalism seems to be removed once one departs from flat space-times. However all the structure on C is not removed since the invariant structure on C which remains is that which is left invariant by the group $\mathscr{F}$ of transformations of twistor coordinates. An infinitesimal twistor transformation is

$$\delta Z^{\alpha} = i\,\frac{\partial H(Z^{\alpha}, Z_{\alpha})}{\partial Z_{\alpha}},$$

where H is real and separately homogeneous of degree unity in Z^{α} and in Z_{α}. Using the Poisson Bracket

$$[\phi, \psi] \equiv i\,\frac{\partial \phi}{\partial Z^{\alpha}}\,\frac{\partial \psi}{\partial Z_{\alpha}} - i\,\frac{\partial \phi}{\partial Z_{\alpha}}\,\frac{\partial \psi}{\partial Z^{\alpha}}$$

the infinitesimal transformation may be written,

$$\delta Z^{\alpha} = [Z^{\alpha}, H].$$

Hence more generally if we may find those functions ψ which satisfy $[\psi, H] = \delta\psi = 0$ then they are part of the invariant structure of C. Examples of such quantities are: —

$$Z^\alpha Z_\alpha;\ Z^\alpha dZ_\alpha;\ dZ^\alpha \wedge dZ\alpha;\ Z^\alpha \frac{\partial}{\partial Z^\alpha}.$$

These expressions of the invariant structure of C have also a significance in the M picture, for example the vanishing of $Z^\alpha Z_\alpha$ indicates that Z^α represents a real null line in M even in the presence of conformal curvature, the division of C into regions above and below N is also invariant. The considerations of the structure of C in the presence of curvature in M suggests that Z^α and Z_α be regarded as canonical variables, that is a gravitational wave in M induces canonical transformations in C. If the passage to a quantized theory is considered then this is a very suggestive state of affairs since the twistor variable Z_α may be regarded as a multiple of the operator $\partial/\partial Z^\alpha$. The canonical transformations induced by curvature may now be interpreted as unitary transformations in an appropriate Hilbert space. (A convenient space to use is that of the analytic functions $f(Z^\alpha)$, used in the discussion of the solution of zero mass fields together with the definition of a scalar product.)

This investigation has uncovered a physical connection between space-time structure and the quantum mechanical formalism; for suppose that one has position and momentum quantum mechanical operators defined initially for some system, if the system is to pass through a region of curvature or large masses in the neighbourhood are to rearrange themselves then one has a situation in which the interpretation of the quantum mechanical operators is changed. The argument may of course be reversed to state that it is the constant shifting of the interpretation of the quantum mechanical operators which manifests itself as curvature of space-time. Therefore within the twistor picture it is possible to see, in quantum mechanical terms, the effects of space-time curvature. In the presence of mass, the field equations of general relativity or conformal invariance breaking interaction terms, the twistor formalism runs into severe difficulties. It is not clear at present as to how the formalism may be extended into these areas of wider physical interest, whether a modification of an extension of the twistor is necessary.

Finally it may be noted that the close connection between the twistor geometry and canonical transformations may give some indication as to the structure underlying quantum mechanics and the geometry of space time.

ACKNOWLEDGEMENTS

Part of this article was written while the author was visiting the Mathematics Department of Birkbeck College, University of London, and the author wishes to thank Professor Penrose and the College for their kind hospitality.

The author wishes to thank Professors D. Bohm and R. Penrose and Dr. D. Schrum for many extremely interesting discussions. He also wishes to thank Dr. D. Goswami for some helpful critical comments.

National Research Council of Canada, Ottawa

NOTES

[1] This is the view of Husserl and the Phenomenologists, for example see, E. Husserl, *Phenomenology and the Crisis of Philosophy*, Harper Torchbooks, New York, 1965.

[2] The criticism of 'Humeian' ideas of perception and their effect upon science was made by A. N. Whitehead in a number of books, for example *The Concept of Nature*, Cambridge University Press, London, 1957; *Symbolism: Its Meaning and Effect*, McMillan, New York, 1959.

[3] N. Bohr, *Atomic Physics and Human Knowledge*, John Wiley, London, 1958; *Essays 1958–1962 on Atomic Physics and Human Knowledge*, John Wiley, London, 1963.

[4] J. A. Wheeler, *Geometrodynamics*, Academic, London, 1962.

[5] The work of Bohm is directed by a holistic approach, see for example, *Foundations of Physics* 1 (1971), 359.

[6] A curious connection was made between relativity and quantum theory as a result of Weyl's attempt at a Unified Field Theory. (H. Weyl, *Sitzber Preuss. Akad. Wiss.*, Berlin (1918), 465.) It will be recalled that this theory employed an affine connection which did not leave the length of vectors invariant under parallel transport, the deviation being given by a four vector field which Weyl identified with the electromagnetic vector potential. The theory was rejected by Einstein, and others, on the grounds that the length of a vector would depend upon the path of its world line. Since the frequencies of atomic spectra observed from the surface of the sun are not broadened as the result of the past histories of the atom he concluded that vectors must indeed be invariant under parallel transport.

London however demonstrated that for certain closed orbits of a vector transported within a Coulomb field, invariance of length could be maintained, it turned out that these orbits were formally identical with the Bohr orbits of the 'Old Quantum Theory'. (F. London, *Zeitschrift für Physik* 42 (1927), 375.) More recently Adler has employed a spinor formulation in a Weyl geometry which leads to the recovery of the Dirac equation for an electron coupled minimally to the electromagnetic field. (R. J. Adler, *Jour-*

nal of Mathematical Physics **11** (1970), 1185.) It is of interest to note that Weyl transforms survive in Zumino's prescription for conformally invariant non-polynomial Lagrangians. (B. Zumino, *1970 Brandeis University Summer Institute in Theoretical Physics, Lectures on Elementary Particles and Quantum Field Theory*, Vol. 2, Massachusetts Institute of Technology, Cambridge, 1970.)

[7] Isospin appears to have no direct correlation with the properties of space-time, rather it is connected with the so-called 'internal-space' of particle physics. However in some gravitational-electromagnetic theory it may turn out that isospin has a geometrical significance.

[8] Realizations of the Conformal group and group of General Coordinate Transformation are discussed by Zumino, Note 6 and by Salam and his co-workers: C. J. Isham, A. Salam, and J. Strathdee, *Annals of Physics* **62** (1971), 98; I. C. T. P., Trieste, preprint, IC/71/14.

[9] C. J. Isham, A. Salam, and J. Strathdee, *Physical Review* **3** (1971), 867.

[10] An introduction into the use of spinors in Relativity Theory is to be found in F. A. E. Pirani, *Brandeis Summer Institute in Theoretical Physics, 1964 Lectures on General Relativity*, Vol. 1, Prentice-Hall, Englewood Cliffs, New Jersey, 1965. The work on twistors is to be found in R. Penrose, *Journal of Mathematical Physics* **3** (1967), 345; *International Journal of Theoretical Physics* **1** (1968), 61; *Magic without Magic: John Archibald Wheeler* (ed. by J. Klauder) to be published.

MENDEL SACHS

ON THE NATURE OF LIGHT AND
THE PROBLEM OF MATTER

I. INTRODUCTION

A very old, yet unresolved problem in physics concerns the basic nature of
light. Through the various periods in the history of science, since the days
of antiquity, new insights have been gained and earlier concepts rejected.
At the present stage of contemporary physics we have learned a great deal
about phenomena concerned with the nature of light; still, logical dichotomy and mathematical inconsistency remain in the usual answers to the
question: What, precisely, is light? It will be my thesis in this paper that
the persistence of these difficulties standing in the way of answering the
question satisfactorily have something to do with present-day problems in
providing a fundamental description of matter – that, indeed, only a resolution of the problem of matter may lead to a resolution of the problem
of light.

Before presenting some of the logical arguments for my contention,
some of the history of physics that concerns the nature of light will be reviewed in order to place the problem in its historical perspective.

II. A CLASSICAL DEBATE ON THE NATURE OF LIGHT

The historic 17th century debate between the adherents of Newton's
corpuscular theory of light, on the one hand,[1] and on the other, those who
supported the approach in terms of a propagating wave (Hooke, Huygens),
dealt respectively with a choice between a discrete (atomistic) model and
one in which it was assumed that light is a continuously propagating disturbance of a continuous medium.[2]

Newton asserted that the dispersion of a beam of light into the different
colors was in fact due to its actual composition as an assemblage of
corpuscles of light characterized by the different colors. In matter-free
space, these were all asserted to move at the same speed – continuously
interacting with a special sort of aether medium which continually scat-

*Hooker (ed.), Contemporary Research in the Foundations and Philosophy of Quantum
Theory, 346–368. All rights reserved*
Copyright © 1973 by D. Reidel Publishing Company, Dordrecht-Holland

ters these corpuscles so as to maintain their motion in a straight line path. Traveling together in this way, such a collection of light corpuscles then gave the view of 'white light'. In a medium of ordinary material, however, such as water vapor or glass, the 'aether' is said to be displaced, and the different types of light corpuscles were then said to travel at different speeds, depending on the internal constitution and the geometry of the propagating medium. This model accounted, for example, for the observation of the dispersion of the sun's light when passing through a region of water vapor in the earth's atmosphere, or through a triangular glass prism, thereby giving the impression of the rainbow. It also explained the blueness of the sky in terms of the preferential scattering of blue light corpuscles toward the earth, as the sun's light is scattered through our atmosphere.

Newton also claimed to explain the spectrum of colors of light observed in the reflections from a solid surface, when coated with a thin liquid film, in terms of a model that involved only reflective and refractive processes. He also explained the refraction of light, when passing from one sort of medium into another, by taking account of the dependence of the velocities of the light corpuscles on the density of the conducting medium. This had to do with the forces acting between the atoms of the respective material media and the light corpuscles. In this application, however, there were some unanswered questions. For example, Newton wondered why a beam of light should bend in one way (say, toward the normal to the surface of the light conducting medium) and not the other, when passing from one medium into another. Then there was a question about *partial* reflection and *partial* refraction. That is, why should some of the light corpuscles be transmitted through the medium and others be reflected from it? Newton assumed that the cause of the reflection of light from a shiny surface is the existence of a repulsive force between the light corpuscles and the atoms of the reflecting surface. This model correctly predicted the equality of the angle of incidence and the angle of reflection of the light. But why should only some of the light corpuscles be reflected from the surface and not the others? Newton speculated that there may be a mechanism to control 'fits of easy transmission' (refraction) and 'fits of easy reflection', as due to a feature of the vibrations of the aether in the neighborhood of a reflecting surface. His conjectures on this question, however, were quite inconclusive.

There was one optical phenomenon that was known in the 17th century that Newton's corpuscular model could not quite explain. This was the observed diffraction of light – making it appear that light can bend around corners. Such a phenomenon may be seen by shining a beam of light through a small hole in a screen, or past a sharp edge or a long thin object (such as a hair – an example that Newton himself studied). When this is done, it is observed that there is no *sharp image* produced (by any of these objects) on a second screen some distance away. Instead, one observes a *continuous change* in the intensity of the 'diffracted light', from the central image of the objects observed, outward. It was further observed that fringes appear in the shadow of the observed object – these are alternate bright and dark regions, with diminishing intensities as the distance from the central image increases. It is also observed that if the size of the diffracting object (say the hole in the screen) is decreased, the diameters of the central image on a second screen, and the fringes, correspondingly increase.

If light were indeed a collection of corpuscles that behave like Newton's mechanical particles, they should move along discrete trajectories. Then why should there be alternate regions on the second screen where these particles are permitted and not permitted to fall (the bright and dark fringes)? Secondly, how could such a model explain the observed *continuous* change in the brightness of the central image of the hole and the light and dark rings that surround it? It appeared to many that the 'diffraction of light' – an effect that makes it appear that light can bend around corners – was not at all compatible with Newton's corpuscular model of light. The experimental evidence for this effect was supposed to have been found long before Newton's day, by Leonardo da Vinci.[3] It was then rediscovered in Newton's day by F. M. Grimaldi and by R. Hooke.[4]

Of course, Newton knew about these phenomena, but he believed that he could explain them in terms of a purely refractive effect, caused by the forces between his light corpuscles and the atoms of the aether medium in the vicinity of the refracting substance. His model explained the observed lack of a sharp image and *some* of the fringes – the ones that he had observed in his own optical experiments on the diffraction of light (from a hair). Newton observed fringes (outside of the geometrical shadow of the diffracting object) which his atomic model could explain. However, his model could not explain the presence of fringes *within* the geometrical shadow of the hair, which, it has been argued, he did not see.

A different explanation of the diffraction phenomenon came from a study of Robert Hooke, who took light to be a set of vibrations of the aether – thus a continuous entity rather than atomistic. Another continuum view was that of Christiaan Huygens, which also took light to be the propagation of continuous waves. With his model, the 'rays' of light with which one described the geometrical optics of lenses, mirrors, etc., were interpreted in terms of a directed wave front. According to *Huygens' principle*, the means of light propagation involves each of the continuum of points on the wave front acting as a new source of light which propagates away from this point in a spherical wave. The sum of spherical wave fronts, from any given region where the wave had been, then propagate together as a common wave front.

Huygens' model was able to explain the diffraction phenomenon in the following way: Consider the beam of light approaching, for example, a small hole in a screen. Each point of the wave front that covers the hole acts as a separate source, each then sending spherical waves radially toward the second screen. One then observes an interference of the continuum of different waves that move spherically from the different points in the hole in the first screen, when, superposed, they reach the second screen a distance away. The effect of spherical waves emanating from the hole then yields the fuzzy image of the hole at the second screen, with the maximum intensity at the center, and the interference of the superposition of waves continuously changes between constructive and destructive at the absorbing screen, thereby giving the observation of bright and dark regions. Thus, Huygens' continuous wave model explained the diffraction phenomenon, as well as the known 'ray' properties of light associated with reflection and refraction, and transmission through focusing devices.

Further enumerating the known properties of light in the 17th century, Huygens' model did not explain the fact that it is polarized. He viewed the continuous wave propagation in terms of longitudinal pulses (as in sound waves). Yet, there was experimental evidence in that day that light is not polarized longitudinally. The Danish physicist, Bartholinus, found that certain transparent crystals transmit light in different directions with different velocities (irrespective of color). Since two directions suffice for the resolution of component velocities, the phenomenon was called 'double refraction'. In 1690, Huygens noted the possibility of associating polarization with the observed doubly refracted rays of light. To explain this effect, Newton imagined his light corpuscles to come in two varieties, each of which was to be refracted differently by the doubly refracting crystal, or by a reflecting surface. On the other hand, Newton did not see how Huygens' continuous wave model, as longitudinal pulses of only one variety, could explain the polarization effect.

III. THE ELECTROMAGNETIC THEORY OF LIGHT

Taking account of all the known properties of light, during the two centuries between Newton and Maxwell, there was no all-encompassing theory of light that could incorporate all of the data. It is interesting to note that the two competing models of light – Newton's atomistic view versus Huygens' continuum view – continued a debate that had persisted since the days of antiquity on the fundamental nature of matter.

Toward the middle of the 19th century, Faraday discovered that the unification of the electric and magnetic manifestations of matter had to do with the relative motion between an observer and the observed charged matter. Maxwell then expressed Faraday's unified field of 'electromagnetism' in terms of partial differential equations that represented the variation of the three components of the electrical field of force and the three components of the magnetic field with respect to the space and time coordinates. According to Faraday's interpretation, the solutions of these equations are the continuous distribution in space and time of the potential effect of charged matter on a test charge (at one place or another and at one time or another).

To ensure that the form of Maxwell's equations would have the same form to relatively moving observers, Lorentz discovered that a quantity (with the dimension of speed) that appears in these equations had to be taken to have the same value in all reference frames. (He took this to be a feature of the lightconducting aether. This result was one of the primary hints that led Einstein to the theory of special relativity not too many years later.) Maxwell then made the important discovery that among the various solutions of his equations, the ones that describe the effect of charged matter on a test charge that is very far away are indeed a mathematical description of continuous waves that propagate at the universal speed that appears in these equations. The latter speed was found (from the ratio of electric and magnetic field amplitudes that would produce the same force on a test charge) to be numerically equal to the speed of light in a vacuum. Maxwell discovered that light is none other than a particular manifestation of electromagnetic forces exerted over a very large distance – called 'electromagnetic radiation'. This new discovery then led to the correct prediction of all of the features of light that were known in the 19th century.

Maxwell then validated Huygens' conceptual model of light, thereby refuting Newton's corpuscular model, even though there were details that differed from Huygens' view. For example, it was now known that light (as any other form of electromagnetic radiation, such as radio waves, X-rays, γ-rays) is a propagating wave that vibrates in a direction perpendicular to its direction of propagation – it is a 'transverse wave'. This was contrary to Huygens' view of light as longitudinal waves. The transverse character of light waves was experimentally verified, some time before Maxwell, by Fresnel.[5] Thus, the 19th century scientist learned that light must be the transverse vibrations of an aether – the medium they assumed must be present to conduct the electromagnetic radiation. The later analysis by Einstein, in the early part of the 20th century, in his development of the theory of special relativity, as well as the negative experimental results of the Michelson-Morley study, then implied the aether to be a superfluous ingredient in the fundamental description of light propagation.

Toward the end of the 19th century, then, there seemed to be no question about the continuous nature of light, as a propagating set of transverse vibrations of the aether, with each component frequency corresponding to a different color response. Together with the dependence of the refraction of light waves on their frequencies and the effect of the coupling between single crystals and the transverse vibrations of a light wave, to produce the double refraction effect, as well as the correct prediction of all other known optical phenomena, including refraction, reflection, diffraction (and interference phenomena generally) and polarization, it was felt that indeed the nature of light had finally been understood in terms of the electromagnetic field theory. It must indeed have been a pleasant feeling for the physicists and philosophers of the late 19th century to have finally settled this age-old dispute about the nature of light, and also to acquire the aesthetically pleasing bonus of unifying the basic descriptions of optics and electromagnetism into one conceptual scheme.

IV. THE QUANTUM THEORY OF LIGHT
AND THE ROLE OF MATTER

The state of content about the 'true' nature of light did not last for too many years after Maxwell. What happened next, in the early decades of the 20th century, seems to have been more devastating to those who sought a logically consistent description of the basic nature of light than were any of the difficulties of the earlier periods. For it now appeared that under certain experimental circumstances light was a continuous field of vibrations, characterized by such wave properties as wavelength, frequency, etc. (exhibited e.g. in interference phenomena), while under other experimental conditions, light appeared to be a collection of 'atoms', each characterized by particle variables, such as momentum and position. This dual set of seemingly dichotomous features of light, incorporated into one theory, was then referred to as the 'wave-particle duality'. It led to one of the two revolutions in 20th century physics – the quantum theory of measurement.

Recalling some of the first experiments that led to this (seeming) dichotomy in the basic description of light, the observations (around 1900) on the spectral distribution of blackbody radiation implied, firstly, that one can view electromagnetic radiation as a 'thing-in-itself' – i.e. as independent of the dynamical properties of the charged matter that consitutes the walls of the cavity that contains the observed radiation. This experimental finding was contrary to Faraday's interpretation of electromagnetic radiation as a particular (far away) manifestation of charged matter. Secondly, Planck discovered that the observed spectral distribution of the blackbody radiation, as a function of temperature, made it appear that one was observing a gas of individual things, each characterized by an energy value that is linearly proportional to its characteristic frequency. Further, since the radiation was observed in a finite cavity (i.e. with walls impenetrable to the radiation contained therein) it follows that the possible modes of vibration of this radiation lie in a discrete spectrum. Thus, Planck concluded that each of the vibrational modes of the radiation must be characterized by a discrete value of energy, called a 'quantum'.

Later on, Bose and Einstein (independently) showed that the Planck spectral distribution function could be derived from a model of the radiation in terms of particles of light that are considered, statistically, as a gas of *indistinguishable* quantum mechanical objects, without restriction as to the number of such particles that can have a particular state of motion at a particular place. This was referred to as 'Bose-Einstein' statistics.

Along with this discovery by Bose and Einstein, however, Planck himself pointed out that one could equally derive his spectral distribution formula from the classical statistics (called 'Maxwell-Boltzmann' statistics) of a set of *distinguishable* vibrational modes – provided that the energy in each of these modes should be taken to be linearly proportional to its frequency.[6]

At this early stage of 20th century physics, then, the question arose once again: Is light, fundamentally, a bundle of discrete particles, or is it a system of continuously distributed waves? To resolve the difficulty, the revolutionary approach of the *quantum theory of measurement* asserted that, indeed, light is both kinds of entity at the same time! It was N. Bohr who first expounded the philosophical basis of this approach when he asserted that these seemingly contrary aspects in the nature of light (and, as later discovered, matter) do in fact *complement* each other to present the entire 'true' description. (This is the *principle of complementarity*).

The latter principle was logically tenable to Bohr and Heisenberg, and the other followers of this view, primarily because of their re-interpretation of the continuous aspect of light, not as a *real* vibration, such as the mechanical oscillation of a sound wave, or a propagating ripple on a pond, but rather in terms of a continuous probability function. According to their approach (called the 'Copenhagen school') the latter was to be identified with the *chance* that a *macroscopic* measuring apparatus can discern that a photon can be found at one place or another throughout space. That is to say, the continuous wave function that describes light was taken to be not more than a *language element* of a macroscopic apparatus – to be used in probability statements reporting its findings on the values of the various physical properties of elementary entities. According to this philosophy, there is nothing more that can be said about the observed elementary entities other than such probability statements of a large scale (macroscopic) measuring device.

According to this new view of light, its wave aspect is not at all the type of intrinsic physical feature as Huygens, Fresnel and Maxwell had supposed (such as the vibrations of a real continuous medium). It rather relates, in this theory, to a 'field of probability' that a device will be able to detect a 'particle of light' (as Newton had assumed to be its basic ingredient) at one place or another. It is important to emphasize at this point that while this is a particle theory of light, it is unlike Newton's theory on the question of determinism. According to Newton's approach, the particles of light are not unlike any other type of mechanical particle – they have *predetermined* trajectories that are independent of the way in which they are observed. The trajectories of light were taken only to depend on the external forces that govern their motions. On the other hand, the quantum theory of measurement is based on a nondeterministic philosophy in its denial of the existence of predetermined, arbitrarily precise values for the properties of the particles of light. Their values, here, are taken to be intimately related to the nature of the observation by a macroscopic measuring apparatus.

The wave-particle theory for matter that was proposed by de Broglie, and the subsequent observations by Davisson and Germer and by G. P. Thomson of electron diffraction by crystals, showed that the wave-particle duality was also a valid description of electrons. Later it was shown to fit the description of the other elementary particles of matter. Thus, Bohr's complementarity principle was asserted to be a valid universal concept.

In this regard, however, it is important to keep in mind some basic physical differences between matter (i.e. particles with non-zero rest mass) and light. Firstly, if light is indeed a collection of corpuscles (photons), then in free space each of them must travel at the same universal speed. They cannot be slowed down or speeded up because they have no inertial mass. They can only be stopped, by annihilating them (by means of absorbing them in 'quantized matter'). Or, they can be started, by creating them (by means of the de-excitation of 'quantized matter'). But once created as particles, they have the peculiar un-particle-like feature of being without inertia since no external force can make them change their speed by arbitrary amounts. In contrast, inertial particles, such as electrons, can move at any speed (up to the speed of light) and indeed can be slowed down or speeded up by external forces by continuously varying amounts.

Secondly, if a given photon has a particular energy, then it cannot be said to be anywhere in particular. If, indeed, the photon is a 'thing', then according to the usual view, one should be able to specify its location, at least in some limited region of space. That is, an observer should be able to say: I know that a photon is in a particular box in the laboratory and nowhere else! – just as he can say he knows that a valence electron of some metal is somewhere within the confines of that piece of metal. But a *single photon*, which, by definition, has a precise energy, is described mathematically in terms of a plane wave – a function that has an equally weighted value at all points in space at any given time. With this description, then, one would have to say that the single photon is everywhere, rather than somewhere – although it can be annihilated somewhere by looking for it at that particular place! Along with this spatial description of the single photon, it is specified to be continually traveling at the speed of light. To the (perhaps naive) inquirer, the logical difficulty appears in trying to answer the question: if the photon is everywhere at the same time, and is traveling continually on its own at the speed of light, where is it going to?

It seems to me that these logical difficulties in the present-day conceptual view of the nature of light, in terms of the photon model, also persist to some degree in the general description of matter – but not so severely. For one thing, as I have mentioned above, material particles can change their speeds continuously in response to external forces, while maintaining their individualities as material particles. For this reason, actual observations always entail matter alone – one quantity of matter responds to another quantity of matter by means of a transfer of energy and momentum between them. On the other hand, the atom of light (photon), even if it is a separate, distinguishable entity, cannot be involved in any direct observation because it cannot change its state of motion in a continuous fashion – it can only disappear! Thus, the existence of a photon can only be *deduced* from the observed facts, as an experimental and a logical necessity to understand these facts. The question then arises: Is it possible that all of the experimental data, that are conventionally interpreted in terms of the existence of photons, may be re-interpreted with a model of light that does not at all entail photons? Aside from 'experimental necessity', it must also be asked if the photon concept is logically necessitated by the theoretical structure of the most basic description of light. Of course, this takes us back to the original question: What is the most basic description of light?

It also seems to me that a second difficulty has to do with the fact that at the present stage of physics, there is not yet an accepted satisfactory theory of matter. There is no difficulty in describing elementary particles, such as electrons, when they move at speeds which are small compared with the speed of light. That is, non-relativistic quantum mechanics seems to me to be at least a *mathematically* consistent formalism. But when one considers these particles to move at speeds that approach the speed of light, it becomes necessary to fully relativize the Schrödinger description. When this is done in the proper way, one arrives at a formalism (quantum field theory) that has no solutions! Indeed, this difficulty has persisted since the onset of quantum mechanics in the 1920's.[7]

One may then argue that at least we do have a good theory of matter in the non-relativistic domain. But even at low energies, this is not good enough! The reason, once again, has to do with the nature of light. For the measurable properties of matter in the microscopic domain are intimately connected with the excitation and de-excitation of atoms, nuclei, elementary particles, molecules, etc. – that is, whenever there is energy-momentum transfer between microscopic matter and the (large quantity of matter that constitutes the) measuring apparatus. The latter entails discrete quantum jumps between the energy levels of microscopic matter, *along with the absorption and emission of photons.* Thus, to *consistently* describe *any* observed property attributed to the characteristics of microscopic matter, it is necessary (both logically and mathematically) to include the light that is emitted and absorbed. Now, while the theoretical description of the matter part of this system does have a non-relativistic limit, no such limit exists for the theoretical description of the light! It then follows that to describe matter, with the light that it emits and absorbs, *in one theoretical scheme*, the fundamental theoretical description must *necessarily* be relativistically covariant. However, as I mentioned above, there is no demonstrably mathematically consistent relativistic quantum field theory of matter (at least at the present stage of physics).

Further logical difficulties remain in the description of the 'quantum jump' itself.[8] For example, when an excited atom de-excites itself by emitting a photon, precisely what is the physical mechanism that causes this de-excitation? Are we to be contented with merely asserting that this physical effect *happens*, without any physically related cause other than a statement that events occur with varying degrees of uncertainty – simply by virtue of their having particular intrinsic energies? Is this not analogous to the Aristotelian view that planets move the way they do only because of their spatial locations, but independent of any underlying physical force (such as the action of the sun's gravitational force)? Then once a photon would be emitted in a de-excitation process, if the atomic electron is still in transit toward the lower energy level, how could one explain some physical occurrence that might transform the atom so as to eliminate this lower energy level? If one should reply that the process of photon emission entails the electron being in the excited and de-excited states simultaneously in the quantized system (even though the precise time when this is so cannot be specified), how might one reply to the objection of relativity theory that simultaneity is only relative to a particular space-time frame – that in other Lorentz frames the electron would be seen to appear in these two energy states at different times? It appears at this stage (to the perhaps naive inquirer) that the concept of the discrete quantum jump to describe the measurement of a physical property of microscopic matter, and the concept of relative simultaneity that is implicit in relativity theory, are indeed logically dichotomous. Is it not possible, then, that the mathematical difficulties in constructing a consistent quantum field theory are, in fact, rooted in the logical inconsistency of a theory that would simply adjoin the concepts of the quantum theory of measurement and the theory of relativity?

V. A POSSIBLE RESOLUTION

A resolution of this problem of light may be rooted in a suggestion that was made, originally, by G. N. Lewis.[9] This was the idea that, indeed, light may not be a 'thing-in-itself' at all, but rather it may relate only to the *process* of propagation of the electromagnetic interaction between electrically charged matter. That is to say, a given quantity of electrically charged matter has associated with it a continuously distributed electromagnetic field of force, according to Faraday's original interpretation. At large distances from this matter, its force field *appears* to a test charge (in terms of the resulting motion of the latter) as a plane transverse wave – such as is described by the homogeneous solutions of Maxwell's equations (i.e. the 'radiation solutions'). This interpretation of light, as Lewis asserts, requires us to consider the process as a perfectly symmetric one, so that we can no longer regard one atom as an active agent and the other as an accidental and passive recipient, but both atoms must play coordinate and symmetrical parts in the process of exchange.

My own research program, which incorporates Lewis' view, is a field theory based on three essential axioms. The first is the principle of general relativity – the assertion that the laws of nature must be in one-to-one correspondence when compared in different frames of reference, irrespective of their motions relative to each other. This axiom implies some interesting consequences in regard to the nature of light. First, the general definition of motion in terms of *continuous changes* of any of the four (space or time) coordinates of one frame, with respect to any of the space or time coordinates of another, implies that the most primitive symmetry group to underlie a theory of light must be without any of the discrete symmetry elements, such as spatial or temporal reflections. This feature implies that the basic language elements with which to describe such a theory of light must necessarily be the functions of the space and time coordinates that vary continuously with respect to these parameters. (The space-time coordinates, in turn, are not more than a continuous set of numbers chosen to facilitate a mapping of the fundamental field variables.) In general relativity theory, the metric relations between the space-time points, in turn, are prescribed by one of the sets of underlying field variables that have been found to relate to the gravitational force (from Einstein's field equations).

Thus, the first axiom of my approach necessarily implies that the theory of light (and matter) is based on the *continuous field concept*. This takes us back to the continuum view of light, as it was assumed by Hooke and by Huygens, in the 17th century, and then concluded by Faraday and Maxwell in the 19th century. The 'particle features' of light do not emerge here as fundamental aspects, as they do in the wave-particle view in contemporary physics. They are rather asymptotic features of a continuous field that reveal *apparent* peaked distributions within a continuum, analogous to the distribution of ripples on a slightly perturbed pond.

A further consequence of the axiom of the principle of general relativity is that the theory of special relativity plays the role of a special limit, where the relative motion of different space-time frames in which the physical laws are compared, is uniform (i.e. constant rectilinear speed). In general relativity – a starting point of this theory – the special relativity limit corresponds to a limit of Euclidean geometry for the space-time (zero curvature) which, in turn, relates to a matterless universe! But all observations, according to this approach, entail a coupling between matter and matter – as components of a closed system. Thus, the successes of special relativity theory, according to this view, are not more than an indication of how good the approximation of a flat space representation is for an actually curved space-time, in those special applications where special relativity has been verified. This is analogous to the consideration of the Galilean transformations (in classical physics) as a good asymptotic approximation (for the Lorentz transformations) where they have been found to be accurate (e.g. to describe a block sliding down a plane). This result then implies that the propagation of light, as an interaction between matter and matter, is necessarily described here in terms of a propagating disturbance in a curved space-time – even though this description can be mathematically approximated with the description of the propagation of such a disturbance in a plane space-time that is tangent to the point of observation. Such an approximation is accurate when sufficiently 'local' observations are being made.

The second axiom of my approach is the assertion of a *generalized Mach principle*. This means that any physical system is necessarily *closed* – none of the physical attributes of a component of the system can be independent (i.e. dynamically uncoupled) from the remaining components of the system. This feature, when applied to the inertial manifestation of matter in particular, is generally called 'the Mach principle'. I have found in earlier work[10], from the incorporation of this axiom with the principle of general relativity, that one can *derive* relations between the inertial properties of elementary particles and the field properties of the space-time continuum. This derivation led to the prediction that (1) purely gravitational forces can only be attractive, (2) in the asymptotic limit, when the components of the closed system become arbitrarily weakly coupled (i.e. when there is sufficiently small energy-momentum transfer between them), the average values of the possible innertial masses of microscopic matter *approaches* a discrete spectrum, (3) the inertia of matter described by spinor variables occurs in mass doublets, and that one can derive the masses of the electron and muon as a particular doublet, when the background gas of electron-positron pairs, that necessarily follows in my theory to permeate all (ordinary) matter, is the primary contribution to the curvature of space-time at the site of one of these observed particles,[11] and (4) in accordance with the Mach principle, as the environment of an observed electron should be continually depleted of other matter, the electron mass itself should correspondingly vanish (i.e. in the limit of a 'free' electron state, its inertial mass would be zero).

The generalized Mach principle, upon which I am basing my field theory, refers, in addition to the inertial manifestation of matter, to all of the other manifestations, including the electromagnetic effects, which in turn are responsible for the phenomenon of light. An implication of such a generalization is that *any* atomistic quantity, such as the electron charge, e, is not a fundamental constant. It is rather the *coupling constant*, e^2, that is the fundamental quantity in this theory. Thus, the quantity of electric charge, Q, assigned to any particular component of an interacting system, is wholly arbitrary – so long as the strength of the coupling for the *closed system* (in principle, the universe) is measured in terms of the fundamental constant e^2. The apparent discreteness of the charge of matter, in multiples of e, then follows from the asymptotic behavior of the field variables that weight the constant e^2, in the various experimental conditions that entail electromagnetic coupling.[12]

The third axiom of my theory is the assertion of a correspondence principle. This requires that the formal structure for the mathematical expression of the theory should be such that its field equations, which are basically nonlinear, should approach the linear eigenfunction form of quantum mechanics in the appropriate limit of sufficiently small energy-momentum transfer between interacting matter, and that the formalism must also incorporate the standard expression of the electromagnetic field theory, in this limit.[13]

Within my conceptual approach, based on the three axioms discussed above, Lewis' view is extended by requiring that the theoretical description of light must necessarily be in terms of a fully relativistic field theory of a closed system. This is a system, without actual parts, in which the formalism is symmetric with respect to the interchange of the variables associated with its interacting components. With this view, the notation 'observer-observed' is only a convenient language for a single elementary entity (that I have called 'elementary interaction'). That is, the general field description is insensitive to which component of the closed system is called 'observer' and which is called 'observed' – in contrast with the approach of the quantum theory of measurement.

Thus, starting with the closed system as the basic existent from which the properties of matter are to be derived, the nature of light is not at all in accord with the previous theories that assumed it to be a thing-in-itself – whether described by particles, waves, or the wave-particle model of the quantum theory. In my theory, which continues the view of Lewis, light is not more than a manifestation of coupled matter. When matter is described most primitively, that is in terms of its microscopic properties, these have a peaked distribution of values – appearing almost to be discrete (i.e. quantized). It is nevertheless significant that, intrinsically, there is a finite width for all of the measured values of the properties of microscopic matter. With the conventional quantum mechanical interpretation, this is due to the Heisenberg Uncertainty Principle. In my theory, it is due to the nonlinear features of the formalism whose presence can in principle never be 'turned off' and which destroys the eigenfunction character of the field equations.

Since the energies of the atomic electrons appear to be weighted most heavily at special (discrete) values, atoms and molecules can only transfer correspondingly discrete quantities of energy between each other. Such discrete quantities of energy transferral then relate to the 'atoms of light' that were originally postulated by Einstein. (It was G. N. Lewis who originally called Einstein's light quanta 'photons'[14].) Nevertheless, Lewis' 'photons' are not the *independent*, individual atoms of light that were supposed by Einstein to explain the data. For one thing, Einstein assumed there to be an equilibrium between the emission and absorption of both induced and spontaneous radiation, when a system is in thermodynamic equilibrium. With Lewis' view, on the other hand, light is not emitted on its own, spontaneously. It is only an expression for a propagating interaction between distant quantities of matter.

Many have commented that this idea appears to defy common sense since the emitting atom would have to 'know' the entire future of the universe before it would emit an electromagnetic signal (say, some quantity of light). This is because it would not be absorbed until the passage of time equal to R/c (R being the spatial separation between the emiter and the absorber, and c is the speed of light) – and, certainly, anything could happen to destroy the absorbing atom before this time had passed! Nevertheless, Einstein himself has emphasized that much of our scientific understanding has indeed come from giving up 'common sense' notions – which, in fact, are not more than a reliance on interpretations of ordinary sense perceptions. Rather, we have learned (as in the case of relativity theory) to base our physical knowledge on (a) logical analyses of particular theoretical hypotheses, (b) a rigorous and consistent mathematical structuring of this theory and (c) a comparison of the mathematical predictions of the theory with the observations of the physical world. In this regard, then, most experiments that are supposed to entail a photon model of light can equally be viewed strictly in terms of the direct electromagnetic coupling between microscopic quantities of matter. This includes such phenomena as the photoelectric effect, the Compton effect, bremsstrahlung, etc., since all of these phenomena entail matter and radiation simultaneously within the same system.

It seems to me that there are only two classes of experimental observations that are usually interpreted in terms of the presence of photons at times when there is no matter in the system. One of these is the observations of the spectral distribution of blackbody radiation, as discussed earlier. The second is 'pair annihilation'. In the latter process, an electron and a positron are supposed to annihilate each other, simultaneously creating a pair of photons. If one wishes to dispose of the photon concept of light, and to replace it with the direct electromagnetic coupling between charged matter, it is then necessary to *derive* all of the experimental facts relating to blackbody radiation and pair annihilation, from a theory that does not involve free (classical) electromagnetic radiation or photons.

Indeed, such predictions have followed from the field theory that I have been pursuing. As discussed earlier, the description of electrodynamics starts here with a single closed system – without truly separable parts! Mathematically, this idea is expressed by saying that there are no 'free field' solutions of the formal expression of the theory. However, in the asymptotic limit, when there is sufficiently small energy-momentum transfer between the components of the system, the formalism approaches a functional structure that would describe a collection of separate electrons and positrons, etc., that appear to be perturbing each other. *In this limit*, the solutions of the field equations are the elements of a Hilbert space, and the predictions of observables follow from the weighted linear operators, as it is usually formulated in the quantum theory. However, to use this limit is only to use an approximation for a formalism that is non-linear, in principle, under all conditions, and is not generally described by solutions that span a Hilbert function space. The 'observation' of electrons as independent 'things', in this approximation, is analogous to the observation of ripples on a pond as separate things. The analogy is a close one since even though the motion of the ripple can be viewed as the dynamical properties of a single thing, it cannot actually be removed from the pond while maintaining its individuality as a ripple! It is, rather, a special manifestation of the entire pond, just as the electron is, in this theory, a manifestation of a closed system of matter.

When this theory was applied to the electron-positron system – *as a closed system* – it was found that the equations, in their exact form, have a bound state solution whose analytical form *predicts* all of the physical observations that are customarily attributed to 'pair annihilation'.[15] In the observation of the converging tracks in a cloud chamber, for example, the tracks disappear at the vertex (according to this theory) because the electron and positron go into a deeply bound state, no longer capable of giving up energy and momentum to the ions in the chamber – thereby making it appear that they are gone! But there is actually no annihilation of matter in this case. It then follows that if sufficient energy $(2\,mc^2)$ should be supplied to the pair, when it is in this deeply bound state (which actually corresponds to its minimum energy – maximum binding), it can be excited into the state that once again allows the freedom to these components of the system to ionize the atoms in the cloud chamber. The latter observation is conventionally interpreted as 'pair creation'.

According to the explicit features of this solution of the coupled nonlinear field equations for the pair, it follows from the field formalism (Noether's theorem) that the mutual energy and the components of momentum, separately, are all zero. Thus, this ground state for the pair is an invariant property of the pair with respect to space-time transformations of relativity theory (in its special or general form). That is to say, this solution of the field equations corresponds to a null energy-momentum vector in all Lorentz frames (or the relative frames of general relativity theory). It should nevertheless be noted that even though this ground state energy is zero (relative to the energy of $2\,mc^2$, when the electron and positron would be (asymptotically) free), each of the particle components, separately, has a non-zero inertial mass. For example, the system would still 'weigh' 2 mg dyne near the earth's surface. The zero energy of the electron-positron bound system in this particular state, along with the non-zero weight, is a feature of the *nonlinear* field description of the theory.

According to the dynamical features of this ground state solution for the pair, it would appear to other charged matter as a pair of plane polarized currents that are correlated with a 90° phase difference, and oriented in a plane that is perpendicular to the direction of propagation of its electromagnetic interaction with other matter. The latter dynamical properties are indeed observed (indirectly) when a pair is supposed to annihilate – the two correlated currents are conventionally associated with the two photons that are supposedly created in the process. Nevertheless, there are no photons in this theory and there is no actual annihilation of matter, while the same experimental observation is predicted.

To explain blackbody radiation, one can consider that the cavity which is supposedly populated with a photon gas is instead populated with an ideal gas of pairs, each in their ground state of null energy-momentum. The analysis of such an ideal gas, according to this nonlinear field theory, then leads to the Planck distribution function.[16] But this is not related here to a gas of quanta (that obey Bose-Einstein statistics). It is rather related to a set of *distinguishable* interactions between the observing apparatus and the ideal gas of pairs in the cavity. Essential features of this theory that led to this result were (a) the special dynamical features of the pairs, exhibited in the *nonlinear* solutions for their states of maximum binding, and (b) the finiteness of the cavity that is taken to contain the gas of such pairs.

With these theoretical results, it is then contended that indeed one can dispense with the photon model of light and that all of the known properties associated with the phenomenon of light can be derived from the nonlinear, relativistic field theory of a *closed system* of matter. The apparent 'quantization' of light – leading to its 'particle' features in the current wave-particle model – follow here from an apparent quantization of the emiting and absorbing matter, that are components of the closed material system. The continuous nature of light (as observed, for example, in diffraction phenomena) follows here from the continuum property of the field of interaction of electrically charged matter.

With these results, it is felt that the logical difficulties of imposing wave-particle dualism and Bohr's principle of complementarity on a theory of light have been eliminated. The nature of light is that it is a particular manifestation of the coupling of the material components of a closed system (in principle, the universe). It is purely a continuum because the fundamental description of matter itself is necessarily based, according to this theory, on the continuous field concept.

VI. CONCLUDING REMARKS ON THE PHILOSOPHY
OF A UNIFIED FIELD THEORY

The philosophy that underlies the theory I have been discussing, which attempts to fully unify the manifestations of light and matter, bears some resemblance to the Leibnizian view of the universe as fundamentally one – without actually separable parts[17]. This is so if one is permitted to interpret his monads, which Leibniz referred to as the different 'reflections' of the single Universe, to mean the independent physical manifestations of this single closed system.

It is interesting to extrapolate from this view of the material aspects of the universe to include man himself. Assuming (with Spinoza) that such an extension is logically necessary,[18] one might then view man's inquiry about the nature of the Universe (that includes himself) as based on an *approximation* that man's consciousness has the freedom to *get into*, thereby *reflecting* the underlying abstract relations about nature. Man then extrapolates from these relations toward a more complete understanding of the entire Universe – that fundamental existent of which man is, in principle, an inseparable component. Thus, it seems to me that man's investigations of the world are not a matter of his looking in, as an impartial outsider, tabulating its physical characteristics according to the responses of his particular sense perceptions. It appears to me to be, rather, a matter of man's reflection, introspection and deduction on the nature of a single, abstract underlying reality – the basic existent that is the Universe, and from which one can derive particulars to be correlated with the physical observations of his senses.[19] Still, there are infinitely many more particulars that fall within the scope of man's reasoning, but are not within the domain of direct responses of the human sensing apparatus. Nevertheless, the latter (unobservable) particulars can play the important role of leading, by logical implication, to tests of directly observable particulars.

There are critics of such a unified approach who argue (with Gödel) that one cannot start at the outset with a complete system because the investigator himself must have the freedom to decide how, where and when to explore along one direction or another – i.e. he has the freedom to freely exercise a decision making process. But this view tacitly assumes that man is indeed a separable entity – independent of the rest of the Universe. Should one accept this assumption as a fundamental axiom, then certainly Gödel's theorem would be *logically* true. On the other hand, the theorem itself is not an *a priori* truth of *nature*. It is only as true as the set of axioms upon which it is based!

According to the Spinozan concept of oneness, and the Leibnizian view of the Universe as a single existent without parts, the tacit assumption that separates man from the universe must be rejected. That is, rather than the atomistic view in which the Universe is said to be composed of many coupled, though separable, individual entities – some of which are the collection of human consciousnesses with their respective free wills – the proponents of a fully unified field theory must view the Universe (with Spinoza) as a fully deterministic existent, that may exhibit an infinite manifold of intrinsic manifestations; yet, where 'free will' (actual individuality) is then only an apparent (illusory) feature that is not more than a *particular approximation* for the oneness of the Universe – just as the motion of a 'discrete ripple' on a pond, as an apparent independent thing, is not actually separable, and is in fact an approximation for a particular manifestation of one continuous pond.

Such a view of the Universe, as a truly closed system, not only serves the purpose of providing an important heuristic function in the metaphysical approach toward a general theory of nature. It also has mathematical and logical consequences that do not follow from the type of universe that is an open set – a sum of individual parts. Thus, the differences in these two metaphysical approaches – one in terms of a closed system, and the other in terms of an open system – imply at least some mutually exclusive features in the physical predictions that, in principle, can be tested in order to further verify one theory or the other. An important mathematical difference is the nonlinear versus linear laws for the closed versus open systems.

In this paper, I have tried to emphasize how the phenomenon of light must be intimately connected with the fundamental features of matter. I have suggested that such connection may be strongly related to the metaphysical basis for a general unified approach in terms of a continuum field theory. Such a unified approach has been adopted, philosophically, by many previous generations of scholars. It is well known that in this century Einstein pursued the view of a unified field theory. In the 19th century, it was suggested by (extrapolations from) Faraday's views, when he introduced the field concept to explain electrical and magnetic phenomena, and in his attempts to further extend it to other types of force fields (gravitation). It is a view that was strongly suggested in the philosophical writings of the 17th century scholars, Spinoza and Leibniz. Aspects of these philosophical works, in turn, can be traced back to influences of scholars in much earlier periods, such as Moses Maimonides in the 12th century[20] and Hasdai Cresas in the 14th century.[21]

I believe that these scholars in the previous periods saw the oneness of the Universe to include man. Should man be able to accept this view, it must lead him to a fully rational approach to science as well as to a higher ethical behavior in regard to his interaction with his fellow constituents of the world – for it is a philosophy that implies humanism and a oneness of man with nature. My conclusions have been drawn from the results of a theoretical study that started with a question about the nature of light. Perhaps this is, in part, behind E. O'Neill's comment, 'there is more to light than meets the eye'.

Department of Physics,
State University of New York at Buffalo

NOTES

[1] A bibliography on the classical works in optics is given in the first chapter of the book by M. Jammer, *The Conceptual Development of Quantum Mechanics*, McGraw-Hill, New York, 1966. Newton's original researches in optics are reported in his treatise, *Opticks, or a Treatise of the Reflexions, Refractions, Inflexions and Colours of Light*, S. Smith, London, 1704.

[2] An account of Huygens' researches is given in his book, *Treatise on Light* (transl. by S. P. Thompson), Macmillan, London, 1912. Hooke's researches on the wave theory of light are expounded in his book, *Micrographia*, 1665.

[3] G. Libri, *Histoire des sciences mathematique en Italie*, J. Renouard, Paris, 1838–1841, Vol. 3, p. 54 (quoted in Jammer, *ibid.*).

[4] F. M. Grimaldi, *Physico-Mathesis de luminie coloribis et iride aliisque adnexis libri duo*, Benatii, Bologna, 1665. An extensive discussion of the different approaches to the diffraction phenomenon, including Newton's explanation, is given in the article by R. H. Stuewer, *ISIS* **61** (1970), 2, No. 207, 188.

[5] A. Fresnel, *Annales de Chimie et de Physique* 11 (1819), 246, 377. *Oeuvres Completes d'Augustin Fresnel*, Imprimerie Imperiale, Paris, 1866.

[6] See, for example, R. C. Tolman, *The Principles of Statistical Mechanics*, Oxford, 1950, pp. 378, 382.

[7] For a detailed analysis of this point, see the book by P. A. M. Dirac, *The Principles of Quantum Mechanics*, Oxford, 1958, 4th ed., p. 306.

[8] For further discussion of this logical point, see E. Schrödinger, *British Journal for the Philosophy of Science*, 3 (1952), 109, 233.

[9] G. N. Lewis, *Proceedings for the National Academy of Sciences*, (*U.S.A.*) 12 (1926), 22.

[10] M. Sachs, *Il Nuovo Cimento* 53B (1968), 398.

[11] M. Sachs, 'The Electron-Muon Mass Doublet from General Relativity', *Il Nuovo Cimento* 78 (1972), 247.

[12] This concept of arbitrary charge of the components of interacting matter, but fixed e^2, was discussed in an earlier publication of the author, *British Journal for the Philosophy of Science* 15 (1964), 213. It is an idea that is also consistent with more recent speculations that nucleons (i.e. neutrons and protons) are composites of constituent fractionally charged particles, that have been called 'quarks'.

[13] A review of the development of this theory, up to 1972, has recently been published in a series of four papers by the author, *International Journal of Theoretical Physics* 4 (1971), 433, 453; 5 (1972), 35, 161.

[14] G. N. Lewis, *Nature* 118 (1926), 874.

[15] M. Sachs and S. L. Schwebel, *Supplemento del Nuovo Cimento* 21 (1961), 197; M. Sachs, *International Journal of Theoretical Physics* 1 (1968), 387.

[16] M. Sachs, *Il Nuovo Cimento* 37 (1965), 977.

[17] Leibniz' approach is discussed in his essay, *Monadology*. A translation appears in M. C. Beardsley, *The European Philosophers from Descartes to Nietzsche*, Modern Library, New York, 1960.

[18] B. Spinoza, *Ethics*, Hafner, New York, 1960, p. 128.

[19] It seems to me that the first stage, in which man becomes aware of the Universe through the acts of his consciousness, is similar to Buber's *I-it* relation. The next stage, where man uses the effects of this awareness as hints in reasoning toward a full understanding of the Universe, in which man is an inseparable component, is an approach toward an understanding similar to Buber's *I-Thou* relation. Buber's philosophy is expressed in his book, *I and Thou*, Scribner's, New York, 1958. Further discussion on this point is given by the author in *Philosophy and Phenomenological Research* 30 (1970), 403.

[20] Maimonides' major philosophical work is his treatise, *The Guide of the Perplexed* (transl. by S. Pines), University of Chicago Press, Chicago, 1963.

[21] Cresas' philosophical ideas are expressed in his book, *Or Ad-o-nai*. Part of this work is translated in the book by H. A. Wolfson, *Cresas' Critique of Aristotle*, Harvard University Press, 1929.

EUGENE P. WIGNER

EPISTEMOLOGICAL PERSPECTIVE
ON QUANTUM THEORY

One can discuss the epistemology of quantum mechanics from two points of view, and there is some confusion in the literature because the writers (including myself) do not always state clearly on which point of view the discussion is based. The first point of view accepts the observable consequences of quantum mechanics as valid, valid accurately and universally; its objective is the determination of the epistemology on which these consequences can be based. The second point of view from which epistemology can be discussed in the quantum mechanical era is based on the realization of the problems of the epistemology which is based on the acceptance of quantum mechanics as a definitive and final theory. These problems – one of them epitomized by a reference[1] to 'Wigner's friend' – lead one to wonder in what respects quantum mechanics may be modified when the interest of science is extended to a larger set of phenomena characterizing complex living beings. It is justified to speculate in this direction because, clearly, present quantum mechanics is based solely on phenomena involving inanimate objects. It is justified and also interesting to speculate on the extension of our theories to the realm of life and consciousness,[2] even if such speculations do not bear fruit in the form of definite, precise conclusions – as they probably will not.

The confidence in the final nature of quantum theory's epistemology will depend, naturally, on the roundedness and inner coherence of that epistemology. It may be of some relevance for this reason, and also because it is interesting anyway, to mention the blemishes on the otherwise beautiful structure of quantum mechanics. These impair also the roundedness and inner coherence of quantum mechanics' structure and may render us more willing to admit that the epistemology which is based on that structure may be subject to modifications. The weaknesses of the struc- will be, therefore, the first subject of my discussion.

Hooker (ed.), Contemporary Research in the Foundations and Philosophy of Quantum Theory, 369–385. All rights reserved

I. RESERVATIONS ABOUT THE STRUCTURE OF
QUANTUM MECHANICAL THEORY

I trust that the remarks on reservations concerning the structure of quantum mechanical theory which follow will not obscure the fact of my profound admiration concerning the practical successes of this theory. I'll group my reservations – there are quite a few of them – into two categories. The problems that arise even in non-relativistic theory form the first, those which render a union between the concepts of quantum mechanics and those of relativity theory difficult form the second category.

(A-1) *Conceptual problems of non-relativistic quantum mechanics – the superposition principle.* The superposition principle – the existence of states $a\psi_1 + b\psi_2$ where a and b are arbitrary constants, ψ_1 and ψ_2 arbitrary states – is such a general principle that one feels there should be some operational procedure to realize it. This would consist in a universal prescription to produce the state $a\psi_1 + b\psi_2$, given in terms of the prescriptions for the realizations of the states ψ_1 and ψ_2. Dr. Gerjuoy has discussed cases in which superpositions can be realized and such realization do constitute some justification of the general principle. However, Dr. Gerjuoy's discussion was confined to highly specialized cases and did not resolve the basic question. It is, in fact, hard to see how the basic question could be approached.

It is surely unnecessary to point out that all the so-called paradoxes of quantum mechanics involve superpositions of classically interpretable states, the superpositions themselves being, however, not interpretable in the naive, classical fashion. This applies to the Einstein-Podolsky-Rosen paradox[3], to Schrödinger's cat[4], and also to the singlet state discussed more recently by J. S. Bell[5]. It is unnecessary to enlarge on this. Let me mention, instead, that the so-called superselection rules[6] do limit the absolute generality of the rule of superposition – they do limit it, however, just enough to impair the mathematical beauty of the general, single and uniform Hilbert space as a frame for the description of all quantum mechanical states. They do not seem to alleviate significantly the conceptual question raised.

(A-2). *The second conceptual problem of non-relativistic quantum mechanics – the problem of measurement.* We learn and teach, respectively, in courses on quantum mechanics that the measurable quantities, or in the words of

Dirac, the observables, are hermitean operators[7]. It can indeed be proved, by means of the theory of measurement, that only hermitean operators can represent measurable quantities[8]. Some books, and some lecturers, go further and claim that *all* hermitean (or more precisely, all self-adjoint) operators can be observables[9]. However, if we ask how the measurement of a given self-adjoint operator should be carried out, the books and the lecturers remain most secretive. One has, of course, no idea how a quantity such as $p + q$ or $pq + qp$ or pqp could be measured – in fact, clearly, most operators cannot. Still, many can. Araki and Yanase[10] furnished the most concise proof that the measurement of almost any self-adjoint operator requires a macroscopic measuring device or, more precisely, a device that is, with comparable probabilities, in a great many of the states of every additive conserved quantity.

There is, however, no rule which would tell us which self-adjoint operators are truly observables, nor is there any prescription known how the measurements are to be carried out, what apparatus to use, etc. In a theory with a positivistic undertone, this is a serious gap.

It may be well if our discussion is a bit more cursory with respect to the second set of difficulties, arising from conflicts with the theory of relativity. I will not mention the difficulties inherent in the relativistic formulation of the quantum mechanics of interacting systems – many of us believe that these are amenable to solution[11], some of us believe that they have been solved already by Isham, Salam, and Strathdee[12] and by Lehman and Pohlmeyer[13], and few of us attribute these difficulties to the conceptual structure of the underlying theory. I will not speak about them when discussing our next subject.

(B-1). *Conceptual problems of relativistic quantum mechanics – the instantaneous nature of the measurement.* Clearly, measurement cannot be an instantaneous process and it is hard to think of one being instantaneous even if one assumes an infinite signal velocity. Much less can it be instantaneous if no signal can travel faster than the velocity of light.[14] Dr. M. Sachs emphasized this point at our meeting.

Actually, the paradoxes which are often found alarming, such as Schrödinger's cat, have little to do with this problem. It is important to realize this point, nevertheless, because it also detracts from the simplicity and hence beauty of the mathematical formulation of the theory.

(B-2). *Conflict between successive measurements.* The relativistic transformation of space-time also leads to a conflict between successive measurements, carried out by observers in different states of motion. The
$t =$ const surfaces of such observers necessarily intersect so that neither
measurement can be considered to occur *after* the other had been carried
out. This is a very serious – though, as we shall see, not insurmountable –
difficulty since what quantum mechanics is supposed to provide us with
are probability connections between subsequent observations. If all these
observations have to be carried out by observers at rest with respect to
each other, the significance of relativistic invariance is much diminished.
The situation is further aggravated by the fact that there is no *operational*
translation of measurements between observers in motion with respect to
each other: a measurement by a moving observer does not appear to be
a measurement for an observer at rest.

It is well known how relativistic field theory overcomes, or tries to overcome, this difficulty.[15] In its original form, it restricted measurements to
the determination of field strengths at space-time points. Such measurements at different points could be compatible (even though, at least in the
simplest case, that of the electromagnetic field, this is questionable). Bohr
and Rosenfeld,[16] in their well known and very ingenious paper, specified
methods for measuring the electromagnetic field at a given point. Their
arguments, I fear, just because of their ingenuity and the equipment they
postulate, lead most readers to the conclusion that the electromagnetic
field *cannot* be measured at points in space-time – not even in arbitrarily
small volumes. The former conclusion is quite in accord with the fact that
an operator which corresponds to the field intensity at a given point does
not exist in the true mathematical sense: only operators which correspond
to integrals of the field intensity over finite domains can be defined in a
mathematically rigorous fashion.[17]

It is for this last reason that modern field theorists postulate the measurement of fields averaged over finite domains.[17] However, if these domains extend only in spatial directions, and are infinitely thin in the
direction of the time axis, at least some of the objections voiced earlier
remain valid. If the domains are bona fide volumes in space-time, one
cannot help feeling that the operator attributed to them can hardly be the
one which is supposed to be measured since, when attributing a field operator to a space-time point in the midst of the domain, it does not take the

effect of the apparatus into account – the apparatus which has interacted with the field at earlier points of the space-time domain.

In order to overcome this last difficulty, and in order to avoid the non-relativistic nature of a domain which is infinitely thin in the time-like direction, one is tempted to use, instead of the traditional constant time cut, the light cone of a point in space-time. This would avoid the difficulties mentioned above but, even though I have tried to implement this, I did not get very far.

(B-3). *Probabilities in a changing world*. Probabilities can be defined only if the process can be repeated many times. This presupposes at least displacement invariance and this is not truly present in a changing world, in an expanding universe.[18]

This point was brought up only for the sake of completeness – few will attribute great significance thereto. The world changes very slowly and the observations take very little time. Nevertheless, it seems to me, the problem does exist.

Our discussion so far concerned the blemishes on the beauty of the conceptual structure of quantum mechanics. Some of these are blemishes because they reduce the mathematical simplicity of the theory – and, as Einstein said, it is easy to believe a theory only if it is truly simple and beautiful. Other blemishes are more closely connected with the present status of the theory, the difficulties encountered in the endeavor to unite quantum and relativistic theories. These are less closely connected with the basic structure of quantum mechanics but are surely not independent therefrom. When concluding this part of our discussion, let me repeat that these are blemishes on a structure which appears to me, in spite of them, truly magnificent and *is*, of course, of immense significance in the whole structure of physics. This should be forgotten even less than the existence of blemishes should – though I believe these should be remembered also.

Let us now turn to the more philosophical problems with which we want to come to grips in spite of the blemishes enumerated.

II. QUESTIONS WITHIN THE FRAMEWORK OF QUANTUM MECHANICS AND TENTATIVE ANSWERS

Before embarking on the truly puzzling questions relating to our subject, let me say a few words on another question which must have bothered many of us. The epistemology of quantum mechanics is supposed to provide the interpretation of the laws of this theory in terms of our observations. It would seem that without such an interpretation the laws of quantum mechanics must be entirely meaningless. How is it possible then that the widespread controversy on the epistemology hardly affects the practical applications of the theory? Most of those working on one of the branches are hardly affected by the controversies and pay very little attention to them. The answer is, in my opinion, that we hardly ever use quantum mechanics in the fashion we use classical mechanics: to predict events. Rather, we use it, as a rule, either to determine material constants, or the possible values of essentially only one observable, the energy. Material constants which we determine by means of the theory are densities, heats of transformation, viscosity, transition probabilities, and some others. These are then inserted into macroscopic equations and we verify the values obtained by means of macroscopic experiments. In this role, quantum mechanics is a servant of macroscopic theories and renders these more definite. The other very common function of quantum mechanics is the determination of the possible energy values of systems, such as hydrogen or helium atoms – and energy is the most important microscopic quantity that we know how to measure. It is perhaps disappointing that so few other conclusions of the general microscopic theory are commonly put to experimental test. However, this is unavoidable since we live in a macroscopic world. It is not surprising, therefore, that the most useful function of quantum mechanics is the providing of material constants for macroscopic equations – a function which at the end of the last century was feared to be the only remaining function of experimental physics.

Let me now take up our subject proper: the epistemology which the complete acceptance of quantum mechanics forces on us. Quantum mechanical theory has two parts: the equations of motion, and the theory of observations. The equations of motion give the change in time of the quantum mechanical determinant of the state of a system: of the state vector. The theory of observations gives the probabilities of the outcomes of observations on systems in terms of the state vector. All is well as long as we do not ask how the observation takes place. However, we do run into difficulties when we try to describe the process of observation by means of the equations of motion, i.e., try to eliminate the second part of the theory by means of the first. Since the equations of motion should be able to describe all events, such an elimination should be possible.[19]

When we try to describe the process of observation by means of the equations of motion, we encounter a contradiction at once: the equations of motion are deterministic, the outcomes of the observation are subject to stochastic laws. Several possible ways can be proposed to eliminate this paradox; they will be discussed next.

(A) The simplest and most natural explanation of the indeterminate outcome of the measurement on an object with completely determined state vector is that the state vector of the measuring apparatus was indeterminate: in one case it had one, in other cases other, directions and the outcome of the measurement is, therefore, different in all these cases.

The possibility just sketched is certainly present. Most measuring apparata, if not all of them, are macroscopic and it would be impossible to determine their state vector. One cc of air can be, at room temperature, with roughly equal probabilities in 10^{20} states, and it would be impossible to ascertain in which one of these it is.

The trouble with this explanation is that, if one discusses this reason for the probabilistic outcome of the measurement – this has been done by von Neumann, myself, and most recently by d'Espagnat[20] – one soon finds that it cannot give the probabilities for the various outcomes of the measurement which are postulated by the second part of quantum theory, by its theory of observations. One therefore has to abandon this explanation for the probabilistic nature of the measurement process, though one may abandon it only reluctantly.

(B) If we could not reduce the theory of observations to the theory of motion, i.e., describe the process of observation in terms of the equations of motion, we can try the opposite: we can eliminate the equations of motion and express all statements of quantum mechanics as correlations between observations. This is indeed possible by attributing the operator[21]

$$Q(t) = e^{iHt/\hbar} \, Q(0) e^{-iHt/\hbar} \tag{1}$$

to the same measurement, carried out at time t, to which we attribute the operator $Q(0)$ if carried out at time 0. We shall denote by P_j the projection operator which leaves the state vectors of outcome j unchanged, annihilates the state vectors of all other measurement outcomes. It is not difficult to see then that an equation similar to the preceding one holds for these P_j

$$P_j(t) = e^{iHt/\hbar} \, P_j(0) e^{-iHt/\hbar} \tag{1a}$$

Quantum mechanics can be, then, reformulated in terms of the projection operator of the successive measurements. Let us denote the projection operator for the outcome i of the first measurement by P_i, for the outcome of j of the second measurement by P'_j. The probability that the second measurement yields the result j if the first one's outcome was i is then given by

$$\text{Trace } (P'_j P_i)/\text{Trace } P_i = \text{Trace } (P_i P'_j P_i)/\text{Trace } P_i \tag{2}$$

and similar expressions can be given for the probabilities of the different outcomes of several successive measurements.[22] These expressions incorporate the equations of motion if one uses for the operator of a measurement which takes place at time t, the expression in terms of the operator for the measurement at time 0, given by (1).

The preceding reformulation of the equations of quantum mechanics, eliminating explicit reference to the equations of motion and to state vectors, corresponds to a conceptual reformulation thereof. According to this, the function of quantum mechanics is to give statistical correlations between the outcomes of successive observations.[23] From a very positivistic point of view, such a reformulation is quite satisfactory: it refers solely to observations and establishes relations between these directly. From the point of view of everyday experience, in particular our belief in reality and the abstract characterizability of such reality, it appears to be very disturbing. No description of the state of the system is used, by state vector or otherwise. It appears that our theory denies the existence of absolute reality – a denial which is unacceptable to many. It seems to me, however, that it is not necessary to go that far in our conclusions. By referring only to outcomes of observations one does not necessarily deny that there is something real behind the observation – whatever the word 'real' may mean. There may be any amount of old-fashioned reality behind the scenes; it is only that quantum theory does not deal with it but only with probabilities for the outcomes of observations. As to myself, I should admit that the old-fashioned concept of reality seems indeed old-fashioned to me and that I do not know how one could define operationally the reality of anything. My own thoughts, impressions and perceptions surely exist, the existence of all else is inferred only on the basis of these impressions and perceptions.[24]

In my opinion, the restriction of quantum mechanical theory to the determination of the statistical correlations between subsequent observations reproduces most naturally the spirit of that theory; the alternative just discussed, renouncing the definition of reality, is the most natural epistemology of quantum mechanics. It considers the state vector to be only a mathematical tool, useful for carrying out certain calculations, but only a tool and, as (2) shows, a tool the use of which can be avoided.

It is well to admit, however, that the most common applications of quantum mechanics, mentioned at the beginning of this section, do not follow the pattern here described. It is very likely that the calculation of densities, heats of transformation, transition probabilities, etc., could be brought into the form of observations. However, this has not been done – and hence it is not clear what additional assumptions might be needed – nor is it very natural to do so.

III. MODIFICATIONS AND EXTENSIONS OF QUANTUM MECHANICS WHICH MAY AFFECT ITS EPISTEMOLOGY

The formulation of Quantum Mechanical Theory presented last seems to avoid all conceptual difficulties. Surely, it restricts quantum mechanics in the sense that it does not give a picture of 'reality', (whatever this term may mean), but it seems to answer the question which is operationally meaningful: what are the correlations between subsequent observations or perceptions. If one were satisfied with a theory restricted in this way, and if one believed it to extend to all phenomena, one could be satisfied with the theory as formulated above.

There are several reasons for the lack of satisfaction with the theory as described and these reasons have stimulated modifications – unfortunately in every case vaguely formulated modifications – of the theory. Naturally, the modifications suggested depend on the reasons which cause the lack of satisfaction of the proposer of the modification. The rest of this article will consist of a discussion of the proposed modifications.

(A) *Hidden variables*[25]. This is the most widely discussed proposal. It is motivated by the conviction that there is a true reality, hidden from the eyes of the quantum theorist, a reality in terms of which the events, in particular also the observations, have a deterministic character. This reality should be the proper subject of any physical theory. It is not adequately described by the state vector – much less by a theory which is as unconcerned with reality as is the one discussed at the end of the preceding section. It is proposed to call the determinants of the real situation of a system 'hidden variables'. These obey some deterministic equations – at present unknown – which, if known, would, together with the knowledge of the values of the hidden variables, enable one to predict the behavior of the system under all conditions.

It is not clear whether it is possible, or for some fundamental reasons impossible, to ascertain the values of the hidden variables. If it is impossible, their existence has a great deal of similarity with the existence of ghosts which cannot be observed and which cannot influence any events. If it is possible, one can produce states which do not behave according to quantum mechanical theory – a possibility which can, of course, not be denied *a priori*. We shall, for the present, disregard this possibility.

Let us return, therefore, to the concept of hidden variables which will remain hidden, that is, undeterminable. It is neither surprising nor difficult to show that they can be so defined as to reproduce the probabilities for the outcomes of measurements postulated by quantum mechanics.[26] However, as was known already to von Neumann, if this is to be true for an infinite succession of measurements, their number must be infinite.[27] Kochen and Specker[28] proved another disagreeable property which the hidden variables must have, greatly strengthening von Neumann's published work on this subject. The most important contribution to the subject was made, however, by J. S. Bell[5], who showed that in a certain case the hidden variables characterizing one particle must depend on the quantity which is being measured on another particle. This appeared so absurd that the supporters of the hidden variable theories now doubt the conclusions of quantum mechanics which force the acceptance of such hidden variables and are testing them experimentally. As I said before, it is possible that these laws are not valid but, personally, I do not expect that this will be their finding.

(B) *The apparatus must be described by macroscopic theory.* I am not sure it is fair for me to discuss this proposal which was so eloquently articulated by Rosenfeld,[29] Daneri, Loinger, Prosperi,[30] and also by Fock.[31] I cannot bring myself to agree with it.

The proposal is based on the fact that the measuring apparatus is macroscopic. It is, in practice, always described in terms of the concepts of classical, that is macroscopic, physics. The object is microscopic and the interaction between object and apparatus should be described by the equations of quantum mechanics. However, after the interaction, one should revert to a classical description of the apparatus. This is the point where the stochastic element enters: the macroscopic description of a state given microscopically, although on the whole much more crude, does not give a unique set of values to the macroscopic variables. This is clear in the case of the well-known paradoxes, such as that of Einstein-Rosen-Podolsky, or in that of Schrödinger's cat. But one can produce much simpler examples: if the wave function gives finite probabilities to more than one value of a macroscopic variable, the classical translation of what is in quantum mechanics a uniquely defined state, gives finite probabilities to several values of the macroscopic variable in question.

There is little question in my mind but that the proposal which we now discuss is in no way in conflict with the procedure which we are using in practice. The trouble is only that it postulates the miracle which disturbs us: that after the measurement, or if I use my own interpretation, after our observation, the apparatus *is* in a state which has a classical description. Hence, the explanation covers up rather than solves the problem. The transition to a classical description of the apparatus is an arbitrary step which may obscure, but does not eliminate, the basic fact that the true equation of motion is deterministic.

One is also in doubt concerning the dividing line between microscopic and macroscopic systems and on the degree of detail to be used in the macroscopic description. There is also a question what to do about quantum phenomena, such as laser action or superconducting currents, which may take place in the apparatus. The terms 'classical description' and 'macroscopic theory' do not seem to be clearly defined in such cases.

(C) *Quantum mechanics is not valid in the macroscopic domain.* This is a somewhat older suggestion of Ludwig and may be considered as a variant of the preceding one.[32] It admits more clearly our inability to describe the process of observation by means of the equations of motion of quantum mechanics. That these equations are not valid for macroscopic systems may be true, though few physicists are willing to admit this and there is no evidence for such lack of validity. Nevertheless, the clarity of the underlying hypothesis seems to me to be preferable to the somewhat vague nature of the preceding proposal.

(D) *There are no isolated macroscopic systems.* The enormous density of the energy levels of a macroscopic system was pointed out before. It was Zeh[33] who pointed out first that as a result of this enormous density it is practically impossible to isolate a macroscopic object from its surroundings: even a single electron, at a distance of a mile, can cause transitions between the quantum mechanical states of a macroscopic body.

Zeh's observation is unquestionably correct and it is also alarming. We are used to formulating physical theory in terms of equations of motion of isolated systems and Zeh's observation tells us that this is, for macroscopic systems, unrealistic. The same applies, no doubt, for the formulation in terms of outcomes of observations which was given at the end of the preceding section. It is possible, of course, that the conclusions arrived at there remain essentially valid in spite of the interference of accidentally present objects but, at least, the proofs demonstrating that the stochastic element cannot be due to the indeterminate state of the apparatus should be reconsidered.

Zeh's observation may be of very great importance but, in my opinion, it does not convert quantum mechanical theory to the description of reality in the traditional sense. It does not explain how it comes about that looking at the apparatus we get one definite impression. The apparatus plus object may be in a 'mixture' after the measurement, correlated with many other objects – the question how we pick out one of the alternate states in which it may be remains unclear.

(E) *Physical theory should be extended to the phenomena of life and consciousness.*[34] There is little doubt that it would be desirable to follow this proposal. The question is only whether a deeper understanding of the phenomena of life and consciousness will alter our views on the role of quantum mechanics and the meaning of observations. It is my opinion that it is likely to do so.

The basic concepts, in terms of which the laws of nature were formulated, have repeatedly changed in the course of history. They were positions and velocities in Newtonian mechanics, they were field intensities in the theories which followed, and if we accept the conclusion of the preceding section, they are observations in quantum mechanics. A deeper insight into the nature of the observation and the observer is surely desirable since the theory now uses a most stereotyped and crude picture of the observer.

This last point becomes most apparent if one tried to describe the observation of another person.[1] It is true that one could consider, in the spirit of the second alternative of the preceding section, the receipt of the communication of another person concerning his observation, as an observation, on our part, of his state. This observation of ours, of his state of mind, permits the drawing of inferences concerning the state of the object on which he has made his observation, just as one can draw similar inferences concerning the state of the object by looking at a measuring instrument which was in contact with that object. This is formally possible – few would, however, consider such a procedure reasonable. It presupposes that we know the behavior of a person as well as we do that of a measuring instrument and treat him as a measuring instrument. In addition, even though 'reality' in the physical and particularly microphysical world may be a questionable concept, one cannot accept the observations of another person, and the resulting content of his consciousness, to be less real than those of ourselves. This suggests that one treat all observations, undertaken by oneself or another person, on a more nearly equal basis.

It is maintained by some that the laws of physics which we now know, or almost know, suffice for the description of life and the attendant phenomena of consciousness. It is not difficult to adduce evidence to contradict this view, and this has been done also by me.[35] It is, in particular, difficult to accept the possibility that a person's mind is in a superposition of two states in the one of which he has received one, in the other of which he has received another, signal. We ourselves never have felt we were in such superpositions. It seems unlikely, therefore, that the superposition principle applies in full force to beings with consciousness. If it does not, or if the linearity of the equations of motion should be invalid for systems in which life plays a significant role, the determinants of such systems may play the role which proponents of the hidden variable theories attribute to such variables. All proofs of the unreasonable nature of hidden variable theories are based on the linearity of the equations. Actually, it is more likely that the concepts which will emerge when our science encompasses the phenomenon of life will differ as radically from those of present quantum mechanics as for instance the concepts of field theories differ from those of point mechanics or those of quantum mechanics differ from those of earlier theories.

It may be well to recall here the point made at the beginning of the present section. It was emphasized there that there is no logical contradiction in quantum mechanics, particularly not if it is formulated entirely in terms of observations. The proposals dealt with here are based on hopes that physical theory can be modified and reformulated in a way which renders it more attractive and more satisfactory to the proponent. The present proposal is motivated by the desire for a less solipsistic theory which does not deal solely with the observations of a single observer but attributes reality also to the contents of the minds of other observers. In addition, it would be desirable, of course, to know more about mental processes in general. Finally, it is hard to accept as complicated a process as the entering of some cognition into a mind as the fundamental one in terms of which the theory should be formulated. Needless to repeat, none of this points to inner contradictions of present quantum mechanics.

(F) *Only the whole world's wave function is meaningful.* This again is a theory[36] which I find very difficult to accept. It appears to me to be a complete denial of the fact, undeniable in my opinion, that our impressions form the primitive reality. The state vector of my mind, even if it were completely known, would not give its impressions. A translation from state vector to impressions would be necessary; without such a translation the state vector would be meaningless.[37]

Actually, if an outside observer could ascertain the state vector of the world, he would find that I am not in a pure state but in a linear combination of immensely many states, each correlated with the states of immensely many outside objects. The epistemology under discussion postulates that I feel to be in only one of the immensely many states in the linear combination of which I actually am. This means, however, that the other states, and hence the total state vector of the world, are meaningless.

One could postulate, in a similar vein, that the world is homogeneous and isotropic and it is only I who sees differences between different locations and different directions – the real state vector is invariant under all Poincaré transformations. Such ideas appear to me – perhaps wrongly – to be detached from reality.

Princeton University

NOTES

The notes below refer to the specific points made in the body of the article and are, probably, grossly incomplete even concerning those points. Even though they do not refer to points of his article, the writer does want to mention two papers which provided, in his opinion, most of the foundation of our thinking on the subject. These are, first, Heisenberg's 'Über den anschaulichen Inhalt...' (*Zeitschrift für Physik* 43 (1927), 172) which revealed the incompatibility of the classical position – velocity specification of states with the ideas of quantum and wave mechanics. It justified the state vector characterization of such states. The second is N. Bohr's article on complementarity (*Naturwissenschaften* 16 (1928), 245; see also his book *Atomic Physics and Human Knowledge*, John Wiley, New York, 1958, which made clear that the new theory has deep philosophical implications.

An excellent review of the whole literature was given by B. S. De Witt and R. Neill Graham, *American Journal of Physics* 39 (1971), 734.

[1] Ludwik Bass, preprint, to appear in *Hermathena*.

[2] J. H. Greidanus, *Transactions of the Royal Netherlands Academy of Sciences* 23 (1966); to appear in *Foundations of Physics*.

[3] A. Einstein, B. Podolsky, and N. Rosen, *Physical Review* 47 (1935), 777.

[4] E. Schrödinger, *Naturwissenschaften* 23 (1935), 807, 823, 844; *Proceedings of the Cambridge Philosophical Society* 31 (1935), 555.

[5] J. S. Bell, *Physics* 1 (1965), 195. Cf. also E. P. Wigner, *American Journal of Physics* 38 (1970), 1005.

[6] G. C. Wick, A. S. Wightman, and E. P. Wigner, *Physical Review* 88 (1952), 101; Dl (1970), 3267; A. S. Wightman, *Il Nuovo Cimento* 14 (1959), 81; J. M. Jauch, *Helvetica Physica Acta* 33 (1960), 711; E. P. Wigner, *Physikertagung* Vienna, Physik Verlag, Mosbach/Baden, 1962, p. 1; G. C. Hegerfeldt, K. Kraus, and E. P. Wigner, *Journal of Mathematical Physics* 9 (1968), 2029.

[7] P. A. M. Dirac, *The Principles of Quantum Mechanics* (several editions), Chapter II, Clarendon Press, Oxford.

[8] E. P. Wigner, *Zeitschrift für Physik* 133 (1952), 101.

[9] J. von Neumann, *Mathematische Grundlagen der Quantenmechanik*, Julius Springer, Berlin, 1932. (English transl. by R. T. Beyer, Princeton University Press, 1955.) Also W. E. Lamb, *Physics Today* 22 (1969), 23.

[10] H. Araki and M. Yanase, *Physical Review* 120 (1961), 666. An earlier proof was given by E. P. Wigner, Note 8.

[11] I am referring particularly to the axiomatic system proposed by A. S. Wightman. See A. S. Wightman and L. Garding, *Arkiv für Fysik* 28 (1964), 129. Also A. S. Wightman, *Physics Today* 22 (1969), 53.

[12] C. J. Isham, A. Salam, and J. Strathdee, Trieste Preprint, 1970, International Atomic Energy Agency.

[13] H. Lehman and K. Pohlmeyer, DESY Preprint, 1970.

[14] H. Salecker and E. P. Wigner, *Physical Review* **109** (1958), 571; also E. P. Wigner, *Helvetica Physica Acta Supplement* **4** (1956), 210; and S. Schlieder, *Communications in Mathematical Physics* **7** (1968), 305.

[15] A. S. Wightman, Note 11.

[16] N. Bohr and L. Rosenfeld, *Det Koneglige Danske Videnskabernes Selskab, Mathematisk-fysiske Meddelelser* **12** (1933), No. 8; *Physical Review* **78** (1950), 194; E. Corinaldesi, *Il Nuovo Cimento* **8** (1951), 494.

[17] See, A. S. Wightman and L. Garding, Note 11, especially p. 132–4. This article also has references to the older literature.

[18] E. Teller, in *Physical Review* **73** (1948), 801 discusses some of the relevant problems. A more complete discussion of the physical situation was given at the Trieste Symposium of 1968. See *Contemporary Physics*, International Atomic Energy Agency, Vienna, 1969.

[19] J. von Neumann, Note 9. F. London and E. Bauer, *La theorie de l'observation en mecanique quantique*, Hermann et Cie, Paris, 1939.

[20] J. von Neumann, Note 9; E. P. Wigner, *American Journal of Physics* **31** (1963), 6; B. d'Espagnat, *Il Nuovo Cimento* **4** (1966), 828.

[21] It is remarkable how difficult it is to find an explicit statement of this formula, well known to all physicists. Cf., however, Section 6i of S. S. Schweber's *An Introduction to Relativistic Quantum Field Theory*, Row, Peterson and Co., Evanston, 1961 or Section 16.22 of D. Bohm's *Quantum Theory*, Prentice Hall, Englewood Cliffs, 1951.

[22] R. M. F. Houtappel, H. Van Dam, and E. P. Wigner, *Reviews of Modern Physics* **37** (1965), 595, Section 4.4.

[23] This point was made by the present author in his lectures at the 1970 session of the Scuola Internazionale di Fisica E. Fermi – IL Course. It was made, probably, by many others.

[24] Even though this observation appears to be obvious, it also appears to provoke sharp denials. Cf. this writer's article, 'Two Kinds of Reality' in *Essays on Knowledge and Methodology*, collection of papers presented at the Marquette University Conference, Ken Cook and Co., Milwaukee, 1965.

[25] An eloquent discussion of this proposal is given by D. Bohm in his *Causality and Chance in Modern Physics*, Routledge and Kegan Paul, London, 1958. See also D. Bohm, *Physical Review* **85** (1952), 166, 180; D. Bohm and J. Bub, *Reviews of Modern Physics* **38** (1966), 453; L. de Broglie, *Foundations of Physics* **1** (1971), 5.

[26] Cf. e.g., the first section of this author's article, Note 5. The fact itself was surely known to J. von Neumann.

[27] This is the argument which, in this writer's opinion, motivated von Neumann against accepting hidden variables theories. Cf. footnote 1 of the second article in Note 5.

[28] S. B. Kochen and E. Specker, *Journal of Mathematics and Mechanics* **17** (1967), 59.

[29] Cf. e.g., L. Rosenfeld, *Supplement to Progress in Theoretical Physics*, extra No. (1965), 222; *Nuclear Physics* **A108** (1968), 241. Dr. Rosenfeld expressed his views also in numerous other articles.

[30] Cf. e.g., A. Daneri, A. Loinger, and G. M. Prosperi, *Nuclear Physics* **33** (1962), 297; *Il Nuovo Cimento* **44** (1966), 119. See also G. M. Prosperi's contribution to the *Scuola Internazionale di Fisica E. Fermi*, IL Course 1970.

[31] V. Fock, see e.g. his paper, 'Classical and Quantum Physics', International Centre for Theoretical Physics, Trieste, 1968.

[32] G. Ludwig, article in *Werner Heisenberg und die Physik unserer Zeit*, Vieweg u. Sohn, Braunschweig, 1961.

[33] H. D. Zeh, *Foundations of Physics* 1 (1970), 69; see also K. Baumann, *Zeitschrift für Physik* 25a (1970), 1954.

[34] E. P. Wigner, *Foundations of Physics* 1 (1970), 33. The articles of Greidanus, Note 2, and of Zanstra, among others, also support this view.

[35] E. P. Wigner, *Proceedings of the American Philosophical Society* 113 (1969), 95. Also article in *The Scientist Speculates* (ed. by I. J. Good), William Heinemann, London, 1962. A. E. Cochran, *Foundations of Physics* 1 (1971), 235.

[36] H. Everett III, *Reviews of Modern Physics* 29 (1957), 454; J. A. Wheeler, *ibid.* 29 (1957), 463; B. S. De Witt, *Physics Today* 23 (1970), 30, and lectures given at the *Scuola Internazionale di Fisica E. Fermi*, IL Course 1970.

[37] This is articulated also in the articles of Notes 34 and 35.